Computational methods for integral equations

Computational methods for integral equations

L.M. DELVES &

J.L. MOHAMED

*Department of Statistics and Computational Mathematics,
University of Liverpool*

CAMBRIDGE
UNIVERSITY PRESS

CAMBRIDGE UNIVERSITY PRESS
Cambridge, New York, Melbourne, Madrid, Cape Town, Singapore, São Paulo, Delhi

Cambridge University Press
The Edinburgh Building, Cambridge CB2 8RU, UK

Published in the United States of America by Cambridge University Press, New York

www.cambridge.org
Information on this title: www.cambridge.org/9780521266291

First published 1985
First paperback edition 1988
Reprinted 1992
Re-issued in this digitally printed version 2008

A catalogue record for this publication is available from the British Library

Library of Congress Catalogue Card Number: 84–4275

ISBN 978-0-521-26629-1 hardback
ISBN 978-0-521-35796-8 paperback

CONTENTS

PREFACE

This book considers the practical solution of one-dimensional integral equations. Both integral equations, and methods for solving them, come in many forms and we could not try, and have not tried, to be exhaustive. For the problem classes covered, we have used the 'classical' Fredholm/Volterra/first kind/second kind/third kind categorisation. Not all problems fit neatly into such categories; then the methods used to solve standard classes of problems must be modified and tailored to suit the needs of nonstandard 'real life' problems. It is hoped that the nature of any such modifications will be obvious to the intelligent reader. Not all categories of problems seem equally important (i.e. frequent) in practice; we have tried to spend most time on the most important classes of problems.

We have also been selective in the choice of methods covered. Here, personal likes and dislikes have helped the selection process, but we have also taken particular note of the fact that the cost of solving even a one-dimensional integral equation of Fredholm type can be unexpectedly high. Methods which converge slowly but steadily are therefore *not* very attractive in practice and particular emphasis is placed on the ability of a given method to obtain rapid convergence, to provide computable error estimates and to produce reliable results at relatively low cost.

It is hoped that the book will serve as a reference text for the practising numerical mathematician, scientist or engineer, who finds integral problems arising in his work. It should also serve as a textbook for third year undergraduate and M.Sc. students attending a course on the numerical solution of integral equations. A basic grounding in numerical analysis and linear algebra is assumed at the first year and (in some places) second year undergraduate level, but otherwise the book is reasonably self-contained.

Chapter 0 introduces the reader to the types of problem considered, while Chapters 1, 3 and 6 cover the necessary mathematical background required in the theory of linear second kind integral equations and eigenvalue problems. Chapters 2, 4 and 5 deal with numerical quadrature and with the application of quadrature methods to the solution of second kind Fredholm and Volterra equations. Expansion methods provide an alternative means of solution and these are covered in Chapters 7, 8 and 9. Singular equations, first kind equations and integro-differential equations are presented respectively in Chapters 11, 12 and 13. Each of the chapters contains a number of worked examples; where numerical results are displayed, these calculations were performed on an ICL 1906S computer at the University of Liverpool. In addition a number of exercises are provided at the end of each chapter. Chapter 10 considers the problems inherent in obtaining an objective comparison of specific computer routines which implement the numerical methods described in previous chapters. Finally, the Appendix, which should be used in conjunction with Chapter 11, contains a number of Chebyshev expansions corresponding to particular types of (known) singularities which may occur in the integral equation.

Special thanks are due to Dr I.J. Riddell for producing the graphs which appear in Chapter 10, to Mr F.J. Cummins who provided the remaining illustrations, and to Don Kershaw for his helpful comments on the manuscript. We are also extremely grateful to Miss D.E. Barton and Mrs M.T. Mosquera for typing endless 'final' drafts.

Liverpool   L.M. Delves
1985   J.L. Mohamed

0
Introduction and preliminaries

0.1   This book

The theory of differential equations is an essential ingredient of any undergraduate course in Mathematics and the majority of Numerical Analysis undergraduate courses introduce the student, at an early stage, to the numerical solution of differential equations. For some reason the theory, and, perhaps more so, the numerical solution, of integral equations are deferred to a later stage: in some sense integral equations must be felt to be either more advanced or of less practical interest than differential equations. This reflects the situation in practical calculations and probably in turn helps to perpetuate it; we turn more readily to a differential formulation of a problem than to an integral formalism. Yet the theory of linear nonsingular integral equations is at least as well developed as that of differential equations and it is in many respects rather simpler. The corresponding operators are *bounded* rather than unbounded, leading to a very straightforward existence theory (the Fredholm theory); perhaps, as one consequence of this, there is a much tighter link between the theory and practice of integral equations than is the case with differential equations. Most of the convergence proofs are constructive in nature and all or nearly all of the constructions have been used as the basis of algorithms for the numerical solution of the underlying equations (although not always with any great success!).

In this book we present the elements of the theory of linear integral equations *ab initio* and at the same time develop various numerical methods for their solution. The theory is developed within the framework given by the space of $\mathscr{L}^2$ functions and operators; this framework is both easy to work within and sufficiently general to cover most (although not all) practical problems. The numerical methods themselves may impose much stronger conditions and differ markedly in their requirements. Thus the

Nystrom methods described in Chapter 4 assume that the functions involved are at least piecewise continuous, while the expansion methods described in Chapter 7 require only the existence and mean convergence of the relevant expansions and hence impose conditions similar to the classical $\mathscr{L}^2$ assumptions. Many interesting facets of the subject are omitted for lack of space; we do not deal with nonlinear Fredholm equations, for example, nor with multidimensional problems.

It is hoped that the material covered is enough to impart the basic flavour of the subject; in particular, we have attempted to give a feeling for which methods work best in practice (i.e. are fastest; most rapidly convergent; most stable; lead to computable and realistic error estimates) for the various classes of standard and especially nonstandard problems encountered in practice. Usually of course no one method stands out as best in all respects, even for a restricted class of problems; we have tried to include those methods which are currently the best available in at least one of these respects, and we have excluded those methods of only historical or curiosity value. It is hoped that this book will serve as a useful text for third year undergraduates and M.Sc. students studying the numerical solution of integral equations. No previous knowledge of $\mathscr{L}^2$ theory is assumed and the subject is adequately covered in Chapter 1.

0.2   Other books

For those readers who require more detail of the mathematical theory of integral equations, a number of books are available. We refer you in particular to:

F. Smithies: *Integral Equations*, Cambridge University Press (1958) for a classic introduction to the subject at an elementary level. A rather more informal approach, which may suit some readers, is given in

R.P. Kanwal: *Linear Integral Equations*, Academic Press (1971)

and in C.D. Green: *Integral Equation Methods*, Nelson (1969).

The first two chapters of the latter give an especially nice introduction to $\mathscr{L}^2$ theory; both books contain some material on the numerical solution of integral equations. A fourth book, directed more towards the interests of applied mathematicians and mathematical physicists, is that of

J.A. Cochran: *The Analysis of Linear Integral Equations*, McGraw-Hill (1972).

Other books which you may find useful in connection with the numerical solution of integral equations are listed below. These contain, in some instances, further details of numerical methods described here, or other methods which we have not considered.

S.G. Mikhlin and J.L. Smolitsky: *Approximate Methods for the Solution of*

Differential and Integral Equations, Elsevier (1967)
is a classic but now dated work.

L.M. Delves and J. Walsh (eds.): *Numerical Solution of Integral Equations*, Oxford University Press (1974)
contains a broad survey of numerical methods and applications.

C.T.H. Baker: *The Numerical Treatment of Integral Equations*, Oxford University Press (1977)
contains a particularly detailed analysis of a large number of numerical methods, mainly those based on the quadrature (Nystrom) method.

C.T.H. Baker and G.F. Miller (eds.): *Treatment of Integral Equations by Numerical Methods*, Academic Press (1982)
contains a number of recent analytical and numerical developments in the treatment of integral equations.

0.3   Types of integral equations

The standard methods for integral equations deal with one-dimensional equations of the forms:

(i)

$$y(s) = \lambda \int_a^b K(s,t)x(t)\mathrm{d}t \qquad (1)$$

or

$$x(s) = y(s) + \lambda \int_a^b K(s,t)x(t)\mathrm{d}t, \qquad (2)$$

where λ is a (possibly complex) scalar parameter and $y(s)$ is known (and often called the driving term) but $x(s)$ is not. These are *linear Fredholm* equations of the first and second kind respectively for the unknown function $x(s)$.

(ii) If in equation (1) or (2) the kernel $K(s,t) = 0$ when $s < t$, they take the forms

$$y(s) = \lambda \int_a^s K(s,t)x(t)\mathrm{d}t, \qquad (3)$$

$$x(s) = y(s) + \lambda \int_a^s K(s,t)x(t)\mathrm{d}t, \qquad (4)$$

with a variable upper limit of integration. These are called *Volterra* equations of the first and second kind respectively.

(iii) If in equation (2) $y(s) = 0$, we have the homogeneous equation

$$x(s) = \lambda \int_a^b K(s,t)x(t)\mathrm{d}t \qquad (5)$$

referred to as an *eigenvalue* equation or a *Fredholm equation of the third kind*. For a given value of λ, (5) will normally have only the trivial solution $x(s) = 0$; those values of λ for which nontrivial solutions exist are called *characteristic values* of the equation. There may be no such (finite) values; for example, if $K(s, t) = 0, s < t$, so that (5) becomes

$$x(s) = \lambda \int_a^s K(s, t)x(t)\mathrm{d}t, \tag{6}$$

then under suitable regularity conditions on K, x, there are no finite characteristic values. Each of equations (1)–(6) is *linear* in the unknown function $x(s)$. More generally, we can recognise nonlinear Fredholm and Volterra equations.

(iv) Nonlinear Fredholm equations of the first kind:

$$y(s) = \lambda \int_a^b K(s, t, x(t))\mathrm{d}t, \tag{7}$$

of the second kind:

$$x(s) = y(s) + \lambda \int_a^b K(s, t, x(s), x(t))\mathrm{d}t, \tag{8}$$

of the third kind:

$$x(s) = \lambda \int_a^b K(s, t, x(s), x(t))\mathrm{d}t. \tag{9}$$

(v) Nonlinear Volterra equations of the first kind:

$$y(s) = \lambda \int_a^s K(s, t, x(t))\mathrm{d}t, \tag{10}$$

of the second kind:

$$x(s) = y(s) + \lambda \int_a^s K(s, t, x(s), x(t))\mathrm{d}t. \tag{11}$$

The classification (1)–(11) is that which is standard in the literature. Such equations arise quite naturally from the theory of differential equations. For example, consider the first order equation

$$\frac{\mathrm{d}x(s)}{\mathrm{d}s} = f(s, x), \tag{12a}$$

$$x(0) = x_0. \tag{12b}$$

Under suitable continuity conditions on $f(s, x)$ we can 'solve' this initial value problem to yield

$$x(s) = x_0 + \int_0^s f(t, x(t))\mathrm{d}t, \tag{13}$$

which is an integral equation for $x(s)$; if $f(s, x)$ is linear in x, the integral equation is linear, and otherwise not. Clearly, any solution of (13) satisfies both (12a) and the initial condition (12b); by going over to an integral equation formulation we include both the differential equation and the boundary conditions.

Similarly, consider the second order equation

$$\frac{d^2x}{ds^2} = f(s, x), \tag{14a}$$

$$x(0) = x_0, \quad \frac{dx(0)}{ds} = x_1. \tag{14b}$$

Integrating once, we have

$$x'(s) = x_1 + \int_0^s f(t, x(t)) dt,$$

whence

$$x(s) = x_0 + x_1 s + \int_0^s dt \int_0^t f(u, x(u)) du$$

$$= x_0 + x_1 s + \int_0^s f(t, x(t)) dt \int_t^s du$$

$$= x_0 + x_1 s + \int_0^s (s - t) f(t, x(t)) dt. \tag{15}$$

Again, the integral equation includes the boundary conditions: any solution of (15) satisfies both (14a) and (14b).

Two-point boundary value problems may be converted in the same way. For any solution of (14a) satisfies an equation of the form

$$x(s) = A + Bs + \int_0^s (s - t) f(t, x(t)) dt,$$

as can be checked by differentiating twice with respect to s.

Now if we require that $x(0) = \alpha$, $x(l) = \beta$, we have

$$\alpha = x(0) = A, \quad \beta = x(l) = A + Bl + \int_0^l (l - t) f(t, x(t)) dt.$$

Solving for A and B we find that x satisfies the integral equation

$$x(s) = \alpha + \frac{\beta - \alpha}{l} s + \int_0^s (s - t) f(t, x(t)) dt$$

$$- \frac{s}{l} \int_0^l (l - t) f(t, x(t)) dt.$$

This can be written in the form

$$x(s) = y(s) - \int_0^l K(s,t)f(t,x(t))dt, \tag{16}$$

where

$$y(s) = \alpha + \frac{\beta - \alpha}{l}s,$$

$$K(s,t) = t\frac{(l-s)}{l}, \quad 0 \leq t \leq s, \tag{16a}$$

$$= s\frac{(l-t)}{l}, \quad s \leq t \leq l.$$

The kernel K is then the 'Green's function' for the problem, in the notation of classical mechanics.

If f is linear in x the resulting integral equation is linear. As an example, we consider the equation

$$\frac{d^2x}{ds^2} + p(s)x(s) = g(s)$$

for which we identify $f(t,x(t)) = g(t) - p(t)x(t)$.
The equation (16) takes the form

$$x(s) = y(s) + \int_0^l K(s,t)p(t)x(t)dt, \tag{17}$$

where

$$y(s) = g(s) - \int_0^l K(s,t)g(t)dt.$$

This is a linear Fredholm equation of the second kind with kernel $K(s,t)p(t)$. A kernel $Q(s,t)$ is said to be *symmetric* if $Q(s,t) = Q(t,s)$. We shall see later that equations with real symmetric kernels have especially simple properties. Now $K(s,t)$ is symmetric, but $K(s,t)p(t)$ is not. However if $p(t) > 0$, $\forall t$, we may write $p(t) = [q(t)]^2$ and introduce

$$\bar{x}(s) = x(s)q(s); \quad \bar{w}(s) = w(s)q(s); \quad \bar{K}(s,t) = q(s)K(s,t)q(t).$$

Clearly $\bar{K}(s,t)$ is symmetric and we find that $\bar{x}(s)$ satisfies the equation

$$\bar{x}(s) = \bar{g}(s) + \int_0^l \bar{K}(s,t)\bar{x}(t)dt$$

which has a symmetric kernel. Such equations often arise in problems of mathematical physics.

It is the close connection between the integral and differential formulation of a given problem which led to the classification of integral equations displayed above. Most discussions of techniques for the numerical solution of integral equations relate to one or more of these standard classes of problems and this book is no exception. In particular, we shall pay considerable attention to the linear Fredholm and Volterra equations (1)–(6) and to nonlinear Volterra equations of the form (10), (11). This is because nonlinear Fredholm equations are much harder to analyse and to solve than their linear counterparts, while nonlinear Volterra equations are not significantly more difficult than the linear variety. This distinction is the same as that which obtains with ordinary differential equations: nonlinear differential equations with initial value boundary conditions are relatively easy to solve numerically, while nonlinear differential equations with multipoint boundary conditions (nonlinear boundary value problems) are very much more difficult to solve than linear boundary value problems.

Integral equations involving unknown functions $x(s_1, s_2, \ldots, s_m)$ in two or more space dimensions also occur regularly in practice but lie outside the scope of this book. However, even for one dimension, the 'standard' classification given above fails to reflect the variety of problems which can arise. For example, taking a random (but perhaps biased) selection of research papers in a variety of fields, we find the following integral equation problems forming the major computational task (in each case x is the unknown).

(i) *Population competition* (Downham and Shah, 1976)

$$\mathbf{x}(s) = \mathbf{y}(s) + \int_a^b \mathbf{K}(s - t, \mathbf{x}(s, t))\mathbf{x}(t)\mathrm{d}t, \quad \mathbf{x}^T = (x_1, x_2, x_3).$$

(ii) *Quantum scattering: close-coupled calculations* (Horn and Fraser, 1975)

$$x(s) = y(s) - \lambda \int_0^\infty K_1(s, z) \int_0^\infty K_2(z, t)x(t)\mathrm{d}t\,\mathrm{d}z.$$

(iii) Problem area as (ii), but a different system and an alternative formulation (Chan and Fraser, 1973):

$$\left(\frac{\mathrm{d}^2}{\mathrm{d}s^2} + k^2\right)\mathbf{x}(s) = \mathbf{y}(s) + \int_0^\infty K(s, t)\mathbf{x}(t)\mathrm{d}t,$$

$$\mathbf{x}^T = (x_1, x_2); \quad K = \begin{pmatrix} K_{11} & K_{12} \\ K_{21} & K_{22} \end{pmatrix}.$$

'Suitable' boundary conditions are supplied in addition.

(iv) *Currents in a superconducting strip (Rhoderick and Wilson, 1962)*

$$y(s) = \frac{1}{\pi} \int_0^1 \frac{t - s}{(t - s)^2 + a^2} x(t)\mathrm{d}t.$$

(v) *Flow round a hydrofoil* (*Kershaw*, 1974)

$$x(s) = y(s) + \lambda \int_C K(s, t)x(t)\mathrm{d}t$$

subject to $\int_C x(t)\mathrm{d}t = 0$, C a closed contour.

These examples illustrate a wide variety of equations; only (iv) can be regarded as being of a 'standard' type.

Additional difficulties are apparent when we look at the structure of the equations involved. Problems may be:

   (i) Linear or nonlinear

  (ii) Single or a set of coupled equations

 (iii) Defined on a finite range or an infinite range

 (iv) Well conditioned or illconditioned

  (v) Without or with side conditions

In addition, the kernel, driving term and solution may be

 (vi) Smooth or nonsmooth

in almost any combination; particularly tantalising are those equations which seem fairly common in practice, for which the kernel and driving term are both badly behaved but the solution is smooth.

This variety makes it difficult to provide standard programs for integral equations which will handle more than a small proportion of a potential user's problems. However, it is useful and sufficient to ignore such complications when describing numerical methods and their performance, since the basic methods extend in obvious ways to the nonstandard problems which arise. It is true that the error and convergence analyses may not extend so readily but they will (hopefully) still indicate the strengths and weaknesses of particular methods. Most of this book therefore deals with the standard class of equations. Sufficient of the underlying theory is given to yield a feeling for the nature of the solutions and for the inherent difficulties involved, but most of the emphasis is on numerical methods for their solution. We have made no attempt to give an exhaustive survey of available methods; rather, emphasis has been concentrated on methods selected according to the criteria (in approximate order of importance)

   (i) Ability to obtain rapid convergence

  (ii) Existence of reliable and cheaply computable error estimates

 (iii) Potential for extension to 'nonstandard' equations

 (iv) Personal predilection.

The ability to provide error estimates is an important facet of any numerical method; in real life, faced with a numerical solution to a new problem, such estimates provide the only guide to the likely accuracy of the solution. Not all methods are equally amenable to the provision of error estimates; for some they come naturally and with little extra work, while for other methods they are difficult or clumsy and hence costly to provide. Where this is so, we take the general point of view that the cost of the error estimate must be added to the cost of the method when judging its relative merits and that the reliability and effectiveness of the estimate is also one of the attributes of the method. Note that we specify *estimates* rather than *bounds* on the error. Given two error estimates of equal realism (that is, which are equally likely to be close to the actual error) and equal cost and given that one is also a bound, we would clearly prefer the bound. In practice bounds usually prove expensive and pessimistic. Most users (it is assumed in this book) will then usually prefer a cheaper and more realistic error estimate, recognising and accepting that estimates may sometimes err on the side of optimism.

Equally, the ease with which a given method, described for a 'standard' problem, can be extended together with its error estimate to a (hopefully) wide class of 'nonstandard' problems, is important in view of the common occurrence of such nonstandard problems.

The ability to obtain rapid convergence is also of obvious interest but it may not be obvious why it should lead the list of priorities. After all, the use of low-order quadrature rules (repeated Trapezoid or Simpson's rule) is often convenient in numerical integration and the relatively low-order fourth order Runge–Kutta rule remains popular for initial value ordinary differential equations (o.d.e.s). However this analogy takes no account of the structural differences between an initial value o.d.e. and an integral equation. If we consider algorithms which compute the numerical solution at N points, where N is some discretisation parameter, we find that the time taken by typical algorithms has the following N-dependence:

initial value o.d.e.s: $\sim AN$

Volterra or Fredholm integral equations: $\sim AN^p, p = 2$ or 3.

$$(18)$$

These estimates are illustrated in Figure 0.3.1, which shows the actual time taken to solve a single initial value o.d.e. using a fourth order Runge–Kutta program, together with the time required to solve a second kind Fredholm equation using a quadrature (Nystrom) method, as a function of N. It is clear that *integral equations are relatively expensive* to solve and there is therefore considerable advantage to be gained in using a high-order method and hence reducing the required value of N.[†]

[†] Viewed another way: a low-order method can be convenient to program and can produce satisfactory and steady convergence to the correct solution, when the error is tabulated against N, but the satisfaction will be considerably moderated by plotting error against computer time rather than N.

There is also obviously considerable advantage to be gained in reducing the exponent p in (18). This exponent can depend on the methods used to implement a given basic algorithm and we therefore pay considerable attention to the relevant numerical techniques.

Finally, personal interest has certainly played a part in determining the emphasis laid in this book on expansion methods, and on Galerkin methods in particular. These are the methods with which we have mainly worked; they also have a pretty theory, are very versatile, and perform rather well. Further, the alternative methods, and in particular Nystrom methods for Fredholm equations and marching methods for Volterra equations, are themselves emphasised in other books. However, we have tried to give sufficient detail that at least the methods discussed can be fairly compared.

Figure 0.3.1 Time taken to solve a single differential or integral equation at N points. Times are in ms on an ICL 1906S computer.
× : second order differential equations, fourth order Runge–Kutta routine.
. : second kind linear Fredholm equations, Nystrom (quadrature) method.

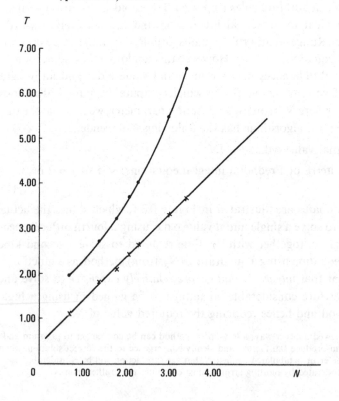

Exercises

1. Classify the following integral equations and show that they have the stated solution.

(a) $\int_0^1 e^{st}x(t)dt = \dfrac{e^{s+1}-1}{s+1}, \quad 0 \le s \le 1$

[solution $x(s) = e^s$].

(b) $x(s) = \cos s - \sin s + 2\int_0^s \sin(s-t)x(t)dt$

[solution $x(s) = e^{-s}$].

(c) $x(s) = -\cosh s + \lambda \int_{-1}^1 \cosh(s+t)x(t)dt, \quad -1 \le s \le 1$

$$\left[\text{solution } x(s) = \dfrac{\cosh s}{\left(\dfrac{\lambda}{2}\sinh 2 + \lambda - 1\right)} \right]$$

(d) $x(s) = e^{-s} + \int_0^s e^{-(s-t)}\{x(t) + \exp(-x(t))\}dt$

[solution $x(s) = \ln(s+e)$].

2. Form the integral equation corresponding to each of the following differential equations.

(a) $\dfrac{dx}{ds} + x = 0; \quad x(0) = 1.$

(b) $\dfrac{d^2x}{ds^2} + x = 0; \quad x(0) = 0, \quad x'(0) = 1.$

(c) $\dfrac{d^2x}{ds^2} + x = 0; \quad x(0) = x(1) = 1.$

(d) $\dfrac{d^3x}{ds^3} + \dfrac{d^2x}{ds^2} + \dfrac{dx}{ds} + x = 0; \quad x(0) = x'(0) = x''(0) = 1.$

1

The space $\mathscr{L}^2(a, b)$

1.1 $\mathscr{L}^2$ functions and kernels

In the convergence analyses which follow in later chapters, and also implicitly in the numerical methods themselves, we shall always make some regularity assumptions on the functions and kernels which appear. Often these assumptions are quite strong: the existence of p continuous partial derivatives in a square, for example. However, especially in those sections concerned with expansion methods, the convergence analysis is concerned with rates of convergence of orthogonal expansions. These rates are determined *implicitly* by the continuity properties of the functions which they represent; see Chapter 9. However, the *existence* and *convergence* of the expansions themselves require some rather minimal regularity assumptions on the functions. We consider throughout complex-valued (actually, almost always real-valued) functions $x(t)$ of a real variable t on an interval (a, b) (which occasionally may be infinite but will usually be finite), and such that in the Lebesgue sense

$$\int_a^b |x(t)|^2 \mathrm{d}t < \infty. \tag{1}$$

The set of all such functions is referred to as the function space $\mathscr{L}^2(a, b)$; we refer to x as an $\mathscr{L}^2$ function.

Two $\mathscr{L}^2$ functions x and y which are equal for 'almost all' values of t, that is, except for values of t being Lebesgue measure zero, are 'equivalent'. Thus, x and y are equivalent if

$$\int_a^b [x(t) - y(t)]^2 \mathrm{d}t = 0. \tag{1a}$$

In the space $\mathscr{L}^2(a, b)$ we are concerned only with 'convergence almost everywhere'; we take sets of measure zero to be of no significance and ignore the difference between 'equality' ($x(t) = y(t), \forall t$) and 'equivalence'. Thus we shall write

$$x(t) = y(t) \Leftrightarrow \int_a^b [x(t) - y(t)]^2 dt = 0$$

while a function $z(t)$ (a 'null function') which is zero almost everywhere will not be distinguished from the zero function:

$$z(t) = 0 \Leftrightarrow \int_a^b z^2(t) dt = 0. \tag{1b}$$

With this convention, the set of $\mathscr{L}^2$ functions forms a complete linear vector space. Furthermore, the space $\mathscr{L}^2$, with an appropriate norm and inner product, is an example of a Hilbert space, i.e. a complete (normed) inner product space, and we shall introduce shortly suitable norm and inner product functions which generate, in a natural way, a metric function on $\mathscr{L}^2$.

Similarly for the two-dimensional kernel function $K(s, t)$ we shall assume that the following condition is satisfied

(i) $\displaystyle\int_a^b \int_a^b |K(s, t)|^2 ds\, dt < \infty. \tag{2a}$

This condition implies that

(ii) For almost all s,

$$\int_a^b |K(s,t)|^2 dt < \infty \tag{2b}$$

(iii) For almost all t,

$$\int_a^b |K(s, t)|^2 ds < \infty. \tag{2c}$$

In the remainder of this chapter, we recall briefly (and very informally) those elementary results for the space $\mathscr{L}^2$, and linear operations on it, which we shall need. For a fuller introduction, see Kreyszig (1978).

1.2 Function norms; inner products

We define the *norm* $\|x\|_2$ (the L_2 norm) of an $\mathscr{L}^2$ function as

$$\|x\|_2 = \left\{ \int_a^b |x(t)|^2 dt \right\}^{\frac{1}{2}}. \tag{1}$$

Unless the context demands it for clarity, we shall omit the suffix henceforth. Note that $\|x\| = 0$ is the condition that x be a null function. It is straightforward to show that this definition yields a suitable norm for the space of $\mathscr{L}^2$ functions. That is, for any two $\mathscr{L}^2$ functions x, y:

    (i) $\| x \| \geq 0$

    (ii) $\| x \| = 0 \Leftrightarrow x = 0$ (2)

    (iii) $\| \lambda x \| = |\lambda| \, \| x \|$

    (iv) $\| x + y \| \leq \| x \| + \| y \|$ (Minkowski triangle inequality).

Proof. (i) and (iii) are obvious from (1); (ii) is a matter of definition (see (1.1.1*b*)). We give a brief proof of the triangle inequality as follows.

    It is sufficient to prove the result when x and y are real and non-negative. Consider

$$\| x + y \|^2 = \int_a^b \{x(t) + y(t)\}^2 dt$$

$$= \| x \|^2 + \| y \|^2 + 2 \int_a^b x(t)y(t)dt$$

$$\leq \| x \|^2 + \| y \|^2 + 2 \left| \int_a^b x(t)y(t)dt \right|.$$

Now it is easy to see that

$$\left| \int_a^b x(t)y(t)dt \right| \leq \| x \| \cdot \| y \| \tag{3}$$

since by (i) we must have

$$\int_a^b [\alpha x(t) + \beta y(t)]^2 dt = \alpha^2 \| x \|^2 + 2\alpha\beta \int_a^b x(t)y(t)dt$$

$$+ \beta^2 \| y \|^2 \geq 0$$

for all real α, β. In order that the quadratic form in (α, β) be non-negative we require

$$\left[\int_a^b x(t)y(t)dt \right]^2 \leq \| x \|^2 \cdot \| y \|^2.$$

Hence

$$\| x + y \|^2 \leq \| x \|^2 + \| y \|^2 + 2 \| x \| \cdot \| y \|$$

and the result follows.

In addition, we define the inner product (x, y) of two $\mathscr{L}^2$ functions as

$$(x, y) = \int_a^b x^*(t)y(t)dt, \tag{4}$$

where the asterisk denotes the complex conjugate.

    It is easy to see (using the properties of the integral) that the inner product is a scalar-valued function which satisfies the following conditions:

(i) $(x, y) = (y, x)^*$,

(ii) $(\alpha x_1 + \beta x_2, y) = \alpha^*(x_1, y) + \beta^*(x_2, y)$, $\hspace{3cm}$ (5)

(iii) $(x, x) \geq 0$,

(iv) $(x, x) = 0 \Leftrightarrow x = 0$.

We define x and y to be orthogonal if $(x, y) = 0$. It follows from (3) that

$$|(x, y)| \leq \|x\| \cdot \|y\| \quad \text{(Schwartz inequality)} \hspace{2cm} (6)$$

and that the norm and inner product functions are closely related by

$$\|x\| = (x, x)^{\frac{1}{2}}. \hspace{5cm} (7)$$

With the norm and inner product functions defined in this way the distance between two $\mathscr{L}^2$ functions x and y may be defined by

$$d(x, y) = \|x - y\| = (x - y, x - y)^{\frac{1}{2}}, \hspace{3cm} (8)$$

where d is a metric which has the following properties:

(i) $d(x, y) \geq 0$,

(ii) $d(x, y) = 0 \Leftrightarrow x = y$, $\hspace{5cm}$ (9)

(iii) $d(x, y) = d(y, x)$,

(iv) $d(x, z) \leq d(x, y) + d(y, z), \quad z \in \mathscr{L}^2$.

We can also define the norm of an $\mathscr{L}^2$ kernel function $K(s, t)$. We use the following theorem, valid for a Lebesgue integral:

Fubini theorem. If $\iint f(s, t)\mathrm{d}s\,\mathrm{d}t$ exists then $\int f(s, t)\mathrm{d}t$ exists for almost all s and

$$\iint f(s, t)\mathrm{d}s\,\mathrm{d}t = \int \mathrm{d}s \int f(s, t)\mathrm{d}t = \int \mathrm{d}t \int f(s, t)\mathrm{d}s.$$

Consequently the order of integration may be interchanged.

Now if $K(s, t)$ is an $\mathscr{L}^2$ kernel, and x an $\mathscr{L}^2$ function, it follows from the definition and (4) that (i) $y(s) = \int K(s, t)x(t)\mathrm{d}t$ is defined for almost all s in $[a, b]$; (ii) $|y(s)|^2 \leq \int |K(s, t)|^2 \mathrm{d}t \int |x(t)|^2 \mathrm{d}t$. Hence by the Fubini theorem, $y(s)$ is an $\mathscr{L}^2$ function, and

$$\int |y(s)|^2 \mathrm{d}s \leq \iint |K(s, t)|^2 \mathrm{d}s\,\mathrm{d}t \int |x(t)|^2 \mathrm{d}t.$$

Hence the operator $\int K(s, t)(\cdot)\mathrm{d}t$ maps an $\mathscr{L}^2$ function into an $\mathscr{L}^2$ function. It is convenient to define the *norm* $\|K\|$ of such a bounded operator K as follows:

$$\|K\| = \sup_{\substack{x \in \mathscr{L}^2 \\ x \neq 0}} \frac{\|Kx\|}{\|x\|}. \hspace{4cm} (10a)$$

It then follows trivially that for any $\mathscr{L}^2$ function x

$$\|Kx\| \le \|K\| \cdot \|x\|. \tag{10b}$$

Note that $y(s)$, defined above, satisfies

$$\|y\|^2 \le \left[\int_a^b \int_a^b |K(s,t)|^2 ds\, dt \right] \|x\|^2,$$

that is

$$\|Kx\| \le \left[\int_a^b \int_a^b |K(s,t)|^2 ds\, dt \right]^{\frac{1}{2}} \|x\|.$$

Thus, by virtue of (10b), $\|K\|$ may be bounded by

$$\|K\| \le \left[\int_a^b \int_a^b |K(s,t)|^2 ds\, dt \right]^{\frac{1}{2}}. \tag{10c}$$

Now define

$$\|K\|_E = \left[\int_a^b \int_a^b |K(s,t)|^2 ds\, dt \right]^{\frac{1}{2}}. \tag{10d}$$

We refer to $\|K\|_E$ as the Euclidean norm of K and it is easy to show that it has the properties of a norm and that it satisfies

$$\|Kx\| \le \|K\|_E \cdot \|x\|.$$

1.3   Sum and product of integral operators

Given two $\mathscr{L}^2$ linear integral operators H, K defined by the relations

$$(Hx)(s) = \int_a^b H(s,t)x(t)dt,$$

$$(Kx)(s) = \int_a^b K(s,t)x(t)dt,$$

we define the sum $H + K$ in an obvious manner as

$$((H+K)x)(s) = \int_a^b (H(s,t) + K(s,t))x(t)dt.$$

It is easy to show (do so) that $H + K$ is $\mathscr{L}^2$, and that

$$\|H + K\| \le \|H\| + \|K\|. \tag{1}$$

Similarly we may define the product operator $L = HK$ as follows:

$$(Lx)(s) = (HKx)(s)$$

$$= \int_a^b H(s,t)(Kx)(t)dt$$

$$= \int_a^b H(s,t) \int_a^b K(t,q)x(q)dq\, dt. \tag{2}$$

But now setting

$$y(t) = \int_a^b K(t,q)x(q)\mathrm{d}q$$

we recall that y is an $\mathscr{L}^2$ function for any $\mathscr{L}^2$ function x. Hence so is $Hy = HKx$; that is, the operator L is $\mathscr{L}^2$. Further, from definition (1.2.10a),

$$\|HK\| = \sup_{\substack{x\in\mathscr{L}^2 \\ x\neq 0}} \frac{\|HKx\|}{\|x\|}.$$

But for any x, from (1.2.10b)

$$\|HKx\| \leq \|H\|\cdot\|Kx\|$$
$$\leq \|H\|\cdot\|K\|\cdot\|x\|.$$

That is,

$$\sup_{\substack{x\in\mathscr{L}^2 \\ x\neq 0}} \frac{\|HKx\|}{\|x\|} \leq \|H\|\cdot\|K\|$$

or

$$\|L\| \leq \|H\|\cdot\|K\|. \tag{3}$$

1.4 The zero operator

Finally, if $\int_a^b K_1(s,t)x(t)\mathrm{d}t - \int_a^b K_2(s,t)x(t)\mathrm{d}t = 0$ for all x, the operator relation $K_1 = K_2$ holds (by definition) and we write

$$K_1 - K_2 = 0. \tag{1}$$

This implies that the kernels are equal except on a set of measure zero; that is,

$$K_1(s,t) = K_2(s,t) \quad \text{(almost everywhere)}.$$

For set $K = K_1 - K_2$. Then for all $\mathscr{L}^2$ functions $x(t)$,

$$\int_a^b K(s,t)x(t)\mathrm{d}t = 0 \quad \text{(almost everywhere)}. \tag{2}$$

For fixed s, take $x(t) = K(s,t)$.

Then (2) yields

$$\int_a^b |K(s,t)|^2\mathrm{d}t = 0 \quad \text{(almost everywhere)}$$

and hence $\int_a^b\int_a^b |K(s,t)|^2\mathrm{d}s\mathrm{d}t = 0$, i.e. $\|K\| = 0$ which in turn implies $K(s,t) = 0$ for almost all s,t.

1.5 Convergence of sequences of functions

In the sequel, we shall construct infinite sequences of functions, and need to discuss their convergence. We recall the two standard function space definitions of convergence.

The sequence of $\mathscr{L}^2$ functions $\{x_n\}$ converges to x

  (i) *strongly* iff $\lim\limits_{n\to\infty} \|x_n - x\| = 0$,

  (ii) *weakly* iff $\lim\limits_{n\to\infty} (x_n, y) = (x, y), \forall y \in \mathscr{L}^2$.

The following theorems, for which we omit proofs, summarise the main properties of these definitions (all limits may be taken either strongly or weakly).

Theorem 1.5.1. (*a*) The sequence $\{x_n\}$ has a (weak or strong) limit in $\mathscr{L}^2$ iff

$$\lim_{m,n\to\infty} (x_n - x_m) = 0. \tag{1}$$

We then say that $\{x_n\}$ is a *Cauchy sequence*; proving that a sequence is Cauchy is the most common way of showing that it has a limit.

  (*b*) If $\{x_n\}$ *converges* to x strongly, it also converges weakly to x. The converse is not true.

    We shall almost always be concerned with strong convergence in this book; with the L_2 norm, this is often referred to as convergence in the mean, so that a (strongly) convergent calculation may not converge at a particular point: a point of discontinuity, for example, or perhaps a badly behaved endpoint. Convergence in other norms (L_∞, L_1) may also be important numerically; see Section 1.8.

    'Weak convergence' is very weak by most computing standards. Consider any orthonormal sequence of functions $\{h_n\} \in \mathscr{L}^2$; then the following are standard results:

Theorem 1.5.2. (Bessel's inequality). For any $x \in \mathscr{L}^2$,

  (*a*) $\sum\limits_{i=0}^{\infty} |(h_n, x)|^2 \le (x, x)$ \tag{2}

  (*b*) As a corollary of (*a*), $\lim\limits_{n\to\infty} (h_n, x) = 0, \quad \forall x \in \mathscr{L}^2.$ \tag{3}

Equation (3) shows that every orthonormal sequence of functions in $\mathscr{L}^2$ converges weakly to zero. For example, on the interval $[0, \pi]$ we may write, correctly:

$$\lim_{n\to\infty} \sin nx = 0 \quad \text{(weakly)}. \tag{4}$$

Theorem 1.5.2 is closely related to the problem of expanding an arbitrary function in terms of an orthonormal set of functions; we meet such expansions frequently in later chapters and the following definitions and theorems summarise their simple convergence properties in $\mathscr{L}^2$:

***Definition* 1.5.1.** The orthonormal sequence $\{h_n\}$ is *complete* if the statement

$$(h_n, x) = 0, \quad \forall n$$

implies

$$x = 0.$$

***Theorem* 1.5.3.** Let $\{h_i\}$ be a complete orthonormal sequence, and for any $x \in \mathscr{L}^2$ define

$$x_n = \sum_{i=1}^{n} (h_i, x) h_i. \tag{5}$$

Then $\{x_n\}$ converges strongly to x (so that we may write without ambiguity

$$x = \sum_{i=1}^{\infty} b_i h_i, \tag{6}$$

where

$$b_i = (h_i, x)). \tag{7}$$

Proof. This theorem is important enough to warrant presenting a proof; note that in passing we prove Theorem 1.5.2. For any n we clearly have (see (5), and note that here x is *not* defined by its expansion but is assumed to be a given element of $\mathscr{L}^2$)

$$0 \leq (x - x_n, x - x_n) = (x, x) - (x_n, x) - (x, x_n) + (x_n, x_n)$$

$$= (x, x) - \sum_{i=1}^{n} \{b_i^*(h_i, x) + b_i(x, h_i)\}$$

$$+ \sum_{i=1}^{n} b_i^* b_i$$

$$= (x, x) - \sum_{i=1}^{n} b_i^* b_i \tag{8}$$

where we have used the orthonormality of the set $\{h_i\}$ and the relation $(h_i, x) = b_i$. Hence

$$\sum_{i=1}^{n} b_i^* b_i = \sum_{i=1}^{n} |b_i|^2 \leq (x, x) \tag{9}$$

and since the left side is a monotone increasing sequence, its limit exists:

$$\sum_{i=1}^{\infty} |b_i|^2 \leq (x, x) \quad \text{(Theorem 1.5.2)}.$$

But now, for any n, and for any $m > n$,

$$x_m - x_n = \sum_{i=n+1}^{m} b_i h_i.$$

Hence

$$\| x_m - x_n \|^2 = \sum_{i=n+1}^{m} |b_i|^2 \tag{10}$$

and therefore, since the series $\Sigma_{i=1}^{\infty} |b_i|^2$ is convergent,

$$\lim_{m,n \to \infty} \| x_m - x_n \|^2 = 0. \tag{11}$$

That is, the sequence $\{x_m\}$ is strongly Cauchy and by Theorem 1.5.1 has some limit y, say.

It remains to prove that $y = x$. But for each i

$$(h_i, y - x) = \left(h_i, \sum_{i=1}^{\infty} b_i h_i \right) - (h_i, x)$$

$$= b_i - b_i = 0. \tag{12}$$

Thus $y - x$ is orthogonal to every element of the sequence $\{h_i\}$. But if $\{h_i\}$ is complete, this implies that $y - x = 0$; that is, $y = x$.

In practice, we often do not know the limit x of a sequence of functions $\{x_n\}$; this sequence may for example represent a sequence of 'approximate solutions' of an equation which may not have an $\mathscr{L}^2$ solution x. However, this possibility is avoided by the following theorem:

Theorem 1.5.4. (Cauchy sequence). Let $\{x_n\}$ be a sequence of $\mathscr{L}^2$ functions and let

$$\lim_{m,n \to \infty} \| x_n - x_m \| = 0.$$

Then there exists an $\mathscr{L}^2$ function x such that

$$\lim_{n \to \infty} \| x_n - x \| = 0.$$

Proof. See Kreyszig (1978), p. 621. The theorem applies generally (and by definition) to complete normed spaces; thus, the proof for the space $\mathscr{L}^2$ involves showing that the space is complete, and this in turn hinges on ignoring the difference between equality and equivalence of two $\mathscr{L}^2$ functions.

1.6   Convergence of sequences of operators

We may also wish to discuss the convergence of sequences of operators. We consider only bounded operators, from $\mathscr{L}^2$ into $\mathscr{L}^2$; that is, operators K defined for all $\mathscr{L}^2$ functions and such that Kx is $\mathscr{L}^2$ whenever x is $\mathscr{L}^2$. Given a sequence $\{K_n\}$ of such operators, one natural extension of the definition of strong convergence for functions is given by that of:

Uniform (operator) convergence. The sequence $\{K_n\}$ converges uniformly to K iff

$$\lim_{n \to \infty} \| K_n - K \| = 0. \tag{1}$$

This definition is in practice rather stronger than required to prove the overall convergence of numerical methods based on sequences $\{K_n\}$. An alternative weaker definition is that of:

Strong (operator) convergence. The sequence $\{K_n\}$ converges strongly to K iff for every y in $\mathscr{L}^2$

$$\lim_{n \to \infty} \| (K_n - K)y \| = 0. \tag{2}$$

That is, $K_n y$ converges strongly to Ky for all $\mathscr{L}^2 y$.

A third definition, which we shall not in fact use, is that of:

Weak (operator) convergence. The sequence K_n converges weakly to K iff for all y in $\mathscr{L}^2$

$$K_n y \text{ converges weakly to } Ky.$$

In the context of expansion methods (see, for example, Chapter 7) we shall often need to manipulate series of the form

$$K \sum_{i=1}^{\infty} b_i h_i = \sum_{i=1}^{\infty} b_i K h_i.$$

Here, the equality between left and right sides may be difficult to establish without a detailed discussion of the operator K involved; we shall always assume that any necessary restrictions on K are satisfied in practice.

1.7   Inverse operators
An equation of the form

$$Tx = y \tag{1}$$

where y is a given $\mathscr{L}^2$ function and T an operator from $\mathscr{L}^2$ onto $\mathscr{L}^2$ may, or may not, have a solution; if it has a solution it may have more than one. If, for any $\mathscr{L}^2$ function y, (1) has an unique solution, then T is said to have an inverse T^{-1}, and this inverse operator is defined by the equation

$$x = T^{-1}y, \tag{2}$$

where for any given y, x is the solution of (1).

Conversely, if there exists an operator T^{-1} such that, for any element y of $\mathscr{L}^2$, $x = T^{-1}y$ satisfies (1), then T^{-1} is the inverse operator to T.

Note that if (1) and (2) are valid for all y, then operating on (2) by T we find

$$TT^{-1}y = Tx = y, \quad \text{for all } y.$$

That is,

$$TT^{-1} = I. \tag{3}$$

The technical requirement that T maps $\mathscr{L}^2$ onto $\mathscr{L}^2$ demands that, for any element x of $\mathscr{L}^2$, there is an element y such that x satisfies (2) for that y. If this is so, then from (1), operating on both sides by T^{-1}, for any element x in $\mathscr{L}^2$

$$T^{-1}Tx = T^{-1}y = x.$$

That is,

$$T^{-1}T = I. \tag{4}$$

Again, in later chapters we shall be concerned with the conditions required for the existence of inverse operators, and hence of solutions of equations of the form (1). These conditions lead to a very simple and beautiful theory which characterises the solution of second kind Fredholm equations with $\mathscr{L}^2$ kernels and driving terms: the 'Fredholm alternative'. We derive this theory, in an elementary manner, in later chapters.

1.8   Other norms

Although the L_2 norm $\|\cdot\|_2$ defined by equation (1.2.1) is very natural in the context of an $\mathscr{L}^2$ theory, other norms are also useful in describing the accuracy and convergence of a numerical method. The two most commonly used norms are the L_p and L_∞ norms:

$$\|x\|_p = \left[\int_a^b |x(t)|^p dt \right]^{1/p},$$

$$\|x\|_\infty = \lim_{p \to \infty} \|x\|_p = \sup_{a \le t \le b} |x(t)|.$$

The cases corresponding to $p = \infty$, $p = 2$ (equation (1.2.1)) and $p = 1$ are the most frequently used.

Exercises

1. Show (by using the triangle inequality) that

$$|\|x\| - \|y\|| \le \|x - y\|.$$

2. Show that the following definitions for the norm of an $\mathscr{L}^2$ operator K are equivalent to (1.2.10a):

$$\|K\| = \sup_{\substack{x\in\mathscr{L}^2 \\ \|x\|=1}} \|Kx\|,$$

$$\|K\| = \inf\{c>0: \quad \|Kx\| \le c\|x\|, \quad \forall x\in\mathscr{L}^2\}.$$

Show, in addition, that the norm defined by (1.2.10a) satisfies (1.2.2).

3. Show that the relation

$$(x, y) = \int_a^b x(t)y'(t)\mathrm{d}t,$$

where x, y are real-valued $\mathscr{L}^2$ functions, does not define an inner product. (*HINT.* Show that property (iv) of (1.2.5) is violated.)

4. Prove Theorem 1.5.1.

2

Numerical quadrature

2.1   Introduction

A numerical quadrature (numerical integration) rule is the basis of every numerical method for the solution of integral equations. We give in this chapter a brief résumé of those aspects of numerical quadrature in one dimension which are most important later on. For a much fuller treatment of the subject, see for example:

P.J. Davis and P. Rabinowitz: *Methods of Numerical Integration*, Acadamic Press (1975).

This gives a very nice introduction to the subject at a reasonable level.

A.H. Stroud and D. Secrest: *Gaussian Quadrature Formulas*, Prentice-Hall (1966)

gives extensive tables of Gauss points and weights for a variety of weight functions.

A.H. Stroud: *Approximate Calculation of Multiple Integrals*, Prentice-Hall (1971)

discusses quadrature in two or more dimensions, with details of many rules; not for the casual reader.

Quadrature rule is a generic name given to any numerical method for evaluating an approximation to an integral If of a function $f(s)$:

$$If = \int_a^b w(s)f(s)ds, \tag{1}$$

where $w(s)$ is a weight function. A method can in principle use any available information about the function $f(s)$: values of its derivative(s) at one or more points, or values of simpler integrals, for example. We consider only the case when the information used is restricted to the value of $f(s)$ at a set of points $\{\xi_i, i = 1, \ldots, N\}$, and the approximation (quadrature rule) Q has the form

$$Qf = \sum_{i=1}^{N} w_i f(\xi_i) = If - Ef, \tag{2}$$

where Ef is the error.

The points $\{\xi_i, i = 1, \ldots, N\}$ are called the *quadrature points* or *abscissae* (or nodes), and the w_i, $i = 1, \ldots, N$ the *quadrature weights*. They are to be chosen so that, for the function $f(s)$ in question, the approximation Qf is as accurate as possible in some sense; what sense depends on how much information we have available about the function, and also possibly on other uses to which the rule Q is to be put. For example, we may need to evaluate many integrals with the same points and weights. We can distinguish two distinct mechanisms for choosing points and weights:

(i) *Points and weights not chosen in advance.* Under this heading are included stochastic (Monte Carlo) rules, and 'automatic' quadrature rules in which the choice of the next few pairs (w_i, ξ_i), and the eventual value of N, is made on the basis of the behaviour of $f(s)$ as sampled previously.

Rules of this type turn out to be not very useful in solving integral equations, because of their expense.

(ii) *Points and weights chosen in advance to be suitable for a given class of function.* We shall be chiefly concerned with this type of quadrature rule. Given a class of functions R in which the integrand $f(s)$ is expected to lie, many criteria could be laid down to specify 'suitability' of the rule. The two most common are:

(a) *'Optimal' rules*: for given N, choose the parameters of the rule to minimise

$$\sup_{f \in R} |If - Qf|.$$

(b) *'Error annihilation' rules*: pick a subclass R_0 from R and arrange the parameters in the rule so that

$$|If - Qf| = 0, \quad f \in R_0.$$

R_0 is then referred to as the *annihilation class* of the rule. The parameters of the rule are the points $\xi_i, i = 1, \ldots, N$, and weights w_i, $i = 1, \ldots, N$, although we may wish to place restrictions on these; for example, that all the weights be equal, or, more commonly, that the points ξ_i be specified in advance. All commonly used rules are of type (ii$b$), although they are in general also 'optimal' (type (ii$a$)), for some (possibly obscure) class of functions; we restrict discussion entirely to this type of rule and in the next section we discuss how the construction can be achieved.

2.2   Method of undetermined coefficients

Suppose that we wish to choose the points ξ_i and weights w_i to make (2.1.2) exact (that is, $Ef = 0$) whenever f belongs to some subclass of functions. This cannot normally be achieved exactly. However, if there is a finite basis $\{h_i, i = 1, \ldots, P\}$ which spans the subclass, the problem may be phrased as follows: Let $f(s) = \Sigma_{i=1}^{P} \alpha_i h_i(s)$. Choose $\{\xi_j, w_j, j = 1, \ldots, N\}$ so that

$$Ef = 0, \quad \forall \alpha_j. \tag{1}$$

Since I is a linear operator, (1) is satisfied for all α_j if it is satisfied whenever $f(s) = h_k(s)$, $k = 1, 2, \ldots, P$. That is, (1) is equivalent to the P conditions

$$\sum_{j=1}^{N} w_j h_1(\xi_j) = m_1 = \int_a^b w(s) h_1(s) ds$$
$$\vdots$$
$$\sum_{j=1}^{N} w_j h_P(\xi_j) = m_P = \int_a^b w(s) h_P(s) ds. \tag{2}$$

Equations (2) are the 'undetermined coefficients' equations for the rule, and the $m_j, j = 1, \ldots, P$ are referred to as *generalised moments* of the weight function $w(s)$ with respect to the set $\{h_i\}$. If we take the points $\{\xi_i, i = 1, \ldots, N\}$ as given, (2) is a set of P linear equations in the N unknowns $w_1, \ldots, w_N$. If we set $P = N$, the system has an unique solution provided that the matrix $\{h_i(\xi_j)\}$ is nonsingular (and the moments $m_1, \ldots, m_N$ exist).

***Definition* 2.2.1.** (The Haar condition). Let $\{h_i(s), i = 1, \ldots, N\}$ be such that, given any N distinct points $\xi_j, j = 1, \ldots, N$, $\xi_L \leq \xi_j \leq \xi_U$, the $M \times M$ matrix $\{h_i(\xi_j), i, j = 1, \ldots, M\}$ is nonsingular for all $M \leq N$. Then the set $\{h_i(s)\}$ is said to satisfy the Haar condition on $[\xi_L, \xi_U]$.

***Theorem* 2.2.1.** Let $\{\xi_i\}_{i=1,\ldots,N}$ be a set of N distinct points with $a \leq \xi_i \leq b$, and let $\{h_i(s), i = 1, \ldots N\}$ satisfy the Haar condition on $[a, b]$. Then (2) with $P = N$ has an unique solution.

Proof. This is just stating the obvious; however, the Haar condition appears in contexts other than this, so that its introduction is not totally unjustified.

***Example* 2.2.1.** Let $h_i(s) = s^{i-1}$; then we seek an N-point rule of degree $N - 1$; that is, a rule exact for all polynomials of degree $N - 1$. This is the most common (but not the only) type of rule used in practice. This basis satisfies the Haar condition for all N. With $[a, b] = [0, h]$, $w(s) = 1$, $N = 2$,

$\xi_1 = 0$, $\xi_2 = h$ we obtain the undetermined coefficient equations

$$w_1 + w_2 = \int_0^h 1 \cdot ds = h,$$

$$0 + w_2 h = \int_0^h s \cdot ds = \frac{h^2}{2},$$

with solution $w_1 = w_2 = h/2$, yielding the usual Trapezoid rule:

$$\int_0^h f(s)ds \approx \frac{h}{2}[f(0) + f(h)].$$

More generally, taking $[a, b] = [a, a + (N - 1)h]$, $w(s) = 1$, $\xi_i = a + (i - 1)h$, $h_i(s) = s^{i-1}$, yields the (closed) Newton–Cotes rule of degree $N - 1$; in particular, when $N = 3$ we obtain Simpson's rule:

$$\int_a^{a+2h} f(s)ds \approx \frac{h}{3}[f(a) + 4f(a + h) + f(a + 2h)].$$

Now note that for any N and P the rule(s) satisfying (2) depend not on the basis $\{h_1, \ldots, h_N\}$ but on the space spanned by this basis. For let $\{\bar{h}_1, \ldots, \bar{h}_N\}$ have the same span. Then any function $f(s)$ in this span has the representation

$$f(s) = \sum_{j=1}^N \alpha_j h_j(s) = \sum_{j=1}^N \bar{\alpha}_j \bar{h}_j(s).$$

But if (2) is valid, then

$$\int_a^b w(s)f(s)ds = \sum_{i=1}^N w_i f(\xi_i).$$

That is,

$$\sum_{j=1}^N \bar{\alpha}_j \int_a^b w(s)\bar{h}_j(s)ds = \sum_{j=1}^N \bar{\alpha}_j \sum_{j=1}^N w_i \bar{h}_j(\xi_i),$$

whence it follows that

$$\int_a^b w(s)\bar{h}_j(s)ds = \sum_{i=1}^N w_i \bar{h}_j(\xi_i), \quad j = 1, 2, \ldots, P.$$

We could therefore construct the Newton–Cotes rules, for example, by considering not the monomials $\{s^{i-1}\}$ but the Legendre polynomials $\{P_{i-1}(s)\}$ or Chebyshev polynomials $\{T_{i-1}(s)\}$ as basis set, and it is sometimes convenient to do so.

2.3   Repeated rules

The Newton–Cotes rules Q_N of degree $N - 1$, with equally spaced points on a fixed interval $[a, b]$, yield a sequence of approximations $Q_N f$ to

If which, however, has generally unsatisfactory convergence properties: the error $|Q_N f - If|$ may not converge smoothly to zero, even for apparently quite well behaved functions f. In the most common case, the weight function $w(s) = 1$, and then the rule is translation invariant and scales linearly; that is, if we compare the two rules

$$Q_N(0, 1): \{\xi_i = (i - 1)/(N - 1); w_i\}$$

and

$$Q_N(a, a + (N - 1)h): \{\xi_i = a + (i - 1)h; w_i'\}$$

then

$$w_i' = h w_i.$$

We can therefore take $Q_N(0, 1)$ as the canonical rule. We obtain the M-panel repeated (Q_N) rule approximation to the integral $\int_a^b f(s)\,ds$ by setting

$$h = \frac{(b - a)}{M};$$

$$If = \int_a^b f(s)\,ds = \sum_{j=1}^{M} \int_{a+(j-1)h}^{a+jh} f(s)\,ds$$

$$= \sum_{j=1}^{M} Q_N(a + (j - 1)h, a + jh)f + E_{N,M}(f). \qquad (1)$$

For fixed N and increasing M these also yield a sequence of approximations to If.

Theorem 2.3.1. With notation as in (1), for every fixed N, and for every (a, b)-Riemann integrable function f:

$$\lim_{M \to \infty} E_{N,M}(f) = 0. \qquad (2)$$

Proof. The M-panel approximation may be viewed as a particular Riemann sum, whose maximum interval decreases linearly as M increases.

2.4   Gauss rules
2.4.1   General

If in (2.2.2) we take the points ξ_i to be variables also, we have $2N$ rather than N variables and we might hope to obtain a (unique?) solution on setting $P = 2N$. The question of the existence and uniqueness of the resulting rule is interesting but in general difficult. We consider only the case of a polynomial basis. In this case, we can show that an N-point rule of degree $2N - 1$ (that is, with $P = 2N$) always exists provided that $w(s) > 0$ (or $w(s) < 0$) for all $a \leq s \leq b$. For, let us take as a basis in (2.2.1) the set of polynomials orthonormal on (a, b) with weight function $w(s)$:

$$h_i(s) = q_{i-1}(s),$$

$$\int_a^b w(s)q_i(s)q_j(s)ds = \delta_{ij}, \quad i,j = 0,1,\ldots,N-1. \tag{1}$$

These polynomials always exist under the stated conditions. Further, the polynomial q_k of degree k has k real, distinct zeros in the interval $[a, b]$ (see e.g., Szego (1939), Section 3.3). We now choose points $\xi_i, i = 1,\ldots,N$ to be the zeros of $q_N(s)$:

$$q_N(\xi_i) = 0, \quad i = 1, 2,\ldots, N \tag{2}$$

and weights $w_j, j = 1,\ldots,N$ to satisfy the first N of equations (2.2.2): that is

$$\sum_{j=1}^N w_j q_k(\xi_j) = \int_a^b w(s)q_k(s)ds, \quad k = 0,1,\ldots,N-1. \tag{3}$$

For $k > 0$, the right side of (3) is zero from the orthogonality relation (1) and the observation that $q_0(x) = $ constant. But (3) holds for $k = N$ also, the left side then being zero since $q_N(\xi_i) = 0, i = 1,\ldots,N$ by construction. For $N < k < 2N$ we note moreover that since q_k is a polynomial of degree k we may write

$$q_k(s) = q_N(s)Q_k(s) + R_k(s),$$

where $Q_k(s), R_k(s)$ are polynomials of degree $< N$. Equation (3) then takes the form

$$\sum_{j=1}^N w_j[q_N(\xi_j)Q_k(\xi_j) + R_k(\xi_j)]$$

$$= \int_a^b w(s)[q_N(s)Q_k(s) + R_k(s)]ds. \tag{4}$$

Using the orthogonality relation (1) and the form (2) for the ξ_i, this reduces to

$$\sum_{j=1}^N w_j R_k(\xi_j) = \int_a^b w(s)R_k(s)ds, \quad k = N+1,\ldots,2N-1. \tag{5}$$

But $R_k(s)$ has an expansion of the form

$$R_k(s) = \sum_{j=0}^{N-1} \alpha_{kj}q_j(s). \tag{6}$$

Inserting (6) into (5) and changing the order of summation, we find that (5) is identically true by virtue of (3). Explicit (if not very useful) relations can be given for the weights w_i; for example, we note that the polynomials q_k satisfy a 3-term recurrence relation

$$q_{k+1}(s) = (\alpha_k s + \beta_k)q_k(s) + \gamma_{k-1}q_{k-1}(s), \tag{7}$$

where α_k is such that $q_{k+1}(s) - \alpha_k s q_k(s)$ is a polynomial of degree k and

$$\beta_k = -\alpha_k \int_a^b s q_k^2(s) w(s) ds,$$

$$\gamma_{k-1} = -\frac{\alpha_k}{\alpha_{k-1}}.$$

(See Exercise 2.) Then after some algebraic manipulation we obtain

$$\sum_{k=0}^N q_k(s) q_k(t) = \frac{q_{N+1}(s) q_N(t) - q_N(s) q_{N+1}(t)}{\alpha_N (s - t)}$$

(Christoffel–Darboux identity). (8)

Now using (7) and (8) we can derive the following expression for the weights w_i:

$$w_i = \frac{-\alpha_N}{q_{N+1}(\xi_i) q_N'(\xi_i)}, \quad i = 1, \ldots, N, \tag{9}$$

where ξ_i, $i = 1, \ldots, N$, are the zeros of $q_N(s)$. Further relations for the weights may be found in Stroud and Secrest (1966), p. 11.

2.4.2   Gauss–Legendre rule

The simplest and probably best-known Gauss rule is that with weight function $w(s) = 1$ on the interval $[-1, 1]$. Then the orthogonal polynomials q_i are the (normalised) Legendre polynomials:

$$q_i(s) = \left(\frac{2i + 1}{2} \right)^{\frac{1}{2}} P_i(s), \quad i = 0, 1, \ldots, N - 1 \tag{10}$$

and the resulting rule is referred to as the Gauss–Legendre rule. For small values of N the points and weights can be calculated algebraically, and the first few are listed in Table 2.4.1.

***Example* 2.4.1.** Let $h_i(s) = s^{i-1}$; we seek a 2-point rule (i.e. $N = 2$) of degree 3 with weight $w(s) = 1$ on the interval $[-1, 1]$. The $2N$ undetermined coefficient equations are:

$$\int_{-1}^1 h_i(s) ds = \sum_{j=1}^2 w_j h_i(\xi_j), \quad i = 1, \ldots, 4,$$

that is,

$$w_1 + w_2 = 2,$$
$$w_1 \xi_1 + w_2 \xi_2 = 0,$$
$$w_1 \xi_1^2 + w_2 \xi_2^2 = \tfrac{2}{3},$$
$$w_1 \xi_1^3 + w_2 \xi_2^3 = 0.$$

Table 2.4.1. *Points and weights for the*
Gauss–Legendre N-point rule

N	1		2		3	
i	w_i	ξ_i	w_i	ξ_i	w_i	ξ_i
1	2	0	1	$-1/\sqrt{3}$	5/9	$-\sqrt{(3/5)}$
2			1	$+1/\sqrt{3}$	8/9	0
3					5/9	$+\sqrt{(3/5)}$

This system has the unique solution

$$w_1 = 1 = w_2; \quad \xi_1 = -\frac{1}{\sqrt{3}}, \quad \xi_2 = \frac{1}{\sqrt{3}}$$

and the 2-point Gauss formula

$$\int_{-1}^{1} f(s)\,ds \approx f\left(-\frac{1}{\sqrt{3}}\right) + f\left(\frac{1}{\sqrt{3}}\right)$$

is exact for all polynomials of degree ≤ 3.

The table demonstrates two obvious properties of these rules:

(i) $\displaystyle\sum_{i=1}^{N} w_i = 2$

which follows because the rule integrates a constant function exactly.

(ii) The rules are symmetric with respect to the point $s = 0$; that is, if (w, ξ) is a member of the rule, so is $(w, -\xi)$.

For larger values of N it is necessary to compute the points and weights numerically. This computation is relatively expensive so that, in practice, tables of Gauss points and weights are stored when required. The need to store them is a (mild) drawback, less importantly because of the space required, and more importantly because of the possibility that, having stored a selection of rules up to a maximum N_{max}-point rule, we will find that we need a rule with $N > N_{max}$. To guard against such cases automatically requires that the program be prepared either to generate a higher degree rule if required, or that it be prepared to use a repeated rule if necessary.

2.4.3 Gauss–Chebyshev open and closed rules

There is one important case for which it transpires that very simple open and closed formulae can be given for the points and weights. We consider the choice

$$(a, b) = (-1, 1); \quad w(s) = (1 - s^2)^{-\frac{1}{2}}. \tag{11}$$

Then the polynomials $q_i(s)$ are the (normalised) Chebyshev polynomials

$$q_i(s) = \begin{cases} \left(\dfrac{2}{\pi}\right)^{\frac{1}{2}} T_i(s), & i = 1, 2, \ldots, N-1, \\ \left(\dfrac{1}{\pi}\right)^{\frac{1}{2}} T_0(s), & i = 0. \end{cases} \tag{12}$$

The nodes ξ_i for the N-point rule are therefore the zeros of $T_N(s) = \cos(N \cos^{-1} s)$, and these are the points

$$\xi_i = \cos\left(\frac{(2i-1)\pi}{2N}\right), \quad i = 1, 2, \ldots, N. \tag{13a}$$

In addition, it follows from (9) that the weights w_i are all equal:

$$w_i = \frac{\pi}{N}, \quad i = 1, 2, \ldots, N. \tag{13b}$$

The rule (13a, b) is of degree $2N - 1$ and referred to as the open Gauss–Chebyshev rule. It is clearly related to the midpoint rule for unit weight function

$$\int_a^b f(s)\,ds \approx h \sum_{i=1}^{N} f\left(a + (2i-1)\frac{h}{2}\right), \quad h = \frac{b-a}{N}. \tag{14}$$

For if we set $(a, b) = (0, \pi)$ and introduce the change of variable $s = \cos\theta$, (14) becomes

$$\int_{-1}^{1} \frac{f(\cos^{-1}s)}{(1-s^2)^{\frac{1}{2}}}\,ds \approx \frac{\pi}{N} \sum_{i=1}^{N} f\left(\frac{(2i-1)\pi}{2N}\right). \tag{14a}$$

That is, it reproduces the rule (13a, b) although this derivation signally fails to show why the rule is of Gauss type! Strictly speaking, the qualifier 'open' is not required, since all Gauss rules for positive weight functions are of open type. However, we can construct modified Gauss rules by imposing conditions on one or more nodes ξ_i and then adjusting the rest of the nodes and the weights, to maximise the degree of the resulting rule. If for the weight function and interval in (11) we insist that the end points $\xi_0 = 1$, $\xi_N = -1$ be nodes, we find that for any N we can construct an $(N + 1)$-point rule of degree $2N - 1$ (the closed Gauss–Chebyshev rule) with points

$$\xi_i = \cos(i\pi/N), \quad i = 0, 1, \ldots, N \tag{15a}$$

and weights

$$w_i = \pi/N, \quad i = 1, 2, \ldots, N-1, \tag{15b}$$

$$w_0 = w_N = \pi/2N,$$

yielding the rule

$$\int_{-1}^{1} \frac{f(s)ds}{(1-s^2)^{\frac{1}{2}}} \approx \frac{\pi}{N} \sum_{i=0}^{N} {}'' f\left(\cos\left(\frac{i\pi}{N}\right)\right), \tag{15c}$$

where the double prime on the summation sign implies that the first and last terms are halved. This closed rule can be obtained from the repeated Trapezoid rule using the mapping in (14); again, the relation fails to indicate why the rule is of high degree. In practice, there is little to choose between the performance of the open and closed rules; each have however other advantages:

    (i) It is sometimes inconvenient, or impossible, to evaluate the endpoint values $f(-1)$, $f(+1)$. The open rule avoids the use of these points.

    (ii) If we carry out a sequence of calculations, successively doubling N, the closed rule has the advantage that the points $\xi_i^{(N)}$ are a subset of the points $\xi_i^{(2N)}$, so that function evaluations can be re-used. This is not true of the open rule.

Example 2.4.2 We may find the exact value of the integral

$$\int_{-1}^{1} \frac{s^2 ds}{(1-s^2)^{\frac{1}{2}}}$$

by applying the open or closed Gauss–Chebyshev rules with $N = 2$.

    (i) *Open Gauss–Chebyshev.* For $N = 2$, $T_2(s) = 2s^2 - 1$ and the zeros of $T_2(s)$ are $\xi_1 = -(1/\sqrt{2})$, $\xi_2 = (1/\sqrt{2})$. The Gauss–Chebyshev formula for $N = 2$ is exact for polynomials of degree ≤ 3, and the integral is exactly

$$\frac{\pi}{2}\left[\left(-\frac{1}{\sqrt{2}}\right)^2 + \left(\frac{1}{\sqrt{2}}\right)^2\right] = \frac{\pi}{2}.$$

    (ii) *Closed Gauss–Chebyshev.* For $N = 2$ the quadrature points are

$$\xi_0 = \cos\frac{0\pi}{2}, \quad \xi_1 = \cos\frac{\pi}{2}, \quad \xi_2 = \cos\frac{2\pi}{2},$$

that is,

$$\xi_0 = 1, \quad \xi_1 = 0, \quad \xi_2 = -1.$$

The integral is thus given exactly by

$$\frac{\pi}{2}[\tfrac{1}{2}(1)^2 + (0)^2 + \tfrac{1}{2}(-1)^2] = \frac{\pi}{2}.$$

2.4.4   Clenshaw–Curtis quadrature

    Provided that the integrand $f(s)$ is reasonably smooth, rules of Gauss type generally give much greater accuracy than low-order repeated rules such as the Trapezoid or Simpson's rule; even if $f(s)$ is not smooth, they usually behave no worse than a low-order rule. These assertions are justified by the error analysis of Section 2.7 and illustrated by the examples in Section 2.8; they imply that the Gauss–Chebyshev rules are of special interest as an easy-to-use sequence of Gauss rule. However, in practice the weight function $w(s) = 1$ occurs much more commonly than the Chebyshev weight function $w(s) = (1 - s^2)^{-\frac{1}{2}}$. It is always possible to write

$$If = \int_{-1}^{1} f(s)ds = \int_{-1}^{1} \frac{\bar{f}(s)ds}{(1 - s^2)^{\frac{1}{2}}}, \tag{16}$$

where $\bar{f}(s) = (1 - s^2)^{\frac{1}{2}} f(s)$.

However, if $f(s)$ is smooth near $s = \pm 1$, $\bar{f}(s)$ is not, and the direct use of a Gauss–Chebyshev rule on (16) will yield results which converge only slowly as N increases. We can make use of the Gauss–Chebyshev rule while avoiding this slow convergence by the following device (Clenshaw and Curtis, 1960; Gentleman, 1972), which, although apparently rather round-about, leads to a very effective quadrature rule and has direct applications to integral equations which are discussed in Chapters 8 and 11.

Suppose that we know the Chebyshev polynomial expansion of $f(s)$:

$$f(s) = \sum_{i=0}^{\infty}{}' b_i T_i(s), \tag{17}$$

where the prime implies that the first term is halved and is conventionally introduced because of the form of equation (12). Then we have formally

$$\int_{-1}^{1} f(s)ds = \sum_{i=0}^{\infty}{}' b_i \int_{-1}^{1} T_i(s)ds$$

$$= \sum_{i=0}^{\infty}{}' b_i \int_{0}^{\pi} \cos i\theta \sin \theta \, d\theta \tag{18}$$

$$= \sum_{\substack{n=0 \\ n \text{ even}}}^{\infty}{}' \frac{2b_n}{1 - n^2}.$$

Now the coefficients b_i are given by (see equation (12)):

$$b_i = \frac{2}{\pi} \int_{-1}^{1} \frac{f(s)T_i(s)}{(1 - s^2)^{\frac{1}{2}}} ds. \tag{19}$$

If we evaluate (19) numerically with the closed Gauss–Chebyshev rule we obtain the approximation

$$b_i \approx a_i^{(N)} = \frac{2}{N} \sum_{k=0}^{N}{}'' f\left(\cos \frac{k\pi}{N}\right) \cos \frac{ik\pi}{N}, \quad i = 0, 1, \ldots, N, \tag{20}$$

where we have truncated the series (17) at $i = N$; the error analysis of the next section indicates that, for $i > N$, (20) yields no significant digits of accuracy. The Clenshaw–Curtis scheme then has the following stages:

(i) Compute $a_i^{(N)}$, $i = 0, \ldots, N$ from (20) (in fact only those coefficients with even suffixes are required).

(ii) Compute $\int_{-1}^{1} f(s)ds = \sum_{\substack{n=0 \\ n \text{ even}}}^{\prime\prime N} 2a_n^{(N)}/(1 - n^2)$. $\qquad$ (21)

In (21) the last term has been halved so that the final result is identical to that given by the original derivation (Clenshaw and Curtis (1960)); the term $a_N^{(N)}$ will normally be very small, and the resulting formula can then be shown to be exact if $f(s)$ is a polynomial of degree $2N - 1$. If we combine the two stages we can present the scheme as a 'standard' integration rule with rather complicated weights:

$$\int_{-1}^{1} f(s)ds = \sum_{\substack{n=0 \\ n \text{ even}}}^{\prime\prime N} \frac{4}{(1 - n^2)N} \sum_{k=0}^{\prime\prime N} \cos\left(\frac{nk\pi}{N}\right) f\left(\cos\frac{k\pi}{N}\right)$$

$$= \sum_{k=0}^{\prime\prime N} w_k f\left(\cos\frac{k\pi}{N}\right), \qquad (22a)$$

where

$$w_k = \frac{4}{N} \sum_{\substack{n=0 \\ n \text{ even}}}^{\prime\prime N} \frac{1}{1 - n^2} \cos\left(n\frac{k\pi}{N}\right). \qquad (22b)$$

The apparent cost of implementing this rule is high; a direct summation of (22b) to compute w_k or of (20) to compute $a_k^{(N)}$, $k = 0, 1, \ldots, N$, involves a total of $(N + 1)^2$ multiplications and additions, compared with only $N + 1$ to actually evaluate the sum (22a) or (21). However (20) has the form of a discrete cosine transformation, and as a result the coefficients $a_i^{(N)}$ can be computed using Fast Fourier Transform (FFT) techniques in $\mathcal{O}(N \ln N)$ operations. The resulting rule is then reasonable in cost and also very stable against rounding errors; for a discussion of the FFT techniques and their implementation in this context, see Gentleman (1972a).

Equation (22b) for the weights w_k can also be viewed as the discrete cosine transformation of the vector $\mathbf{v}$ with entries

$$v_n = \begin{cases} \dfrac{2}{(1 - n^2)}, & n \text{ even} \\ 0, & n \text{ odd}. \end{cases} \qquad (23)$$

The weights w_k can therefore also be computed directly in $\mathcal{O}(N \ln N)$ operations; this will be the faster formulation if a number of (unrelated) functions are to be integrated using the same value of N. (For a situation when many but related integrals are involved, see Chapter 8.)

2.5   Mapped rules

2.5.1   Mapping the independent variable

Most quadrature rules are not used directly in the form in which they are derived but undergo some (usually simple) transformations to put them in a form which is more generally useful. Consider the integral

$$If = \int_a^b w(s)f(s)ds \tag{1}$$

and introduce the new variable $z = z(s)$ where the mapping function $z(s)$ is one to one on the interval $[a, b]$ and hence has an inverse

$$S(z):S(z(s)) = 1, \, a \le s \le b. \tag{2}$$

Then

$$If = \int_c^d \bar{w}(z)F(z)dz,$$

where

$$F(z) = f(s(z)),$$

$$\bar{w}(z) = w(s(z))\frac{ds(z)}{dz}, \tag{3}$$

$$c = z(a); \, d = z(b).$$

Now suppose that we have a rule $Q(w_i, \xi_i, i = 1, \ldots, N)$ for the interval $[c, d]$ with weight function $\bar{\bar{w}}(z)$:

$$\int_c^d \bar{\bar{w}}(z)g(z)dz \approx \sum_{i=1}^N w_i g(\xi_i). \tag{4}$$

Then we may approximate If as follows:

$$If = \int_c^d \bar{\bar{w}}(z)\left(\frac{\bar{w}(z)}{\bar{\bar{w}}(z)}\right)F(z)dz$$

$$\approx \sum_{i=1}^N \left(w_i \frac{\bar{w}(\xi_i)}{\bar{\bar{w}}(\xi_i)}\right)f(s(\xi_i))$$

$$= \sum_{i=1}^N w_{mi}f(\xi_{mi}), \tag{5a}$$

where

$$w_{mi} = w_i\frac{\bar{w}(\xi_i)}{\bar{\bar{w}}(\xi_i)}; \quad \xi_{mi} = s(\xi_i). \tag{5b}$$

Equation (5a) defines a rule $Q_m(\mathbf{w}_m\mathbf{\xi}_m)$ for the interval $[a, b]$ with weight function $w(s)$. We refer to it as a mapped version of the rule Q, and the points $\mathbf{\xi}_m$ and weights $\mathbf{w}_m$ as the mapped points and weights. The properties of the mapped rule depend on those of Q and on the mapping $z \leftrightarrow s$; they can best be summed up in practice by checking the annihilation class of Q_m, that is, the set of functions for which (5b) is exact.

If the annihilation class of Q is $\{h_1(z), \ldots, h_p(z)\}$, (5) is exact whenever

$$\frac{\bar{w}(z)}{\bar{\bar{w}}(z)} F(z) \in \text{span} \{h_1(z), \ldots, h_p(z)\}.$$

That is,

$$f(s(z)) = F(z) \in \text{span} \left\{ \frac{\bar{\bar{w}}(z)}{\bar{w}(z)} h_1(z), \ldots, \frac{\bar{\bar{w}}(z)}{\bar{w}(z)} h_p \right\}$$

and hence the annihilation class of Q_M is the set of functions $\{h_{Mi}(s), i = 1, \ldots, p\}$ where

$$h_{Mi}(s) = \frac{\bar{\bar{w}}(z(s))}{\bar{w}(z(s))} h_i(z(s)). \tag{6}$$

The mapping should therefore be chosen so that this set bears some relation to the function $f(s)$ to be integrated.

2.5.2 Linear maps

The simplest case, and one which is so common that it is usually carried out without comment, occurs when $w(s) \equiv 1$ and the finite interval $[a, b]$ is mapped onto some standard finite interval $[c, d]$ which we take to be $[-1, 1]$. For this case we may take the linear transformation

$$s = \tfrac{1}{2}[(b - a)z + (a + b)] \tag{7a}$$

with inverse

$$z = \frac{1}{(b - a)}[2s - (a + b)]. \tag{7b}$$

Given $w(s) = 1$, we find

$$\bar{w}(z) = \frac{b - a}{2} \tag{7c}$$

and given a rule (4) defined on $[-1, 1]$ with $\bar{\bar{w}}(z) = 1$, the mapped rule has points and weights

$$\xi_{Mi} = \tfrac{1}{2}[(b - a)\xi_i + (a + b)]; \quad w_{Mi} = \frac{b - a}{2} w_i \tag{7d}$$

and annihilation class

$$\left\{ \frac{2}{b - a} h_i(z(s)) \right\}. \tag{7e}$$

In the most common case the annihilation class $\{h_i\}$ spans the space of polynomials of degree $\leq P$, the degree of the rule; the mapped class then also spans this same space.

2.5.3   Nonlinear maps

Suppose, for example, that we want to integrate over $[0,1]$ a function $f(s)$ which tends to zero as $s^{\frac{1}{2}}$ at the origin. Consider the nonlinear transformation

$$z(s) = s^{\frac{1}{2}}.$$

Then

$$\int_0^1 f(s)\mathrm{d}s = 2\int_0^1 F(z)z\,\mathrm{d}z.$$

Now choose a rule with weight function 1 and annihilation class

$$\{z^i\} = \{s^{i/2}\}.$$

The rule is exact if

$$F(z)z \in \mathrm{span}\,\{s^{i/2}; i = 0, 1, \ldots, P\},$$

that is, if

$$F(z) \in \mathrm{span}\,\{s^{(i-1)/2}; i = 0, 1, \ldots, P\}.$$

Notice that the rule will in fact integrate exactly a function with $s^{-\frac{1}{2}}$ singularity.

2.6   Infinite range integrals
2.6.1   Interval truncation

Infinite range integrals of the general form

$$I_1 = \int_a^\infty w(s)f(s)\mathrm{d}s \tag{1}$$

$$I_2 = \int_{-\infty}^\infty w(s)f(s)\mathrm{d}s \tag{2}$$

are relatively common and require special consideration. Usually the product $w(s)f(s)$ vanishes sufficiently rapidly near the 'infinite' end(s) of the range that the integral exists in the Riemann sense. We assume this is true; then

$$I_1 = \lim_{R\to\infty} \int_a^R w(s)f(s)\mathrm{d}s = \lim_{R\to\infty} I_1(R), \tag{3}$$

$$I_2 = \lim_{R\to\infty} \int_{-R}^R w(s)f(s)\mathrm{d}s = \lim_{R\to\infty} I_2(R), \tag{4}$$

and it is tempting to avoid consideration of the infinite range problem by truncating the range at some point R, for example

$$I_1 \approx I_1(R), \quad R \text{ finite and 'large'}.$$

Whether this is economic depends on the accuracy sought and in particular on the rate of decay of $w(s)f(s)$ for large $|s|$. If this decay is slow it is necessary to choose R large; then it will prove necessary to use many points to evaluate the finite range integral accurately. We illustrate with two examples:

Example **2.6.1.** Evaluate

$$I_1 = \int_0^\infty e^{-s}\sin s\, ds = \tfrac{1}{2}, \tag{5a}$$

$$I_2 = \int_0^\infty \frac{ds}{s^2+100} = \frac{\pi}{20}. \tag{5b}$$

In each case we truncate the upper limit at a point R. For fixed R we evaluate the integrals to five significant figures using an N-point Gauss rule; Table 2.6.1 shows the value of N necessary to achieve this. Because of the rapid convergence (for fixed R) as N increases the actual accuracy for this minimum N is usually much greater than this.

We see from this table that this crude technique is reasonably satisfactory for I_1: a modest truncation range and modest value of N lead to satisfactory results. But, for the long-tailed integrand of I_2, the situation is not so satisfactory. As the upper limit R is increased the number of quadrature points required to maintain accuracy rises (only slowly in this example because the integrand is smooth); worse, convergence to the infinite range result is very slow. Clearly better techniques should be sought in general.

2.6.2 Infinite range Gauss rules

An alternative to the brute force truncation of the range is the use of a rule developed directly for the infinite range. The most commonly used of these are the *Gauss–Laguerre* and *Gauss–Hermite* rules which we present below.

The Gauss–Laguerre rule is that with weight function $w(s) = e^{-s}$ on the interval $[0, \infty]$; the polynomials $q_i(s)$ are the (normalised) Laguerre polynomials:

$$q_i(s) = \frac{1}{i!}L_i(s), \quad i = 0,1,\dots,N-1,$$

with

$$L_i(s) = e^s \frac{d^i}{ds^i}(e^{-s}s^i).$$

Table 2.6.1. *Numerical evaluation of $I_1(R)$, $I_2(R)$ for the integrals in Example 2.6.1 with truncated range. The values of N are the number of points needed to evaluate the finite range integral to five significant figures, using an N-point Gauss–Legendre rule*

I_1			I_2		
R	N	$I_1(R)$	R	N	$I_2(R)$
∞	Exact	0.5	∞	Exact	0.157 079 632
5.0	8	0.502 275	5.0	4	0.046 365
10.0	8	0.500 314	10.0	4	0.078 540
15.0	16	0.500 000	15.0	4	0.098 279
			20.0	8	0.110 715
			25.0	8	0.119 029
			50.0	16	0.137 340
			100.0	16	0.147 113
			200.0	32	0.152 084

Thus the formula

$$\int_0^\infty e^{-s} f(s)\,ds \simeq \sum_{i=1}^N w_i f(\xi_i), \qquad (6)$$

where ξ_i are the roots of the Laguerre polynomial $L_N(s)$ with weights given by

$$w_i = \frac{(N!)^2}{\xi_i [L'_N(\xi_i)]^2}, \quad i = 1,\dots,N \qquad (7)$$

is exact for all polynomials of degree $\leq 2N - 1$.

***Example* 2.6.2.** We evaluate the integral

$$\int_0^\infty e^{-s} s^3 \, ds$$

using the Gauss–Laguerre rule with $N = 2$, which is exact for all polynomials of degree ≤ 3. Now for $N = 2$, $L_2(s) = s^2 - 4s + 2$ and the zeros of $L_2(s)$ are $\xi_1 = 2 + \sqrt{2}$, $\xi_2 = 2 - \sqrt{2}$. The weights are given by

$$w_i = \frac{(2!)^2}{\xi_i [L_2'(\xi_i)]^2}, \quad i = 1,2$$

and we find $w_1 = \frac{1}{4}(2 - \sqrt{2})$, $w_2 = \frac{1}{4}(2 + \sqrt{2})$. Thus the Gauss–Laguerre rule yields the value

$$\tfrac{1}{4}(2 - \sqrt{2})(2 + \sqrt{2})^3 + \tfrac{1}{4}(2 + \sqrt{2})(2 - \sqrt{2})^3 = 6$$

which is the exact value of the integral.

More generally, we can evaluate integrals of the form

$$\int_a^\infty e^{-\alpha s} f(s) ds, \quad -\infty < a < \infty; \quad \alpha > 0, \tag{8a}$$

using the so-called *shifted Gauss–Laguerre* rule:

$$\int_a^\infty e^{-\alpha s} f(s) ds \approx \frac{e^{-\alpha a}}{\alpha} \sum_{i=1}^N w_i f\left(\frac{\xi_i}{\alpha} + a\right), \tag{8b}$$

where ξ_i are the roots of $L_N(s)$ and the weights are given by equation (7). Note that (8b) is obtained from (8a) on simply changing the variable to $z = \alpha(s - a)$.

Example 2.6.3. We use the shifted Gauss–Laguerre rule with $N = 2$ to evaluate

$$\int_1^\infty e^{-s} s^3 ds$$

whose exact value is $16/e$. Using the results of the previous example we find that the rule yields the value

$$\frac{1}{e}\left[\tfrac{1}{4}(2 - \sqrt{2})(3 + \sqrt{2})^3 + \tfrac{1}{4}(2 + \sqrt{2})(3 - \sqrt{2})^3\right] = \frac{16}{e}.$$

Consider now the Gauss–Hermite rule with weight function $w(s) = e^{-s^2}$ on the interval $[-\infty, \infty]$; the polynomials $q_i(s)$ are the (normalised) Hermite polynomials:

$$q_i(s) = \frac{1}{\pi^{\frac{1}{4}} 2^{i/2} (i!)^{\frac{1}{2}}} H_i(s), \quad i = 0, 1, \ldots, N - 1,$$

with

$$H_i(s) = (-1)^i e^{s^2} \frac{d^i e^{-s^2}}{ds^i}.$$

Thus the formula

$$\int_{-\infty}^\infty e^{-s^2} f(s) ds \approx \sum_{i=1}^N w_i f(\xi_i), \tag{9}$$

where ξ_i are the roots of the Hermite polynomial $H_N(s)$ with weights given by

$$w_i = \frac{2^{N+1} N! \sqrt{\pi}}{[H_N'(\xi_i)]^2}, \quad i = 1, \ldots, N, \tag{10}$$

is exact for all polynomials of degree $\leq 2N - 1$.

Example 2.6.4. We evaluate the integral

$$\int_{-\infty}^{\infty} e^{-s^2} s^4 ds$$

whose exact value is $\frac{3}{4}\sqrt{\pi}$, using the Gauss–Hermite rule with $N = 3$. Now $H_3(s) = 4s(2s^2 - 3)$ and its zeros are $\xi_1 = 0$, $\xi_2 = -\sqrt{\frac{3}{2}}$, $\xi_3 = \sqrt{\frac{3}{2}}$. The weights are given by

$$w_i = \frac{2^4 3! \sqrt{\pi}}{[H_3'(\xi_i)]^2}, \quad i = 1, 2, 3,$$

and we find $w_1 = \frac{2}{3}\sqrt{\pi}$, $w_2 = \sqrt{\pi}/6 = w_3$. Thus the Gauss–Hermite rule yields the value

$$\frac{2\sqrt{\pi}}{3} \cdot 0^4 + \frac{\sqrt{\pi}}{6} \cdot \left(-\sqrt{\frac{3}{2}}\right)^4 + \frac{\sqrt{\pi}}{6}\left(\sqrt{\frac{3}{2}}\right)^4 = \frac{3\sqrt{\pi}}{4}.$$

We expect the rule to be exact here, since for $N = 3$ the rule is in fact constructed to be exact for all polynomials of degree ≤ 5.

2.6.3   Mapped finite range rules

Both the Laguerre and Hermite rules have exponentially decaying weight functions. We might hope therefore that they are well suited to integrals such as (5a) which have this feature, and less well suited to integrals of the form (5b) with long-tailed integrands. We can produce other infinite range rules by suitably mapping finite range rules and we give here one example of such a mapping which makes use of the Gauss–Legendre rule. The approach can be described in two ways: either we can map the integral to be calculated onto $[-1, 1]$ or we can map a rule defined on $[-1, 1]$ onto the required integration interval. We take the latter viewpoint because we believe it yields a more direct insight into the types of integrand for which the particular mapping used is appropriate. An example of such a mapped rule is given by (15) below.

Consider the integral

$$I = \int_a^{\infty} f(s) ds. \tag{11}$$

Making the change of variable

$$s = \frac{2(a + \alpha)}{(s' + 1)} - \alpha,$$

where α is arbitrary subject to the constraint $a + \alpha > 0$, we find

$$I = 2(a + \alpha) \int_{-1}^{1} \frac{F(s')}{(s' + 1)^2} ds', \tag{12}$$

where

$$F(s') = f(s).$$

Now suppose that $\{\xi_i, w_i, i = 1, \ldots, N\}$ are the points and weights of an N-point Gauss–Legendre rule. The approximation

$$I \approx I_N = 2(a + \alpha) \sum_{i=1}^{N} w_i \frac{F(\xi_i)}{(\xi_i + 1)^2} \tag{13}$$

will in fact be an identity if $F(s')/(s' + 1)^2$ is a polynomial of degree $\leq 2N - 1$ in s'; that is, for the class of functions

$$F(s') = (s' + 1)^k, \quad k = 2, 3, \ldots, 2N + 1. \tag{14}$$

Equation (14) defines the annihilation class for a rule of order $2N - 1$ of the form

$$\int_a^\infty f(s) \mathrm{d}s = \sum_{i=1}^{N} w_i' f(\xi_i'), \tag{15}$$

with

$$\xi_i' = \frac{2(a + \alpha)}{(\xi_i + 1)} - \alpha,$$

$$w_i' = \frac{2(a + \alpha)w_i}{(\xi_i + 1)^2},$$

where w_i, ξ_i are the Gauss–Legendre weights and points (see (13)). Because of the form of the annihilation class (14) this is sometimes called the 'Gauss–rational' rule. We might expect that the rule is better suited to integrands, such as that which appears in (5*b*), which have 'long tails', than is the Gauss–Legendre rule, and that the latter would be better suited to integrals such as (5*a*). We might also guess that the value chosen for the parameter α in equations (8), (15) will affect the results obtained. Both of these guesses prove correct. Table 2.6.2 shows the results obtained for the 'long-tailed' integrand of (5*b*) using a Gauss–Laguerre rule with α varying.

It would seem from this table that the best results are obtained with $\alpha = 0.05$ when an accuracy of about 2×10^{-4} is achieved with $N = 48$. Table 2.6.3 compares this with the results obtained for the Gauss–rational rule (equation (15)) with $\alpha = 10$ and gives similar results for the short-tailed integrand of (5*a*). As expected, the Gauss–Laguerre rule is better suited to (5*a*) and the Gauss–rational rule is better suited to (5*b*). In the case of equation (5*b*) both the Gauss–Laguerre and Gauss–rational rules prove much cheaper than the crude 'truncated interval' approach illustrated in Table 2.6.1.

Table 2.6.2. *Results obtained for integral (5b), using the N-point Gauss–Laguerre rule, defined by equation (8), with various values for the parameter* α. *The exact value is* $I = 0.157\,079\,6$.

N \ α	1.0	0.5	0.2	0.1	0.05
2	0.055\,143\,6	0.090\,962\,4	0.127\,465\,3	0.149\,325\,7	0.147\,944\,5
4	0.092\,764\,9	0.121\,565\,1	0.142\,308\,8	0.150\,119\,0	0.161\,040\,0
8	0.122\,889\,3	0.139\,492\,7	0.149\,990\,6	0.153\,376\,0	0.156\,128\,6
16	0.140\,124\,0	0.148\,541\,6	0.153\,657\,6	0.155\,373\,8	0.156\,047\,1
32	0.148\,784\,5	0.152\,925\,0	0.155\,417\,0	0.156\,248\,3	0.156\,670\,3
48	0.151\,614\,3	0.154\,344\,9	0.155\,985\,5	0.156\,532\,6	0.156\,805\,6

Table 2.6.3. *Evaluation of the integrals (5a), (5b) using the N-point Gauss–Laguerre and Gauss–rational rules, with* $a = 0$ *and values of* α *as shown (see equations (8), (15))*

Integral	(5a)		(5b)	
N	Laguerre, $\alpha = 1$	Rational, $\alpha = 10$	Laguerre, $\alpha = 0.2$	Rational, $\alpha = 10$
∞	0.5	0.5	0.157\,079\,6	0.157\,079\,6
2	0.432\,459\,5	0.245\,840\,8	0.127\,465\,3	0.150\,000\,0
4	0.504\,879\,3	0.594\,976\,8	0.142\,308\,8	0.156\,862\,7
8	0.499\,998\,7	0.495\,531\,6	0.149\,990\,6	0.157\,079\,4
16	0.500\,000\,0	0.499\,987\,6	0.153\,657\,6	0.157\,079\,6
32		0.500\,000\,0	0.155\,417\,0	
48			0.155\,985\,5	

2.7 Error estimates for numerical quadrature
2.7.1 An error expansion

We now turn to the estimation of quadrature errors for a given rule Q. If Q has been constructed to have a given annihilation class $\{h_i, i = 1, \ldots, P\}$, we may proceed systematically as follows. Suppose, as is usually the case, that $h_1, \ldots, h_P$ are members of an infinite sequence $\{h_i\}$, and that f has the expansion (convergent in the mean)

$$f(s) = \sum_{i=1}^{\infty} b_i h_i(s). \tag{1}$$

Then

$$Ef = (I - Q)f = \sum_{i=1}^{\infty} b_i Eh_i$$

$$= \sum_{i=P+1}^{\infty} b_i Eh_i, \tag{2}$$

the last result following since by construction $Eh_i = 0$, $i = 1,\ldots,P$. We therefore have the bound

$$|Ef| \leq \sum_{i=P+1}^{\infty} |b_i||Eh_i|. \tag{3}$$

In principle at least the terms Eh_i can be evaluated (or estimated) for the given annihilation class; they are independent of f.

We expect that, for large i, the exact integral Ih_i bears little relation to the approximate integral Qh_i, and in that case we might as well use the crude bound

$$|Eh_i| \leq |Ih_i| + |Qh_i|$$
$$= \left|\int_a^b w(s)h_i(s)\mathrm{d}s\right| + \left|\sum_{k=1}^N w_k^{(N)}h_i(\xi_k^{(N)})\right|.$$

Now suppose that

$$|h_i(s)| \leq K_i, \quad a \leq s \leq b.$$

Then

$$|Eh_i| \leq K_i\left[\int_a^b |w(s)|\mathrm{d}s + \sum_{k=1}^N |w_k^{(N)}|\right]. \tag{4}$$

We suppose in addition that $w(s) > 0$, and that the weights are positive:

$$w_k^{(N)} > 0, \quad k = 1,\ldots,N.$$

The second assumption is valid, for example, for a Gauss rule with positive weight function $w(s)$. We also assume that the function 1 is in the annihilation class of the rule; that is, that a constant function is integrated exactly. Then

$$\sum_{k=1}^N |w_k^{(N)}| = \sum_{k=1}^N w_k^{(N)} = \int_a^b w(s)\mathrm{d}s = W \text{ say}$$

and we find from (4) and (3):

$$|Ef| \leq 2W \sum_{i=P+1}^{\infty} K_i|b_i|. \tag{3a}$$

Equation (3a) is valid for every expansion set $\{h_i\}$, under the stated assumptions concerning the rule. If however the set $\{h_i\}$ is orthogonal with weight function $w(s)$ and includes the unit function (we then assume $h_1(s) = $ constant), that is

$$\int_a^b w(s)h_i(s)h_j(s)\mathrm{d}s = 0, \quad i \neq j$$

$$h_1(s) = \text{constant}$$

it follows that

$$\int_a^b w(s)h_i(s)\mathrm{d}s \equiv 0, \quad i \neq 1.$$

Then the factor 2 may be removed from (3a), Qh_i representing an approximation to a zero integral.

Equation (3a) yields an error (bound) expansion for the rule Q. For a given class of functions f, the coefficients b_i, $i > P$, can often be estimated closely enough to establish their rate of decrease as i increases. They will be small if (1) converges rapidly; we should try to choose the annihilation class so that this is true.

Note that the effective use of (3) may depend quite crucially on the basis used to represent the annihilation class. For example, if Q is of polynomial degree N (that is, is exact whenever f is a polynomial of degree $\leq N$) the annihilation class may be represented equally by either of the choices

$$h_i(s) = (s - a)^{i-1}, \quad i = 1, 2, \dots, N+1, \tag{5a}$$

or

$$h_i(s) = P_{i-1}(\bar{s}), \quad i = 1, 2, \dots, N+1, \tag{5b}$$

where $P_i(\bar{s})$ is the Legendre polynomial for the interval $[a, b]$. Clearly, there are many ways of representing a polynomial of degree N, and each choice of the set $\{h_i\}$ will lead to a different error estimate. The choice (5b) is particularly appropriate for a rule of Gauss type; more generally, if the set $\{h_i\}$ is chosen to be the set of orthogonal polynomials with weight function $w(s)$, the coefficients b_i may be closely estimated in terms of the analyticity properties of $f(s)$; see Chapter 9. However, we consider first the use of the monomial expansion (5a).

2.7.2   Peano's theorem

In this case, (1) reduces to the Taylor series expansion of $f(s)$ about the point a, and we obtain a simpler result by using not the infinite series but a finite series with exact remainder. The result is embodied in:

***Theorem* 2.7.1** (Peano's theorem). Consider the integral

$$If = \int_a^b w(s)f(s)\mathrm{d}s$$

and an N-point quadrature rule

$$Qf = \sum_{i=1}^N w_i f(\xi_i).$$

Let us suppose that Qf is *exact* for all polynomials P_n of degree $\leq p$:

$$Ef = (I - Q)P_n = 0, \quad n = 0, 1, \ldots, p. \tag{6}$$

Then p is the *degree* of the rule Q. Suppose further that $f(s)$ has continuous derivatives of all orders up to q, and a piecewise continuous derivative of order $q + 1$; let $r = \min(p, q)$. Then

$$|Ef| \leq \frac{M^{(r+1)}}{r!} K_r \tag{7}$$

where

$$K_r = \int_a^b |E_s[(s - t)_+{}^r]| \, dt \tag{8}$$

$$M^{(i)} = \sup_{a \leq s \leq b} |f^{(i)}(s)|$$

and

$$(s)_+{}^m = s^m, \quad s \geq 0$$
$$= 0, \quad s < 0.$$

The integral in (8) is called the Peano kernel and the error operator E_s indicates the error in integrating over s. We may prove the result as follows. Taylor's theorem with remainder yields

$$f(s) = f(a) + f'(a)(s - a) + \cdots + \frac{f^r(a)}{r!}(s - a)^r$$
$$+ \frac{1}{r!} \int_a^s f^{(r+1)}(t)(s - t)^r \, dt. \tag{9}$$

We may write the last term, R, in the form

$$R = \frac{1}{r!} \int_a^b f^{(r+1)}(t)(s - t)_+{}^r \, dt. \tag{10}$$

Now apply the linear operator $I - Q \equiv E$ to both sides of equation (9). The first $r + 1$ terms all vanish since the rule Q is of degree $\geq r$. Hence we have

$$Ef = \frac{1}{r!} E \int_a^b f^{(r+1)}(t)(s - t)_+{}^r \, dt.$$

But the operator E acts on the variable s and hence commutes with t:

$$Ef = \frac{1}{r!} \int_a^b f^{(r+1)}(t) E_s(s - t)_+{}^r \, dt. \tag{11}$$

Hence

$$|Ef| \leq \frac{M^{(r+1)}}{r!} \int_a^b |E_s(s - t)_+{}^r| \, dt. \tag{12}$$

Equation (12) yields a bound on the error; in some cases we can in fact find an identity satisfied by the error. If $E_s(s-t)_+^r$ is single signed for $a \le t \le b$, we can write for some $\xi \in (a,b)$:

$$Ef = \frac{f^{(r+1)}(\xi)}{r!} \int_a^b E_s(s-t)_+^r \, dt \qquad (13)$$

which shows that the bound (12) cannot be too pessimistic.

***Example* 2.7.1.** To demonstrate the use of (11), let us calculate the error for the Trapezoid rule. We set $a = 0$, $b = h$; then the error $E_{\text{Trap}}(f)$ is

$$E_{\text{Trap}}(f) = \int_0^h f(s)ds - \frac{h}{2}[f(0) + f(h)].$$

This is zero if $f(s)$ is a polynomial of degree 1: we may set $p = 1$ in (6). For functions with at least two piecewise continuous derivatives we may therefore set $r = 1$, and (11) yields

$$E_{\text{Trap}}(f) = \int_0^h f^{(2)}(t) E_{\text{Trap}}(s-t)_+ \, dt,$$

where

$$E_{\text{Trap}}(s-t)_+ = \int_0^h (s-t)_+ \, ds - \frac{h}{2}[(0-t)_+ + (h-t)_+]. \qquad (14)$$

We now evaluate the components of (14).

(1) $(-t)_+ = 0$ for all positive t

(2) $(h-t)_+ = (h-t)$, $t \le h$

(3) For $0 \le t \le h$, $\displaystyle\int_0^h (s-t)_+ \, ds = \int_0^t (s-t)_+ \, ds + \int_t^h (s-t)_+ \, ds$

$$= 0 + \int_t^h (s-t)ds = \frac{(h-t)^2}{2}.$$

Thus $E_{\text{Trap}}(s-t)_+ = \frac{1}{2}[(h-t)^2 - h(h-t)] = -(t/2)(h-t) \le 0$, $0 \le t \le h$. Hence we may use the form (13):

$$E_{\text{Trap}}(f) = f^{(2)}(\xi) \int_0^h \left(\frac{-t}{2}(h-t)\right)dt$$

$$= \frac{-f^{(2)}(\xi)}{12}h^3. \qquad (14a)$$

A similar but more tedious calculation yields the error form for Simpson's rule:

$$\int_a^{a+2h} f(s)ds = \frac{h}{3}[f(a) + 4f(a+h) + f(a+2h)] - \frac{h^5}{90}f^{(4)}(\xi). \qquad (14b)$$

Example 2.7.2. We can also use (12) to bound the error for a Gauss–Legendre integration with N points. We give here a crude treatment, as follows. An N-point Gauss rule has degree $v = 2N - 1$, and hence for an analytic function if we span the whole interval (a, b) with a single N-point Gauss rule we find with $(a, b) = (-1, 1)$:

$$|Ef(\text{Gauss})| \le \frac{M^{(2N)}}{(2N-1)!} \int_{-1}^{1} |E_s(s-t)_+^{2N-1}| dt. \qquad (15)$$

How rapidly this converges to zero depends on the three terms on the right hand side, all of which are functions of N. We can easily derive a (very weak) bound on the last term as follows:

$$|E_s(s-t)_+^{2N-1}| = \left| \int_{-1}^{1} (s-t)_+^{2N-1} ds - \sum_{i=1}^{N} w_i(\xi_i - t)_+^{2N-1} \right|$$

$$\le \left| \int_{-1}^{1} (s-t)_+^{2N-1} ds \right| + \left| \sum_{i=1}^{N} w_i(\xi_i - t)_+^{2N-1} \right|.$$

But

$$(s-t)_+ \le 2, \quad \forall s, t \in [-1, 1] \quad \text{and} \quad \sum_{i=1}^{N} w_i = 2,$$

whence we obtain

$$|E_s(s-t)_+^{2N-1}| \le 2^{2N+1},$$

and therefore

$$\int_{-1}^{1} |E_s(s-t)_+^{2N-1}| dt \le 2^{2N+2},$$

whence we find

$$|Ef(\text{Gauss})| \le \frac{M^{(2N)}}{(2N-1)!} 2^{2N+2} \sim 4M^{(2N)} \left(\frac{N}{\pi} \right)^{\frac{1}{2}} \left(\frac{e}{N} \right)^{2N}, \qquad (16)$$

where we have used Stirling's estimate for $(2N)!$.

Now if we integrate a function such as cos, sin, for which $M^{(i)}$ is uniformly bounded for all i, we clearly get very rapid convergence in N. In other cases $M^{(i)}$ is a rapidly increasing function of i. For example:

Theorem 2.7.2. (Cauchy's estimate). Let $f(z)$ be analytic in a circle of radius R and let $M^{(i)}(r) = \sup_{|z| < r} |f^{(i)}(z)|$. We set $\rho = r/R$. Then for any $\varepsilon > 0$

$$M^i(r) \le i!(\rho + \varepsilon)^i M^{(0)}$$

If we insert this bound into (16) we find for functions analytic in a circle of radius R:

$$|Ef(\text{Gauss})| \le \frac{2N!(\rho + \varepsilon)^{2N} M^{(0)}}{(2N-1)!} 2^{2N+1} \sim 4NM^{(0)}(2\rho + 2\varepsilon)^{2N},$$

which is still very rapid convergence provided that $\rho > 2$. If $\rho < 2$ this state merely reflects the weak way we bounded $|E_s(s-t)_+^{2N-1}|$ above. Similarly, if the function f has only a finite number of continuous derivatives, these estimates are not useful and much tighter estimates can be achieved in these cases (see Section 2.7.4).

2.7.3   Error estimates for repeated (M-panel) rules

Error estimates for a repeated rule of fixed order follow very simply from those for the canonical underlying rule. Let Q_M be an M-panel repeated N-point rule based on a rule Q of polynomial degree p; let $f(s)$ have q continuous derivatives and a piecewise continuous derivative of order $q+1$; as before, set

$$r = \min(p, q). \tag{17}$$

Then applying Peano's theorem to each panel, we find from (12):

$$|E_M f| = |(I - Q_M)f| \le \frac{M^{(r+1)}}{r!} \sum_{k=1}^{M} \int_{a+(k-1)h}^{a+kh} |E_s[(s-t)_+^r]| dt, \tag{18}$$

where

$$h = \frac{(b-a)}{M} \tag{19}$$

and

$$E_s[(s-t)_+^r]$$
$$= \int_{a+(k-1)h}^{a+kh} (s-t)_+^r \, ds - \sum_{i=1}^{N} w_i(a+(k-1)h + h\xi_i - t)_+^r. \tag{20}$$

Now for all s, t in (20), $|s - t| \le h$, and for all y, $(y)_+ \le |y|$. We therefore have quite generally for some A dependent on the rule but not on M:

$$|E_s[(s-t)_+^r]| \le Ah^{r+1} \tag{21}$$

and hence

$$|E_M f| \le MA \frac{M^{(r+1)}}{r!} h^{r+2} = \frac{AM^{(r+1)}}{r!} (b-a)h^{r+1}. \tag{22}$$

Equation (22) is equally valid for repeated Gauss or repeated Newton–Cotes rules. It shows that, as h tends to zero, the error for a high-order rule reduces more rapidly than that for a low-order rule, provided that the integrand f is sufficiently smooth (so that $r = p$ in (17)). If the integrand is not smooth, then two rules of degree $p_1 > q$ and $p_2 > q$ are predicted by (22) to converge at the same rate; thus we might as well use the rule of higher degree since we incur no penalty.

These conclusions of course assume that the bounds (22) are realistic. It is possible to identify classes of problems for which they are very pessimistic and then the conclusions are not valid. For example, if $f(s)$ is periodic with period P, the Trapezoid rule error in evaluating the integrand over an entire period:

$$\int_a^{a+P} f(s)\,ds - h\sum_{k=0}^{M}{}'' f(a+kh), \quad h = \frac{P}{M} \tag{23}$$

converges very rapidly to zero with h, for smooth f; this is a consequence of the relation between the Trapezoid rule and the closed Gauss–Chebyshev rule noted in Section 2.4.3.

We illustrate the use of (22) by considering in detail the repeated Trapezoid and Simpson's rules over an interval $[a,b]$ divided into N panels of width $h = (b-a)/N$. These have the familiar form

$$\text{Trapezoid:} \int_a^b f(s)\,ds = h[\tfrac{1}{2}f(a) + f(a+h) + \cdots + f(a+(N-1)h)$$

$$+ \tfrac{1}{2}f(b)] + E_{\text{Trap}}$$

and, for N even, the repeated Simpson rule yields

$$\int_a^b f(s)\,ds = \frac{h}{3}[f(a) + 4f(a+h) + 2f(a+2h) + 4f(a+3h) + \cdots$$

$$+ 4f(a+(N-1)h) + f(b)] + E_{\text{Simp}} \quad (N \text{ even}).$$

Using the error forms (14a, b) for these two rules, we can bound the error in each case by

$$|E_{\text{Trap}}| = \left| -\sum_{i=1}^{N} \frac{h^3}{12} f^{(2)}(\xi_i) \right| \leq \frac{NM^{(2)}}{12} h^3 \equiv (b-a)M^{(2)}\frac{h^2}{12},$$

$$|E_{\text{Simp}}| = \left| -\sum_{i=1}^{N/2} \frac{h^5}{90} f^{(4)}(\xi_i) \right| \leq \frac{Nh^5 M^{(4)}}{180} \equiv \frac{(b-a)M^{(4)}}{180} h^4,$$

where

$$M^{(i)} = \sup_{s\in(a,b)} |f^{(i)}(s)|.$$

2.7.4   Error expansions for the Gauss–Legendre rule

We now return to the direct use of equation (3a) for the estimation of quadrature errors, and consider the case when $[a,b] = [-1,1]$, $w(s) = 1$, and the Gauss–Legendre N-point rule of degree $2N - 1$ is used. Then it is natural to take as basis for the annihilation class the set of normalised Legendre polynomials

$$h_n(s) = \left(\frac{2n-1}{2}\right)^{\frac{1}{2}} P_{n-1}(s). \tag{24}$$

Now (Szego, 1939, p. 163) there exists K such that for all n and $|s| < 1$

$$|P_n(s)| \leq K. \tag{25}$$

We may therefore set in (3a)

$$K_i = Ki^{\frac{1}{2}}. \tag{26}$$

Moreover, we can illustrate the way in which the coefficients b_i depend on the properties of the function $f(s)$ by quoting the following lemma which follows from the results in Bain and Delves (1977).

Lemma 2.7.1. Let $f(s) \in C^p[-1, 1]$; that is, $f^{(i)}(s)$ is continuous on the closed interval $[-1, 1]$, $0 \leq i \leq p$, and suppose $f^{(p+1)}(s)$ exists and is continuous in $[-1, 1]$ except possibly for a finite number of finite discontinuities. Then for some K_f and all $n > 0$

$$|b_n| \leq K_f n^{-p-1}. \tag{27}$$

Substituting (26), (27) into (3a) and recalling that $P = 2N - 1$ we find

$$|Ef| \leq WKK_f \sum_{i=2N}^{\infty} i^{-p-\frac{1}{2}}$$

$$\leq WKK_f (2N)^{-p+\frac{1}{2}} \frac{(p+\frac{1}{2})}{(p-\frac{1}{2})}, \tag{28}$$

where we have used the following relation

$$\sum_{j=N+1}^{\infty} (j-i)^{-r} \leq (N+1-i)^{-r+1} \frac{r}{r-1}, \quad \text{provided } r > 1 \tag{28a}$$

(see Delves and Freeman, 1981, Lemma A.8). For a smooth function, $p \gg 1$ and (28) predicts very rapid convergence as N increases. For functions which are not very smooth, (28) is pessimistic for two reasons:

(i) The bound leading to (3a) is itself rather weak, and this becomes important when convergence is slow.

(ii) Lemma 2.7.1 is also weaker than can be obtained by a more careful analysis; see Bain and Delves (1977) for stronger results.

The main advantage of expansion (3a) over the use of the Peano bound (15), lies in its direct dependence on the coefficients b_i. In the context of an expansion method for the solution of an integral equation (see Chapter 7), approximations to these coefficients are calculated as part of the solution

Table 2.7.1. *N-point Gauss–Legendre quadrature errors*
for the function $f_1(s)$, equation (29)

N	2	4	10	16
|Error|	2.6×10^{-3}	$1.1, -4$	$1.6, -6$	$4.5, -7$

process; quite realistic estimates of the quadrature error can then be computed with no additional costs from (3a), whereas the estimate of $M^{(2N)}$ in (15) poses a major numerical difficulty.

As an example of the use of (28) we consider the function

$$f_1(s) = (1+s)^{\frac{3}{2}}; \; If_1 = \int_{-1}^{1} (1+s)^{\frac{3}{2}} ds = \frac{(2)^{\frac{5}{2}}}{5}. \qquad (29)$$

This function has one continuous derivative on $[-1, 1]$; we may set $p = 1$ in (27), (28), which then predict

$$|b_n| \leq \text{const} \cdot n^{-2},$$
$$|Ef_1| \leq \text{const} \cdot N^{-\frac{3}{2}} \text{(predicted)}. \qquad (30)$$

The actual errors obtained using an N-point Gauss–Legendre rule are shown in Table 2.7.1. These errors are fitted approximately by the relation

$$|Ef_1| \leq \text{const} \cdot N^{-4.5} \text{(actual)}, \qquad (31)$$

that is, very much more rapid convergence than is predicted by (28). However, if we replace the weak bound (27) by the tighter bound

$$|b_n| \leq \text{const} \cdot n^{-4.5}$$

which follows from the results of Bain and Delves (1977) (or from the exact expansion of $(1+s)^{\frac{3}{2}}$) we find in place of (30) the tighter estimate

$$|Ef_1| \leq \text{const} \cdot N^{-3} \text{(predicted)}$$

which is closer to the observed convergence rate, but still pessimistic.

2.7.5 An error expansion for Gauss–Chebyshev quadrature

As a second example of the use of equation (2), we consider the case of the open and closed Gauss–Chebyshev rules for the integral

$$If = \int_{-1}^{1} \frac{f(s) ds}{(1-s^2)^{\frac{1}{2}}}. \qquad (32)$$

In this case it is natural to represent the annihilation class of the rule via the Chebyshev polynomials.

To adhere to the standard notation for a Chebyshev expansion, we replace (1) by

$$f(s) = \sum_{i=0}^{\infty}{}' b_i T_i(s) \tag{33}$$

where the coefficients b_i satisfy the relation

$$b_i = \frac{2}{\pi} \int_{-1}^{1} (1 - s^2)^{-\frac{1}{2}} f(s) T_i(s) \, ds. \tag{34}$$

Then for either the open N-point Chebyshev rule (2.4.13) or the closed $(N+1)$-point rule (2.4.15) we find, since both have polynomial degree $2N - 1$:

$$Ef = \sum_{i=2N}^{\infty} b_i E T_i. \tag{35}$$

We now show that the simple properties of the Chebyshev polynomials allow an explicit calculation of the errors ET_i which leads to a remarkably simple identity for the error, avoiding the rather crude bounding process we employed in (4) above. The results are due originally to Riess and Johnson (1969).

 (i) *Open Chebyshev rule.* The open N-point rule has the form

$$Qf = \frac{\pi}{N} \sum_{k=1}^{N} f(\xi_k^{(N)}) = If - E_o f$$

where

$$\xi_k^{(N)} = \cos\left[\frac{(2k-1)\pi}{2N} \right].$$

We now make the split $f = f_1 + f_2$; $f_1 = \sum_{i=0}^{'2N-1} b_i T_i(s)$, and note that $E_o f = E_o f_1 + E_o f_2$. But since the open rule is of degree $2N - 1$, $E_o f_1 = 0$. Further, the orthogonality of the Chebyshev polynomials implies $I T_i = 0$, $i > 0$, whence it follows that

$$E_o f = E_o f_2 = -\frac{\pi}{N} \sum_{i=2N}^{\infty} b_i \sum_{k=1}^{N} T_i(\xi_k^{(N)}) = -\frac{\pi}{N} \sum_{i=2N}^{\infty} S_{Ni} b_i, \tag{36}$$

where

$$S_{Ni} = \sum_{k=1}^{N} T_i(\xi_k^{(N)}) = \sum_{k=1}^{N} \cos\left[(2k-1)\left(\frac{i\pi}{2N} \right) \right]. \tag{37}$$

But, for any x,

$$\begin{aligned}
\sin 2Nx &= (\sin 2Nx - \sin(2N-2)x) + (\sin(2N-2)x \\
&\quad - \sin(2N-4)x) + \cdots + (\sin 2x - \sin 0x) \\
&= 2\sin x [\cos(2N-1)x + \cos(2N-3)x + \cdots + \cos x] \\
&= 2\sin x \sum_{k=1}^{N} \cos(2k-1)x. \tag{38}
\end{aligned}$$

Now set $s = i\pi/2N$; then the sum on the right side of (36) is identically S_{Ni}. Further,

$$\sin 2Nx = \sin i\pi = 0.$$

Hence, if $\sin(i\pi/2N) \neq 0$, that is, i is not a multiple of N,

$$S_{Ni} = 0. \tag{39}$$

And if $i = 2Nj, j = 1, 2, \ldots$ then

$$S_{Ni} = \sum_{k=1}^{N} \cos((2k-1)j\pi) = \begin{cases} -N, j \text{ odd} \\ N, j \text{ even} \end{cases} \tag{40}$$

and we obtain from (36) the remarkable identity

$$E_o f = -\pi \sum_{j=1}^{\infty} (-1)^j b_{2Nj}. \tag{41}$$

(ii) *Closed Chebyshev rule.* The discussion of the closed rule follows the same path. The closed $(N+1)$-point rule has the form

$$Qf = \frac{\pi}{N} \sum_{k=0}^{N}{}'' f(\xi_k^{(N)}), \tag{42}$$

where

$$\xi_k^{(N)} = \cos\left(\frac{\pi k}{N}\right)$$

with error $E_c f$. The analogue of (36) is the identity

$$E_c f = -\frac{\pi}{N} \sum_{i=2N}^{\infty} \bar{S}_{Ni} b_i, \tag{43}$$

where

$$\bar{S}_{Ni} = \sum_{k=0}^{N}{}'' \cos \frac{ik\pi}{N} = \begin{cases} 0, i \neq N_j \\ N, i = N_j \end{cases} \tag{44}$$

and hence we find immediately the identity

$$E_c f = -\pi \sum_{j=1}^{\infty} b_{2Nj}. \tag{45}$$

Equations (41), (45) show clearly that the open and closed rules can be expected to be of about the same accuracy, since both yield the bounds

$$|Ef| \leq \pi \sum_{j=1}^{\infty} |b_{2Nj}|. \tag{46}$$

The expansion coefficients b_k can be estimated using the same techniques as for the Legendre polynomial expansion discussed briefly in Section 2.7.4. If we have available the convergence rate estimate

$$|b_k| \leq C k^{-p} \tag{47}$$

Table 2.7.2. *Errors using the N-point open Gauss–Chebyshev rule for the integral $\int_{-1}^{1}(1+s)^{\frac{3}{2}}(1-s^2)^{-\frac{1}{2}}ds$ $= 8\sqrt{2}/3$*

N	2	4	10	16
Error	1.9×10^{-2}	$1.0, -3$	$2.5, -5$	$3.8, -6$

Table 2.7.3. *Errors using the (N + 1)-point closed Gauss–Chebyshev rule for the integral $\int_{-1}^{1}(1+s)^{\frac{3}{2}}(1-s^2)^{-\frac{1}{2}}ds = 8\sqrt{2}/3$*

N	2	4	10	16
Error	-2.1×10^{-2}	$-1.2, -3$	$-2.9, -5$	$-4.4, -6$

then (46) yields

$$|Ef| \leq \frac{pC\pi}{(p-1)} \cdot (2N)^{-p}. \tag{48}$$

As an example, we consider again the function $f(s) = (1 + s)^{\frac{3}{2}}$. For this function

$$|b_k| \leq \text{const} \cdot k^{-4} \tag{49}$$

and then (48) yields

$$|Ef| \leq \text{const} \cdot N^{-4}. \tag{50}$$

The actual errors obtained using the open and closed Chebyshev rules are shown in Tables 2.7.2 and 2.7.3 respectively. These errors are well fitted by the forms

$$|Ef| \simeq \text{const} \cdot N^{-4.1} \tag{51a}$$

$$|Ef| \simeq \text{const} \cdot N^{-4.2} \tag{51b}$$

for the open and closed rules. We see that for both rules the predicted and observed convergence rates are extremely close to each other. Note that the open rule underestimates the value of the integral while the closed rule overestimates it.

2.8 High-order vs. low-order rules

Faced with a choice of quadrature rules, most users will make a mental balance between convenience and efficiency.[†] In the context of this book, efficiency is the more important of these; then the error analyses of Section 2.7 suggest that the following broad remarks should contain some truth: (N is the number of points in the rule; convergence is measured as $N \to \infty$)

(i) For a well-behaved (smooth) integrand, a rule of high (polynomial) degree will converge more rapidly than one of low degree.

(ii) For an ill-behaved function low degree and high degree rules will converge at about the same rate.

These remarks, if justified, suggest that we should standardise, except when explicit indications to the contrary exist, on rules of high degree. Exceptions can indeed be found, usually when a polynomial expansion of the integrand is not very efficient; the case of a periodic function integrated over a complete period, for which the repeated Trapezoid rule (which is of high trigonometric degree) is very efficient, was quoted earlier. Nonetheless, the conclusion drawn is, we believe, reasonable; we illustrate this with some numerical examples.

2.8.1 Families of rules

In talking about convergence rates we clearly have in mind the use of a family of rules $\{Q_N\}$ with increasing numbers of points N, which we will apply successively until the answers achieved have settled down to the required accuracy; this is a reasonable model for the behaviour of a cautious user, at least at the exploratory stages of a calculation. We consider four types of rule family:

(i) The repeated low-order M-point Newton–Cotes rule with P panels and M fixed; if the rule is doubly closed (uses both endpoints as abscissae) $N = P(M-1) + 1$. Typical (regrettably popular) examples of such rules are the repeated Trapezoid and Simpson's rules.

(ii) As (i), but with a high-order Newton–Cotes rule.

(iii) As (i), but with an M-point Gauss rule.

(iv) A sequence of N-point Gauss rules with increasing N.

The use of high-order Newton–Cotes rules is sometimes avoided on the grounds that they have undesirable properties due to some of the weights

---

[†] and then use the repeated Trapezoid or Simpson's rule because they are familiar and trivial to program.

Table 2.8.1. *Values of the error $\varepsilon_N = |Q_N f - If|$ obtained by approximating the integral in (1) using an N-point closed Newton–Cotes rule of degree $N - 1$ on the interval $[0, 1]$*

N	ε_N	N	ε_N	N	ε_N
2	3.5×10^{-2}	6	7.2, -5	14	1.6, -7
3	2.1, -3	8	3.3, -6	16	3.8, -8
4	7.8, -4	10	1.2, -8	18	2.1, -7
5	1.3, -4	12	3.8, -8	20	1.6, -7

associated with the rules being negative. A statement of this kind requires qualification. It is true that a sequence of Newton–Cotes rules of increasing order, analogous to family (iv) above, is dangerous in practice: the sequence cannot be guaranteed to converge without quite strong restrictions on the class of integrands considered, and in practice erratic convergence or lack of convergence can easily be displayed. However, the use of a repeated rule of fixed high order does not have this danger. The negative weights do lead to some cancellation error but the loss is not very large and does not increase rapidly with the degree of the rule. We illustrate this with an example:

Example 2.8.1. We evaluate the integral

$$If = \int_0^1 \frac{1}{s^2 + 1} ds = \frac{\pi}{4} \tag{1}$$

using a sequence of N-point closed Newton–Cotes rules of degree $N - 1$. This sequence of rules is that *not* recommended above, but the results given in Table 2.8.1 show that convergence *is* achieved for this integral.

Further, the convergence is quite rapid for $N \leq 10$; then, no further improvement is obtained, the results fluctuating in accuracy. This fluctuation is due to the presence of the 'undesirable' negative weights in the rules, but the results also show that the cancellation errors do not build up rapidly for N in this range.

2.8.2   Numerical performance

We now put these families of rules through a miniature test. To be specific, we test the following sequences of rules:

Rule 1: Repeated Trapezoid rule with a total of N points

Rule 2: Repeated Simpson's rule with a total of N points

These rules are examples of family (i).

Rule 3: A repeated 8-point Newton–Cotes rule with a total of N points

Rule 4: A repeated 4-point Gauss–Legendre rule with a total of N points

These rules are examples of family (ii) and (iii) respectively, and are each of degree 7.

Rule 5: An N-point Gauss–Legendre rule of degree $2N - 1$

This rule exemplifies family (iv).

For each rule we evaluate the following integrals:

Smooth:

$$I_1 = \int_0^{2\pi} \cos^2 s \, ds = \pi,$$

$$I_2 = \int_0^1 \cos s \, ds = 0.841\,470\,98.$$

Oscillatory:

$$I_3 = \int_0^{2\pi} [s \sin(30s) \cos(50s)] ds = 0.117\,809\,72.$$

Non-smooth:

$$I_4 = \int_0^1 s^{\frac{3}{2}} \, ds = 0.4,$$

$$I_5 = \int_0^1 |s^2 - 0.25|^{\frac{1}{2}} ds = 0.464\,742\,50.$$

Results are given in Tables 2.8.2–6.

We make the following remarks about these results.

(i) I_1: *Table* 2.8.2. This is the integral of a periodic function over a complete period; the repeated Trapezoid rule is therefore expected to be rapidly convergent, and is in fact exact for $N > 2$. Simpson's rule is also exact for any even number of panels P for $P > 2$. However, the other rules also converge extremely rapidly and the example gives little support for a suggestion that such cases are worth catering for especially.

(ii) I_2: *Table* 2.8.3. The integrand is again periodic, but not over the interval of integration; it is also very smooth. The high-order rules converge extremely fast and certainly faster than the Trapezoid or Simpson's rules.

(iii) I_3: *Table* 2.8.4. The integrand is 'mathematically smooth', indeed analytic, but contains many oscillations within the interval of integration.

Tables 2.8.2 to 2.8.6. *Achieved accuracies* ε_N *for integrals* I_1 *to* I_5 *using rules 1 to 5 and various values of N.*

Table 2.8.2. I_1

Rule 1		Rule 2		Rule 3		Rule 4		Rule 5	
N	ε_N	N	ε_N	N	ε_N	N	ε_N	N	ε_N
2	3.14	3	3.14	8	2.3, −1	4	−3.9, −1	2	−2.8
4	5.8×10^{-11}	5	−1.05	15	−2.2, −3	8	3.4, −3	4	−3.9, −1
8	−2.9, −11	9	5.8, −11	29	0.0	16	1.4, −10	8	−1.2, −5
16	2.9, −11	17	5.8, −11					16	0.0

Table 2.8.3. I_2

Rule 1		Rule 2		Rule 3		Rule 4		Rule 5	
N	ε_N	N	ε_N	N	ε_N	N	ε_N	N	ε_N
2	-7.1×10^{-2}	3	3.0, −4	8	5.2, −9	4	4.3, −9	2	−2.0, −4
8	−1.4, −3	5	1.8, −5	15	4.8, −9	8	4.8, −9	4	4.3, −9
32	−7.3, −5	9	1.1, −6					8	4.8, −9
128	−4.3, −6	17	7.6, −8						
		33	9.3, −9						
		65	5.1, −9						

Table 2.8.4. I_3. *Results for rule 5 marked with an asterisk were obtained with a repeated 64-point rule*

Rule 1		Rule 2		Rule 3		Rule 4		Rule 5	
N	ε_N	N	ε_N	N	ε_N	N	ε_N	N	ε_N
4	-1.2×10^{-1}	5	−1.2, −1	8	−4.5	4	2.1, −1	4	2.1, −1
16	−1.2, −1	17	7.0, −1	15	1.6	16	−1.1, −1	16	6.8, −1
64	−1.5, −1	65	−1.4, −1	57	1.3, −1	64	3.3, −3	64	−2.4, −1
128	6.0, −2	129	1.3, −1	113	1.7, −1	128	1.2, −1	128*	1.8, −1
256	1.0, −2	257	−5.6, −3	225	−6.1, −2	256	−1.2, −3	256*	3.3, −9
512	2.4, −3	513	−2.25, −4	449	7.9, −4	512	2.4, −6	512*	4.8, −9
1024	5.9, −4	1025	−1.28, −5	897	−1.6, −6	1024	1.0, −8		

Such functions are very difficult to deal with and all the rules are clearly unsatisfactory for $N \leq 18$, when each rule has about one point per half-cycle of the integrand. After this, they behave much as (the authors) expected: the Trapezoid and Simpson's rules converge slowly, and the high-order rules

Table 2.8.5. I_4

Rule 1		Rule 2		Rule 3		Rule 4		Rule 5	
N	ε_N	N	ε_N	N	ε_N	N	ε_N	N	ε_N
2	1.0×10^{-1}	3	2.4, -3	8	6.7, -5	4	-5.0, -5	2	-1.2, -3
4	1.2, -2	5	4.3, -4	15	1.2, -5	8	-8.8, -6	4	-5.0, -5
8	2.4, -3	9	7.7, -5	29	2.1, -6	16	-1.6, -6	8	-1.9, -6
16	5.3, -4	17	1.4, -5	57	3.7, -7	32	-2.7, -8	16	-6.8, -8
32	1.3, -4	33	2.4, -6	113	6.6, -8	64	-4.8, -8	32	-2.3, -9
64	3.1, -5	65	4.3, -7	225	1.2, -8	128	-8.6, -9	64	-6.9, -11
128	7.6, -6	129	7.6, -8						

Table 2.8.6. I_5

Rule 1		Rule 2		Rule 3		Rule 4		Rule 5	
N	ε_N	N	ε_N	N	ε_N	N	ε_N	N	ε_N
2	2.2×10^{-1}	3	-2.4, -1					2	6.7, -2
4	3.4, -2	5	-2.0, -2			4	2.6, -2	4	2.6, -2
8	8.5, -3	9	-7.2, -3	8	8.0, -4	8	8.2, -4	8	9.7, -3
16	2.5, -3	17	-2.5, -3	15	-2.6, -3	16	2.9, -4	16	3.6, -3
32	8.1, -4	33	-9.0, -4	29	-9.2, -4	32	1.0, -4	32	1.3, -3
64	2.6, -4	65	-3.2, -4	57	-3.3, -4	64	3.6, -5	64	4.6, -4
128	9.1, -5	129	-1.1, -4	113	-1.2, -4	128	1.3, -5		
				225	-4.1, -5				

rapidly. The results illustrate the rule of thumb that, unless the oscillatory part of an integrand is dealt with explicitly (via a suitable choice of $w(s)$ in equation (2.1.1), for example) there is no substitute for 'putting points in the wiggles'.

(iv) I_4 *and* I_5: *Tables* 2.8.5, 2.8.6. These integrands are mildly ill-behaved, having respectively one and zero continuous derivatives on the closed interval $[0, 1]$. This difference is reflected in the results: for I_4, the rules of higher order converge a little faster than Simpson's rule (and much faster than the Trapezoid rule) while for I_5, all the rules converge about equally slowly; rule 4 in fact is about a factor of 10 better than the others, but this factor is independent of N for $N > 8$ and cannot be given much significance. We sum up the results as evidence for a rule that 'high-order rules work as well as, or better than, low-order rules'. Note that this applies to both Newton–Cotes and Gauss rules and that the higher the order of the rule, the better its performance overall.

The order dependence is particularly striking if we compare the fixed order repeated 8-point Newton–Cotes rule (rule 3) with the fixed order repeated 4-point Gauss rule (rule 4): the Newton–Cotes rule performs about as well as the Gauss rule which uses half as many points.

2.9 Singular integrals

Many integrals arising in practice have integrands which are in some sense unpleasant. The commonest causes of unpleasantness seem to be:

   (i) Rapid oscillations.
  (ii) A singularity either in the integrand or in a low-order derivative of the integrand.
(iii) Discontinuities in the integrand or in a low-order derivative.

Difficulties of type (iii) are best overcome by integrating separately over subregions in which the integrand is continuous. Difficulties of type (ii) are rather common in the field of integral equations and we discuss them briefly here. Difficulties of type (i) are not discussed at all here.

2.9.1 Gauss rules

Often we can factor the singularity out of the integrand, in the sense that the problem may be formulated as

$$If = \int_a^b w(s)f(s)\mathrm{d}s \tag{1}$$

where $w(s)$ is singular or has some other unpleasantness, and $f(s)$ is smooth or relatively smooth. If a rule tailored to the weight function $w(s)$ is available, the difficulties associated with (i) effectively vanish. In particular, if a Gauss rule (see Section 2.4) for the weight function $w(s)$ is available, its use is strongly recommended; for tables of available Gauss points and weights, see Stroud and Secrest (1966). If no Gauss rule is available it may be worthwhile to construct either a Gauss or a Newton–Cotes type rule. This is straightforward in practice given the moments m_i:

$$m_i = \int_a^b s^i w(s)\mathrm{d}s \tag{2}$$

but the expense of the calculation must be offset against the subsequent gains that we make in performing the integration: clearly, the more integrals we need to compute for a given $w(s)$, the more worthwhile the construction of the rule will be.

2.9.2 Singular Chebyshev rules

Alternatively, rules of generalised Gauss–Chebyshev or Clenshaw–Curtis type can be constructed (see Section 2.4). Suppose for convenience that $[a,b] = [-1,1]$, and that $f(s)$ has the known Chebyshev expansion:

$$f(s) = \sum_{i=0}^{\infty}{}' a_i T_i(s). \tag{3}$$

Then

$$If = \sum_{i=0}^{\infty}{}' a_i \int_{-1}^{1} w(s) T_i(s) ds = \sum_{i=0}^{\infty}{}' a_i \bar{m}_i. \tag{4}$$

Provided the modified moments $\bar{m}_i$ are available (4) can be made the basis of an effective integration scheme by following the development of Section 2.4.4: we truncate (3) at $i = N$ and replace the a_i by approximations given in terms of the known function values $f(\cos(k\pi/N))$, $k = 0, 1, \ldots, N$. A procedure of this type proves particularly effective for dealing with singular integral equations; see Chapter 11 for details.

2.9.3 Subtraction of the singularity

Alternatively, the singularity can be 'subtracted out'. Suppose that the 'singularity' (that is, source of difficulty) occurs at a single point $s_0 \in [a, b]$ and that (1) is rewritten in the form

$$If = \int_{a}^{b} w(s)[f(s) - f(s_0)] ds + f(s_0) \int_{a}^{b} w(s) ds \tag{5}$$

$$= \int_{b}^{a} w(s) g(s) ds + f(s_0) m_0. \tag{5a}$$

Provided that m_0 is known, If can be estimated by integrating $g(s)$; this is usually easier than the original problem because $g(s_0) = 0$.

It is easier to see the advantages and limitations of this approach by looking at a particular singularity. Suppose that

$$w(s) = s^{-\frac{1}{2}}; \quad If = \int_{0}^{1} s^{-\frac{1}{2}} f(s) ds.$$

Then the subtraction identity (5) takes the form

$$If = \int_{0}^{1} s^{-\frac{1}{2}}[f(s) - f(0)] ds + 2f(0).$$

Now usually $f(s) - f(0)$ will approach zero linearly as $s \to 0$, and hence $s^{-\frac{1}{2}}[f(s) - f(0)]$ will be finite (indeed zero) at $s = 0$; the subtracted integral is indeed 'easier' than the original and we can choose to evaluate it without taking further account of the singularity. However, the singularity has not completely disappeared; the new integral still has a singular first derivative so that we have weakened, rather than removed, the difficulty.

The subtraction process can however be repeated, successively if required, to weaken the singularity further. An appropriate formulation for doing this, and numerical examples, are given in Chapter 11.

2.9.4   Ignoring the singularity

Finally, singularities can be ignored. Weak singularities (that is, integrands which are regular everywhere but which have singular derivatives of some low order) rarely preclude convergence, but merely delay it. Even if the integrand is singular at some point s_0 it is usually possible to choose a sequence of rules which avoid using s_0 as an abscissa, and which will converge to the exact result. Procedures of this type are described, for example, in Davis and Rabinowitz (1975, Section 2.12). They are *not recommended* here, except as a method of last resort.

Exercises

1. Find constants a, b such that the formula

$$\int_0^1 e^{-s} f(s)\,ds \approx af(0) + bf(1)$$

   is exact when f is a polynomial of degree ≤ 1.

2. Let $q_i(s)$ be a polynomial of degree i in s. If the sequence $(q_i(s))$ is such that

$$\int_a^b w(s)q_i(s)q_j(s)\,ds = 0, \quad i \neq j,$$

(a) show that its members satisfy the recurrence relation

$$q_{i+1}(s) = (\alpha_i s + \beta_i)q_i(s) + \gamma_{i-1}q_{i-1}(s), \quad i = 1, 2, \dots.$$

   where α_i is the ratio of the leading coefficients of q_{i+1} and q_i;

(b) derive the expressions

$$\beta_i = -\frac{\alpha_i}{g_i} \int_a^b w(s)s q_i^2(s)\,ds,$$

$$\gamma_{i-1} = -\frac{\alpha_i}{\alpha_{i-1}} \frac{g_i}{g_{i-1}},$$

   where

$$g_i = \int_a^b w(s)q_i^2(s)\,ds.$$

   (*HINT* (a). Write $q_{i+1}(s) - \alpha_i s q_i(s)$ in the form

$$\beta_i q_i(s) + \sum_{r=0}^{i-1} \gamma_r q_r(s)$$

   and take the inner product of this equation with $q_r(s)$, $r = 0, 1, \dots, i-2$.)

3. Let $f(s) = \sum_{i=0}^{\prime\infty} a_i T_i(s)$. The Clenshaw–Curtis method uses the approximation

$$f(s) \approx \sum_{i=0}^{N}{}'' \alpha_i T_i(s), \quad \alpha_i = (2/N) \sum_{j=0}^{N}{}'' \cos(\pi i j/N) f(\cos(\pi j/N)).$$

Show that the method can be expressed as

$$\int_{-1}^{1} f(s)\,ds \approx \sum_{j=0}^{N} h_j f(s_j)$$

where $s_j = \cos(\pi j/N)$ and, when N is even, that is, $N = 2n$,

$$h_j = h_{N-j} = \frac{2}{n} \sum_{k=0}^{n}{}'' \frac{1}{1-4k^2} \cos\frac{k\pi j}{n}, \quad j = 1, \dots, n,$$

$$h_0 = h_N = \frac{1}{N^2 - 1}.$$

4. Suppose f has an expansion of the form

$$f(s) = \sum_{i=0}^{\infty}{}' a_i T_i(s)$$

and an approximate expansion given by

$$f(s) \approx \sum_{i=0}^{N}{}'' \alpha_i T_i(s),$$

where the α_i are defined as in Exercise 3.
Use the following summation orthogonality property:
if $i \le N$, $j > N$ and $p = 0, 1, 2, \dots$

$$\sum_{k=0}^{N}{}'' T_i(x_k) T_j(x_k) = \begin{cases} 0, & j \ne 2pN \pm i, \\ N, & j = pN, \\ \dfrac{N}{2}, & j = 2pN \pm i, \quad i \ne 0 \text{ or } N, \end{cases}$$

to show that

$$\alpha_i = a_i + a_{2N-i} + a_{2N+i} + a_{4N-i} + a_{4N+i} + \cdots.$$

5. Show, using Peano's Theorem, that the error in Simpson's rule is given by

$$E_{\text{Simp}}(f) = \int_{-1}^{1} f(s)\,ds - \tfrac{1}{3}[f(-1) + 4f(0) + f(1)]$$

$$= -\frac{f^{(4)}(\xi)}{90}, \quad -1 < \zeta < 1.$$

3

Introduction to the theory of linear integral equations of the second kind

3.1 Regular values and the resolvent equations

In this chapter we introduce the elements of the theory of linear integral equations. The aim is twofold; first, to introduce some necessary notation, and second, to study the interesting questions – does a linear integral equation have any solutions? If so, is the solution unique? We can give quite a detailed answer to this question for linear Fredholm and Volterra equations of the second kind. To do so, we introduce a parameter λ, which we consider as a complex variable, into the kernel and write the Fredholm equation (0.3.2) in the form

$$x(s) = y(s) + \lambda \int_a^b K(s, t)x(t)\,dt \tag{1}$$

or

$$Lx = (I - \lambda K)x = y \tag{2}$$

where we have introduced for brevity the operator $L = I - \lambda K$. We then study the question: for which values of λ does a solution of (2) exist (within the space $\mathscr{L}^2$)? If a solution exists, is it unique? If, for some value of λ, no unique solution exists, are there no solutions or a multiplicity of solutions? We give a partial answer in this chapter; the existence and uniqueness theory is completed in Chapter 6.

If an unique solution of (2) exists, for all $\mathscr{L}^2$ functions y, then (by definition) an operator inverse to $I - \lambda K$ exists and we can formally write

$$x = (I - \lambda K)^{-1}y = L^{-1}y \tag{3}$$

where L^{-1} is an $\mathscr{L}^2$ operator. It is by no means clear a priori that L^{-1} exists. If it does exist it will depend on the parameter λ; where it is necessary to make this dependence explicit, we shall write L_λ, L_λ^{-1}. Now suppose that L^{-1} does exist. Then if the function x defined by (3) satisfies (2), it follows that

$$y = Lx = LL^{-1}y. \tag{4}$$

If this is valid for arbitrary y, then (by definition), the operator relation

$$LL^{-1} = I \tag{5}$$

holds. An operator L^{-1} satisfying (5) is said to be a *right inverse* of L. But, on the other hand, from (3) and (2)

$$x = L^{-1}y = L^{-1}Lx.$$

Now, if we assume that (3) has an unique solution y for arbitrary x, this implies the operator equation

$$L^{-1}L = I. \tag{6}$$

An operator L^{-1} satisfying (6) is said to be a *left inverse* of L. If there exists an operator satisfying both (5) and (6):

$$L^{-1}L = LL^{-1} = I \tag{7}$$

then L^{-1} is both a left and right inverse of L and is called the *inverse operator*. We shall refer to (7) as the *resolvent equations*; see Section 3.6 below where this term is introduced again.

If, for a given value of λ, an operator L^{-1} satisfying (7) exists, then it is unique and so is the solution x of (2), since any solution satisfies (3). We can show this formally:

Definition 3.1.1. If for a given $\lambda = \lambda_0$, an $\mathscr{L}^2$ operator L^{-1} exists satisfying (7), λ_0 is a *regular value* of the operator K.

Note that $\lambda = 0$ is a regular value of every kernel, with inverse $L_0^{-1} = I$.

Theorem 3.1.1. If for a given λ, L^{-1} exists, then it is unique.

Proof. Suppose that L_1^{-1}, L_2^{-1} are two $\mathscr{L}^2$ operators satisfying (7) and let $\Delta = L_1^{-1} - L_2^{-1}$. Then from (7)

$$L_1^{-1}L = LL_1^{-1} = I,$$
$$L_2^{-1}L = LL_2^{-1} = I,$$

and subtracting we find

$$\Delta L = L\Delta = 0.$$

Hence, multiplying by the operator L_1^{-1} we find from the second of these that

$$L_1^{-1}L\Delta = 0.$$

But $L_1^{-1}L\Delta = I\Delta = \Delta$ and hence $\Delta = 0$.

Note that the proof has in fact only used property (6). The distinction between left and right inverses, although important for some types of operator, is not important for linear Fredholm operators of the second kind (that is, operators of the form in (2)).

Theorem 3.1.2. If λ is a regular value of K, with inverse operator L^{-1}, then for any $\mathscr{L}^2$ function y, equation (2) has an unique $\mathscr{L}^2$ solution x satisfying

$$x = L^{-1}y. \tag{8}$$

Proof. This theorem is more or less obvious from the way we have defined L^{-1}, but we give the formal steps anyway. Substituting (8) in (2) we find

$$LL^{-1}y = y, \tag{9}$$

and since $LL^{-1} = I$, (9) is satisfied; that is, the function x defined by (8) is a solution of (2). Conversely, let x_1, x_2 be any two solutions of (2). Then from (2)

$$L(x_1 - x_2) = 0$$

and hence

$$L^{-1}L(x_1 - x_2) = 0.$$

That is, $I(x_1 - x_2) = x_1 - x_2 = 0$. Hence $x_1 = x_2$ and the solution of (2) is unique.

Theorem 3.1.2 is obviously very convenient, since equations with non-unique solutions are much more awkward than ones with unique solutions. Remember however that both y and x have been restricted to be $\mathscr{L}^2$ functions and the operators L, L^{-1} are from the space $\mathscr{L}^2$ into $\mathscr{L}^2$ (that is, they produce an $\mathscr{L}^2$ function from an $\mathscr{L}^2$ function). Equations of the form (2) may have non-unique solutions if we accept as 'solutions' non-square-integrable functions; see Section 3.7 below for an example.

3.2   The adjoint kernel and the adjoint equation

If $K(s, t)$ is $\mathscr{L}^2$, so is the adjoint (or Hermitian adjoint, or Hermitian conjugate) kernel $K^\dagger$ defined by the relation

$$K^\dagger(s, t) = [K(t, s)]^*.$$

It follows at once from this definition that

$$(K^\dagger)^\dagger = K,$$
$$\|K^\dagger\| = \|K\|,$$
$$(\lambda K)^\dagger = \lambda^* K^\dagger,$$
$$(K_1 + K_2)^\dagger = K_1^\dagger + K_2^\dagger. \tag{1}$$

Associated with the kernel $K(s, t)$ are the integral operator K and its adjoint $K^\dagger$. These operators also satisfy relations (1).

Moreover,

$$
\begin{aligned}
(K_1 K_2)^\dagger(s, t) &= [K_1 K_2(t, s)]^* \\
&= \int K_1^*(t, u) K_2^*(u, s) du \\
&= \int K_2^\dagger(s, u) K_1^\dagger(u, t) du \\
&= K_2^\dagger K_1^\dagger(s, t),
\end{aligned}
$$

i.e.

$$
(K_1 K_2)^\dagger = K_2^\dagger K_1^\dagger. \tag{2}
$$

Relations (1) and (2) are the same as those for square matrices and their Hermitian adjoints. The analogy is made even closer if we compute the inner products (x, Ky), $(x, K^\dagger y)$ of the operators K and $K^\dagger$ with two arbitrary $\mathscr{L}^2$ functions x, y:

$$
\begin{aligned}
(x, K^\dagger y) &= \int_a^b x^*(s) \left[\int_a^b K^\dagger(s, t) y(t) dt \right] ds \\
&= \int_a^b x^*(s) \int_a^b K^*(t, s) y(t) dt \, ds \\
&= \left[\int_a^b x(s) \int_a^b K(t, s) y^*(t) dt \, ds \right]^* \\
&= \left[\int_a^b y^*(t) \left\{ \int_a^b K(t, s) x(s) ds \right\} dt \right]^* \\
&= (y, Kx)^*. \tag{3}
\end{aligned}
$$

Equation (3) should be compared with the relation for a square matrix $\mathbf{A}$ and its Hermitian adjoint:

$$
(\mathbf{A}^\dagger)_{ij} = (A_{ji})^*.
$$

If $K = K^\dagger$ we say that K is *Hermitian*; if it is real then it is also symmetric (see Chapter 0), while obviously the unit operator I is symmetric; that is,

$$
I^\dagger = I.
$$

The integral equation

$$
u = y^* + \lambda^* K^\dagger u
$$

is called the *adjoint equation* of

$$
x = y + \lambda K x.
$$

There is a close connection between a given equation and its adjoint, as the following theorem shows.

Theorem 3.2.1. If K is an $\mathscr{L}^2$ kernel, λ is a regular value of K if and only if λ^* is a regular value of $K^\dagger$. Further, if $(I - \lambda K)^{-1} = L^{-1}$, then $(I - \lambda^* K^\dagger)^{-1} = (L^{-1})^\dagger$.

Proof. If λ is a regular value of K, then the operator L^{-1} exists. Then, from (3.1.7) and (3.1.2), with $L = I - \lambda K$,

$$I^\dagger = I = (LL^{-1})^\dagger = (L^{-1})^\dagger L^\dagger,$$
$$I^\dagger = I = (L^{-1}L)^\dagger = L^\dagger (L^{-1})^\dagger.$$

These relations identify $(L^{-1})^\dagger$ as the inverse operator to $L^\dagger = I - \lambda^* K^\dagger$. *Since this operator exists, λ^* is a regular value of $K^\dagger$.* Conversely, if $L^\dagger$ has inverse $(L^{-1})^\dagger$ the argument can be repeated to show that λ is a regular value of K.

3.3   Characteristic values and characteristic functions

If in (3.1.1) we set $y = 0$, we obtain the homogeneous equation

$$x(s) - \lambda \int_a^b K(s, t) x(t) \mathrm{d}t = 0 \tag{1}$$

with operator form

$$(I - \lambda K)x = 0.$$

Equation (1) always has the solution $x = 0$, referred to as the trivial solution. If it has a nontrivial solution $x_0 \neq 0$, then there cannot exist an unique solution of (3.1.1). For if x satisfies (3.1.1), so does $x + \alpha x_0$ for any value of α. Thus λ cannot then be a regular value of K; it is called a *characteristic value* and $x_0(s)$ the corresponding *characteristic function*.

If x_1 and x_2 are both characteristic functions for the same value of λ, then clearly so is $\alpha x_1 + \beta x_2$; thus, all such functions form a linear subspace of $\mathscr{L}^2$ – the characteristic subspace belonging to λ.

We will sometimes rewrite the characteristic equation in the form

$$Kx = \gamma x,$$

$$\gamma = \frac{1}{\lambda}.$$

γ is then an *eigenvalue* of K, and x can equally be called an *eigenfunction*. We prove in Chapter 6 that every λ is either a regular value or a characteristic value.

3.4   The Neumann series

If λ is a regular value of K, the equation

$$x = y + \lambda Kx \tag{1}$$

has the unique solution

$$x = (I - \lambda K)^{-1}y = L^{-1}y. \tag{2}$$

We can show that any given value λ is a regular value if we can construct explicitly either the inverse operator $(I - \lambda K)^{-1} = L^{-1}$ or an unique solution.

Since the value $\lambda = 0$ is always a regular value it is natural to seek a representation for L^{-1} as a power series in λ. Purely formally the binomial theorem yields the representation

$$L^{-1} = (I - \lambda K)^{-1} = I + \lambda K + \lambda^2 K^2 + \cdots + \lambda^n K^n + \cdots$$

$$= I + \sum_{n=1}^{\infty} \lambda^n K^n. \tag{3}$$

Alternatively we can try to find a representation for x by solving (1) by successive approximations. For $\lambda = 0$ the (unique!) solution is $x = y$. We now define a sequence of approximations x_n to x as follows. We set

$$x_0 = y,$$
$$x_1 = y + \lambda Kx_0 = y + \lambda Ky,$$
$$x_2 = y + \lambda Kx_1 = y + \lambda Ky + \lambda^2 K^2 y,$$
$$\vdots$$

and in general

$$x_{n+1} = y + \lambda Kx_n. \tag{4a}$$

The recurrence relation (4a) has the solution

$$x_n = y + \sum_{i=1}^{n} \lambda^i K^i y. \tag{4b}$$

Equation (4b) is meaningful for arbitrary λ, K and any finite n. It defines a sequence of functions x_n and this sequence may or may not have a limit as $n \to \infty$. However, again proceeding formally, it suggests the following representation for x

$$x = \sum_{i=0}^{\infty} \lambda^i K^i y. \tag{4c}$$

Equation (4c) is the *Neumann series* for the solution x of (1). The same series is obtained if we formally apply the series (3) for the operator L^{-1} to (2), and (3) is also called the Neumann series for the inverse operator L^{-1}. If for some fixed λ and all y the series (4c) is convergent to a solution of (1), or if

equivalently the series (3) is convergent to an operator satisfying relations
(3.1.7), then λ is a regular value of K. It is quite easy to give sufficient
conditions for this to be so.

***Theorem* 3.4.1.** The Neumann series (3) for L^{-1} converges strongly if
$\|\lambda K\| < 1$.

Proof. The simplest proof of convergence is to treat (1) as an operator
equation in the Banach space of $\mathscr{L}^2$ functions with norm $\|\cdot\|$ as defined in
Chapter 1; then a standard contraction mapping theorem yields the stated
result. We give a direct proof here.

Define

$$H_n = \sum_{j=0}^{n} \lambda^j K^j \tag{5}$$

and take $n > m$.

Then

$$H_n - H_m = \sum_{j=m+1}^{n} \lambda^j K^j$$

and hence

$$\|H_n - H_m\| \leq \sum_{j=m+1}^{n} \|\lambda K\|^j = \frac{\|\lambda K\|\{\|\lambda K\|^m - \|\lambda K\|^n\}}{1 - \|\lambda K\|}. \tag{6}$$

Now if $\|\lambda K\| < 1$, then

$$\lim_{n \to \infty} \|\lambda K\|^n = 0$$

and hence

$$\lim_{m,n \to \infty} \|H_n - H_m\| = 0.$$

Thus under the conditions of the theorem, the sequence $\{H_m\}$ is Cauchy and
therefore $\lim_{n \to \infty} H_n$ exists.

We should also show that this limit, which we may call L^{-1}, satisfies the
relations (3.1.7). The second of these may be written as

$$I - (I - \lambda K)L^{-1} = 0.$$

Now consider the sequence R_n:

$$R_n = I - (I - \lambda K)H_n.$$

Inserting the definition (5) we find

$$R_n = I - \sum_{j=0}^{n} \lambda^j K^j + \sum_{j=0}^{n} \lambda^{j+1} K^{j+1}$$

$$= \lambda^{n+1} K^{n+1}.$$

and hence

$$\| R_n \| \le \| \lambda K \|^{n+1}$$

and

$$\lim_{n \to \infty} \| R_n \| = 0$$

which implies

$$\lim_{n \to \infty} R_n = 0.$$

The operator L^{-1} is therefore a right inverse of L; a similar proof shows that it is also a left inverse.

The condition $\| \lambda K \| < 1$ is sufficient for λ to be a regular value, but by no means necessary. As an example, consider a kernel of the product form

$$K(s, t) = u(s)v(t) \tag{7}$$

where $u(s)$, $v(s)$ are real and orthogonal; that is,

$$(u, v) = \int_a^b u(s)v(s)\mathrm{d}s = 0. \tag{8}$$

Then (see Chapter 1) the operator K^2 has kernel

$$K^2(s, t) = \int_a^b K(s, r)K(r, t)\mathrm{d}r$$

$$= \int_a^b (u(s)v(r))(u(r)v(t))\mathrm{d}r = 0.$$

That is, $K^2 = 0$. Hence $K^n = 0$, $n \ge 2$ and the Neumann series (3) converges for all λ since it terminates. Hence, every finite λ is a regular value of the kernel (7) provided that u, v satisfy (8).

3.5   Generalisation of the Neumann series

Theorem 3.4.1 shows that for any kernel K, all (real or complex) values of λ are regular values provided that

$$|\lambda| < 1/\| K \|. \tag{1}$$

Equation (1) defines a disc in the complex plane, with centre the origin. We now generalise this result and show that if some value λ_0 is known to be a regular value, then all values λ sufficiently close to λ_0 are also regular values.

Theorem 3.5.1. If λ_0 is a regular value of K with inverse operator $(I - \lambda_0 K)^{-1} = L_0^{-1}$, then every value of λ such that

$$|\lambda - \lambda_0| < \frac{1}{\| L_0^{-1}K \|} \tag{2}$$

is also a regular value of K, with inverse operator

$$L^{-1} = P + \sum_{i=1}^{\infty} (\lambda - \lambda_0)^i Q^i P \tag{3}$$

where $P = L_0^{-1}$, $Q = PK$.

Proof. We sketch a constructive proof which follows closely on the development leading to the series (3.4.4c).

Equation (3.4.1) can be written in the form

$$[I - \lambda_0 K]x = y + (\lambda - \lambda_0)Kx$$

or on multiplying by the operator P:

$$x = Py + (\lambda - \lambda_0)Qx. \tag{4}$$

Now, as in Section 3.4, we seek an iterative solution of (4). We set $x_0 = Py$; then as for the Neumann series we obtain the obvious set of successive approximations x_n to x satisfying

$$x_{n+1} = Py + (\lambda - \lambda_0)Qx_n. \tag{5}$$

The solution of the recurrence relation (5) is given by

$$x_n = \sum_{i=0}^{n} (\lambda - \lambda_0)^i Q^i Py \tag{6}$$

and taking the limit as $n \to \infty$ yields the formal relation

$$x = Py + \sum_{i=1}^{\infty} (\lambda - \lambda_0)^i Q^i Py. \tag{7}$$

If (7) is valid for arbitrary y, equation (3) follows. The convergence of (7) can be investigated in exactly the same way as that of the Neumann series (3.4.3), convergence to L^{-1} following under the sufficient restriction (2).

3.6   The resolvent operator and the resolvent kernel

In some texts (for example, Smithies, 1958) the discussion of the existence of solutions of (3.1.1) is given not in terms of the inverse operator $L^{-1} = (I - \lambda K)^{-1}$, but in terms of the *resolvent operator H*, defined by the relation

$$L^{-1} = I + \lambda H. \tag{1}$$

Alternatively, but completely equivalently, H may be defined by the *resolvent equations*

$$H - K = \lambda KH = \lambda HK \tag{2}$$

which follow from (3.1.7) on inserting (1).

If L^{-1} exists, so (trivially) does H, and vice versa; the development of the theory, in terms of either L^{-1} or H, is a matter of taste and not content.

However, the reason why it is sometimes convenient to introduce H is illustrated by the Neumann series (3.4.3) for L^{-1}. This may equivalently be written as a series for H:

$$H_\lambda = K + K \sum_{i=1}^{\infty} \lambda^i K^i \tag{3}$$

from which two results follow:

(i) $\lim_{\lambda \to 0} H_\lambda = K$ $\qquad\qquad$ (4)

(ii) Considering the ith term in (3)

$$H_i \equiv \lambda^i K^{i+1}.$$

We note that K^i is, for any i, an integral operator of Fredholm form with an $\mathscr{L}^2$ kernel $K_i(s, t)$:

$$(K^i u)(s) = \int_a^b K_i(s, t) u(t) dt \tag{5}$$

where the kernel $K_i(s, t)$ satisfies a recurrence relation which follows from the operator relation

$$K^i = KK^{i-1}$$

and is

$$K_i(s, t) = \int_a^b K(s, r) K_{i-1}(r, t) dr. \tag{6}$$

The resolvent operator therefore has an associated kernel, the *resolvent kernel* $H_\lambda(s, t)$, which we may represent formally as the series

$$H_\lambda(s, t) = K(s, t) \left[I + \sum_{i=1}^{\infty} \lambda^i K_i(s, t) \right]. \tag{7}$$

The numerical calculation of the inverse operator L^{-1} is usually reduced to the calculation of this resolvent kernel, although not necessarily via the series (7), which is possibly only convergent for small $|\lambda|$. For other regular values λ_0 of K there exists an equivalent series based on (3.5.2) for $H_\lambda(s, t)$, and from Theorem 3.5.1 we obtain the following

Corollary 3.6.1. In the range $\rho(K)$ of regular values of K, the resolvent kernel $H_\lambda(s, t)$ is an analytic function of λ for each s, t.

For in the neighbourhood of any point λ_0 of $\rho(K)$, $H_\lambda(s, t)$ is given by an absolutely convergent power series in $(\lambda - \lambda_0)$.

3.7 Linear Volterra equations of the second kind

A linear Volterra equation of the second kind has the form

$$x(s) = y(s) + \lambda \int_a^s Q(s, t) x(t) dt \tag{1}$$

where $Q(s, t)$ is the kernel. If we define the kernel $K(s, t)$:

$$K(s, t) = \begin{cases} Q(s, t), & t \leq s, \\ 0, & t > s; \end{cases}$$

(2a)
(2b)

then for any finite $b > a$, (1) takes the standard Fredholm form (0.3.2)

$$x(s) = y(s) + \lambda \int_a^b K(s, t)x(t)dt$$

and the discussion of this chapter is directly applicable. A kernel $K(s, t)$ satisfying (2b) is said to be a Volterra kernel, and the corresponding integral operator, a Volterra operator. The product of two $\mathscr{L}^2$ Volterra operators is also an $\mathscr{L}^2$ Volterra operator. For, let $H(s, t)$, $K(s, t)$ be two Volterra kernels. Then the product $L = HK$ has kernel

$$L(s, t) = \int_a^b H(s, u)K(u, t)du.$$

(3)

But $H(s, u) = 0$ when $s < u$ and $K(u, t) = 0$ when $u < t$.

Hence $L(s, t) = 0$ when $s < t$. Indeed, (3) can be written in the form

$$L(s, t) = \int_t^s H(s, u)K(u, t)du, \quad a \leq t \leq s \leq b.$$

(3a)

One major feature of a Volterra operator is that it has no non-zero eigenvalues: the set $\rho(K)$ of regular values is the entire complex plane, and the Neumann series converges for all (real or complex) λ. The proof of this assertion is straightforward but lengthy. We need two lemmas:

Lemma 3.7.1. Let K be an $\mathscr{L}^2$ Volterra kernel and x an $\mathscr{L}^2$ function, and $b > a$. If

$$x_n(s) = \int_a^s K^n(s, t)x(t)dt$$

then

$$|x_n(s)| \leq \frac{k_1(s)\|x\|}{[(n-1)!]^{\frac{1}{2}}} \left\{ \int_a^s [k_1(t)]^2 dt \right\}^{\frac{1}{2}(n-1)}, \quad a \leq s \leq b$$

(4)

where

$$k_1(s) = \left[\int_a^s |K(s, t)|^2 dt \right]^{\frac{1}{2}}.$$

Proof. See Smithies (1958), p. 32.

Lemma 3.7.2.

$$|K^{n+1}(s,t)| \le \frac{\|K\|_E^{n-1}}{[(n-1)!]^{\frac{1}{2}}} k_1(s) k_2(t) \tag{5}$$

where

$$k_2(t) = \left[\int_t^b |K(s,t)|^2 ds \right]^{\frac{1}{2}}. \tag{5a}$$

Proof. Since $K(s,t)$ is an $\mathscr{L}^2$ function of s for all t, we may take $x(s) = K(s,t)$ in Lemma 3.7.1.

Then

$$\|x\| = k_2(t); x_n(s) = K^{n+1}(s,t)$$

and

$$\int_a^s [k_1(t)]^2 dt \le \int_a^b [k_1(t)]^2 dt = \|K\|_E^2, \tag{6}$$

whence the inequality follows at once from Lemma 3.7.1.

Theorem 3.7.1. If K is an $\mathscr{L}^2$ Volterra operator, the Neumann series

$$L^{-1} = (I - \lambda K)^{-1} = I + \sum_{n=1}^{\infty} \lambda^n K^n$$

converges strongly for all λ to the inverse operator of K.

Proof. We show that the series is convergent; the proof that it converges to L^{-1} then follows exactly as for Theorem 3.4.1.

Defining H_n as in (3.4.5), we obtain the bound, for $n > m$

$$\|H_n - H_m\|_E \le \sum_{j=m+1}^{n} \|\lambda^j K^j\|_E. \tag{7}$$

But from Lemma 3.7.2, (6) and (1.2.10d) it follows that

$$\|\lambda^j K^j\|_E \le |\lambda|^j \frac{\|K\|_E^j}{[(j-2)!]^{\frac{1}{2}}}$$

and hence that, for all λ,

$$\lim_{j \to \infty} \|\lambda^j K^j\|_E = 0, \tag{8}$$

whence the sequence $\{H_n\}$ is Cauchy for all λ and the theorem follows.

As a consequence of this theorem

(i) Equation (1) has an unique $\mathscr{L}^2$ solution for any y and all λ

(ii) The resolvent kernel $H_\lambda(s,t)$ of K is a Volterra kernel and is an entire function of λ for any s,t

(iii) Since no regular value is a characteristic value, the homogeneous equation

$$x(s) = \lambda \int_a^s K(s,t)x(t)dt \tag{9}$$

has no nontrivial $\mathscr{L}^2$ solution for any λ.

The following example (Smithies, 1958, p. 35) shows that non-$\mathscr{L}^2$ solutions of (9) may exist.

We take

$$K(s,t) = \begin{cases} t^{s-t}, & 0 < t \le s \le 1, \\ 0, & \text{otherwise.} \end{cases} \tag{10}$$

(Note that, in particular, $K(s,0)=0$).

Then K is bounded in $0 \le t \le s \le 1$ and is an $\mathscr{L}^2$ Volterra kernel. It therefore has no finite characteristic value. But the homogeneous equation (9) with kernel (10) and $\lambda = 1$ takes the form

$$x(s) = \int_0^s t^{s-t}x(t)dt, \quad 0 \le s \le 1 \tag{11}$$

and direct substitution verifies that this equation has the non-zero solution $x_0(s)$:

$$x_0(s) = \begin{cases} s^{s-1}, & 0 < s \le 1, \\ 0, & s=0, \end{cases} \tag{12}$$

in apparent contradiction to remark (iii) above.

The contradiction is only apparent since $x_0(s)$ is not square integrable. However, the example serves to illustrate how the details of an existence and uniqueness theory can depend on what is agreed to represent an 'acceptable solution'.

Exercises

1. Let S, T be $\mathscr{L}^2$ operators. Prove that if $(I - ST)$ has inverse operator U then $(I - TS)$ has inverse operator

$$(I - TS)^{-1} = I + TUS.$$

2. If $K(s,t)$ is an $\mathscr{L}^2$ operator with $K = K^\dagger$ then K is said to be self-adjoint or Hermitian; if $K = -K^\dagger$ then K is skew symmetric. Show that if A, B are $\mathscr{L}^2$ operators which are either both self-adjoint or

both skew symmetric, then

(i) $AB + BA$ is self-adjoint

(ii) $AB - BA$ is skew symmetric.

3. If the kernel $K(s, t)$ has characteristic function x corresponding to the characteristic value λ^*, and $K^\dagger$ has characteristic function y corresponding to the characteristic value μ, where $\lambda \neq \mu$, show that x and y are orthogonal.

4. Show that if K is an $\mathscr{L}^2$ Volterra operator then K^n is an $\mathscr{L}^2$ operator with kernel defined by

$$k_n(s, t) = \int_t^s k(s, u)k_{n-1}(u, t)du.$$

Use the results of Theorem 3.7.1 to show that the unique solution of

$$x(s) = y(s) + \lambda \int_0^s e^{(s-t)}x(t)dt, \quad 0 \leq s \leq 1,$$

is given by

$$x(s) = y(s) + \lambda \int_0^s e^{(\lambda+1)(s-t)}y(t)dt.$$

4

The Nystrom (quadrature) method for Fredholm equations of the second kind

4.1   Iteration and the Neumann series

We consider now the numerical solution of Fredholm equations of the second kind:

$$x(s) = y(s) + \lambda \int_a^b K(s,t)x(t)dt; \tag{1a}$$

which we write in the form

$$x = y + \lambda Kx. \tag{1b}$$

Many numerical techniques have been used successfully for such equations and in this chapter we discuss in detail a straightforward yet generally applicable technique: the 'Nystrom' or 'quadrature' method (see Section 4.2 below). However, it is convenient to begin by considering techniques based on the use of the Neumann series (3.4.4c):

$$x = \sum_{i=0}^{\infty} \lambda^i K^i y \tag{2a}$$

in truncated form

$$x \approx x_n = \sum_{i=0}^{n} \lambda^i K^i y. \tag{2b}$$

If the series (2a) converges, then $\lim_{n \to \infty} \| x - x_n \| = 0$. The approximate solution x_n is most easily produced iteratively via the obvious recurrence relation:

$$\left. \begin{array}{l} x_{n+1} = y + \lambda Kx_n \\[4pt] \text{with initial value} \\[4pt] x_0 = y \end{array} \right\} \tag{2c}$$

Provided that the function $Kx_n = \int_a^b K(s,t)x_n(t)dt$ can be computed, $(2c)$ is very convenient to use: we may steadily increase the degree of approximation until convergence is reached 'to a sufficient accuracy'. Two obvious questions are

(i) How do we know what is sufficient accuracy?

(ii) How do we carry out the iterations numerically?

We study these in turn.

4.1.1   A computable error bound

Question (i) has a rather satisfactory answer in terms of an *a posteriori* error bound which we can compute given two successive iterates x_n, x_{n+1}. Let us set

$$e_n = x_n - x.$$

Then on subtracting $(2c)$ from $(1b)$ we find

$$e_{n+1} = \lambda K e_n. \tag{3}$$

Moreover,

$$x_{n+1} - x_n = (x_{n+1} - x) + (x - x_n) = e_{n+1} - e_n,$$

that is,

$$e_n = e_{n+1} - (x_{n+1} - x_n).$$

Hence,

$$\|e_n\| \le \|e_{n+1}\| + \|x_{n+1} - x_n\| \le \|\lambda K\| \cdot \|e_n\| + \|x_{n+1} - x_n\|,$$

that is,

$$(1 - \|\lambda K\|)\|e_n\| \le \|x_{n+1} - x_n\|;$$
$$\|e_n\| \le \|x_{n+1} - x_n\|/(1 - \|\lambda K\|). \tag{4}$$

Inequality (4) represents a computable bound on the error, provided that we know or can estimate $\|\lambda K\|$ and $\|\lambda K\| < 1$. This last condition is also sufficient to guarantee that the Neumann series $(2a)$ converges to x; see Section 3.4. If $\|\lambda K\| > 1$, then inequality (4) is invalid and the series $(2a)$ may not converge. We return to this case below.

4.1.2   Numerical evaluation of the terms in the Neumann series

Given the function $x_n(s)$ the iterative scheme $(2c)$ requires that we produce the function $\lambda(Kx_n)(s)$:

$$\lambda(Kx_n)(s) = \lambda \int_a^b K(s,t)x_n(t)dt.$$

In general we shall not be able to carry out analytically the integrations involved. In this case we naturally turn to numerical quadrature. We introduce a quadrature rule R for the interval $[a, b]$ with weights w_i and points t_i:

$$Rf = \sum_{i=1}^{N} w_i f(t_i) = If - Ef = \int_a^b f(s)ds - Ef. \tag{5}$$

If we first ignore the error, the integral equation (1) is replaced by the approximate equation

$$x(s) = y(s) + \lambda \sum_{j=1}^{N} w_j K(s, t_j)x(t_j) \tag{6}$$

and (2c) by the approximate iterative scheme

$$x_{n+1}(s) = y(s) + \lambda \sum_{j=1}^{N} w_j K(s, t_j)x_n(t_j). \tag{7}$$

Now let us consider how convenient, or inconvenient, the iterations are. The first three iterations have the form

$$x_0(s) = y(s),$$

$$x_1(s) = y(s) + \lambda \sum_{j=1}^{N} w_j K(s, t_j)y(t_j),$$

$$x_2(s) = y(s) + \lambda \sum_{j=1}^{N} w_j K(s, t_j)\left\{ y(t_j) + \lambda \sum_{k=1}^{N} w_k K(t_j, t_k)y(t_k) \right\}.$$

We see that, at least if we only ask for $x_n(t_j)$,

(i) the only function evaluations involved are for $y(t_j)$ and $K(t_j, t_k)$;

(ii) the values of the successive iterates $x_n(s)$ need only be tabulated at the points t_k for the iterations to proceed;

(iii) we may write the iterations (for any value of n) in the form

$$\mathbf{x}_{n+1} = \mathbf{y} + \lambda \mathbf{K} \mathbf{x}_n \tag{7a}$$

where

$$\left. \begin{array}{l} (x_n)_i = x_n(t_i) \\ y_i = y(t_i) \end{array} \right\}, \quad i = 1, \ldots, N, \tag{7b}$$

and

$$\mathbf{K}_{ij} = w_j K(t_i, t_j), \quad i, j = 1, \ldots, N.$$

From (7) we have immediately some interesting consequences:

(iv) Repeating the manipulations leading to (4), it is easy to show that the iterations (7) will converge if in any matrix norm

$$\| \lambda \mathbf{K} \| < 1, \tag{8}$$

a relation with obvious analogy to the conditions for convergence of the original Neumann series.

(v) If the iterations converge, they converge to the solution x_R of the equation

$$(I - \lambda K)x_R = y. \tag{9}$$

We return to this remark later; the suffix R makes explicit the dependence of (9) on the quadrature rule R.

(vi) If we define the 'error vector' e_n

$$e_n = x_R - x_n \tag{10}$$

(note that this 'error' refers only to the error relative to the exact solution of (9), and not to the solution of the integral equation) then an argument identical to that leading to (4) also yields the error bound

$$\|e_n\| \le \|x_{n+1} - x_n\|/(1 - \|\lambda K\|) \tag{11}$$

where $\|x\|$, $\|\lambda K\|$ are any two compatible vector/matrix norms for which $\|I\| = 1$.

4.1.3 Truncation error in the solution of the approximating equation (6)

Let us suppose that we solve (9) exactly. This equation defines an approximation to $x(s)$ only at the 'mesh points' t_j; but we may choose to treat (6) (with $x(t_j)$ replaced by $(x_R)_j$) as defining an approximate solution for all $a \le s \le b$; if we do this, we have an exact solution of (6) everywhere, which we shall call $x_R(s)$ to indicate that it depends on the rule R used. How closely does x_R approximate the exact solution x of (1)?

We set $e_R(s) = x(s) - x_R(s)$ and note that $x_R(s)$ satisfies the integral equation

$$x_R(s) = y(s) + \lambda \int_a^b K(s, t)x_R(t)dt - E_t\lambda K(s, t)x_R(t) \tag{12}$$

where (see equation (5)) E_t indicates the error functional for the quadrature rule R operating on $\lambda K(s, t)x_R(t)$, viewed as a function of t for fixed s. We then have, subtracting (12) from (1):

$$e_R(s) = E_t\lambda K(s, t)x_R(t) + \lambda \int_a^b K(s, t)e_R(t)dt. \tag{13}$$

Thus, the error function $e_R(s)$ satisfies a Fredholm integral equation of the second kind with the same kernel as (1), but with a different driving term.

Equation (13) has the solution

$$e_R(s) = [I - \lambda K]^{-1}E_t\lambda K(s, t)x_R(t) = [I + \lambda H]E_t\lambda K(s, t)x_R(t) \tag{14}$$

where H is the resolvent kernel. Hence taking norms we find

$$\|e_R\| \le (1 + \|\lambda H\|)\|E_t\lambda K(s, t)x_R(t)\|. \tag{14a}$$

This equation is not very useful in practice because of the difficulty of computing $\|\lambda H\|$. However, if $\|\lambda K\| < 1$ we find, on taking norms in (13) and rearranging, the simpler bound

$$\|e_R\| \leq \frac{\|E_t \lambda K(s,t) x_R(t)\|}{1 - \|\lambda K\|}. \tag{14b}$$

It is usually possible to estimate $\|\lambda K\|$ rather simply; the numerator of (14b) can also be bounded in general, given error forms for the quadrature rule used. The condition $\|\lambda K\| < 1$ is a natural one in the current context of iteration; however, it is often violated in practice and for later use we derive a form which does not suffer from this limitation and which avoids introducing the resolvent kernel H. We rewrite (1) and (12) in the equivalent forms

$$x(s) = y(s) + \lambda \sum_{i=1}^{N} w_i K(s,t_i) x(t_i) + E_t \lambda K(s,t) x(t), \tag{15}$$

$$x_R(s) = y(s) + \lambda \sum_{i=1}^{N} w_i K(s,t_i) x_R(t_i). \tag{16}$$

Subtracting (16) from (15), and then setting $s = t_i$, we find that $e_R(s)$ satisfies the equations

$$e_R(t_i) = \lambda \sum_{j=1}^{N} w_j K(t_i,t_j) e_R(t_j) + E_t \lambda K(t_i,t) x(t), \tag{17}$$

that is,

$$\mathbf{e}_R = \lambda \mathbf{K} \mathbf{e}_R + \mathbf{k} \tag{17a}$$

where $(\mathbf{k})_i = E_t \lambda K(t_i,t) x(t)$.
(17a) has the solution

$$\mathbf{e}_R = (\mathbf{I} - \lambda \mathbf{K})^{-1} \mathbf{k},$$

whence, on taking norms:

$$\|\mathbf{e}_R\| \leq \|(\mathbf{I} - \lambda \mathbf{K})^{-1}\| \cdot \|\mathbf{k}\|. \tag{18}$$

In (18), $\|\mathbf{e}_R\|$ and $\|\mathbf{k}\|$ are finite vector norms, $\|\mathbf{e}_R\|$ bounding the error at the mesh points; the norm $\|(\mathbf{I} - \lambda \mathbf{K})^{-1}\|$ can be calculated from a knowledge of the matrix K. The norm $\|\mathbf{k}\|$ has the disadvantage that it contains the unknown solution x but we can perhaps estimate $x(s)$ in some way (for instance by $x_R(s)$) and hence estimate $\|\mathbf{k}\|$. In any case, (18) has the advantage of being directly applicable even if $\|\lambda K\| > 1$.

4.1.4   Total error in the iterates x_n

Finally, if we evaluate the integrals for each iteration numerically and terminate the iterations (2c) at $\mathbf{x}_n$, the total error on the mesh used is given by

$$\|\mathbf{x}_n - \mathbf{x}\| = \|(\mathbf{x}_n - \mathbf{x}_R) + (\mathbf{x}_R - \mathbf{x})\| \leq \|\mathbf{x}_n - \mathbf{x}_R\| + \|\mathbf{x}_R - \mathbf{x}\| \quad (19)$$

and we have bounded both of these terms above.

4.2 The Nystrom or quadrature method

4.2.1 The Nystrom method

If the iterative scheme (4.1.2c) converges, it converges to a solution of (4.1.9). May we not therefore solve this directly? And could we not have written down (4.1.9) directly *ab initio*?

The answer to both these questions is: Yes. Equation (4.1.9) follows directly if we approximate the integral in (4.1.1) by a quadrature rule, ignore the error term, and write the resulting equation for $x(s)$ at the rule nodes t_i. We may solve (4.1.9) directly even if $\|\lambda K\| > 1$; equation (4.1.18) then shows that the approximation can be expected to be sensible: the error remains bounded provided only that the quadrature error is bounded and $\mathbf{I} - \lambda\mathbf{K}$ is nonsingular. This prescription defines the very common *Nystrom* or *quadrature* method for the solution of (4.1.1) and we shall refer to (4.1.9) as the Nystrom equations, and to $\mathbf{x}_R$, and its extension $x_R(s)$, as the *Nystrom solution* (for the rule R).

Why then consider the Neumann iterations at all? To answer this, let us look on our development in reverse and consider (4.1.9) as the basic equation. Then (4.1.7a) can be recognised as a (modified) Jacobi iterative scheme for the solution of (4.1.9).[†] We should use the Neumann series if and only if the iterative solution of (4.1.9) seems to promise to be quicker than a direct solution.

For one-dimensional problems, this is rather often not the case. A typical one-dimensional rule to cope with a smooth kernel might have 10–15 Gaussian nodes and a direct solution of the equations is then so painless that iterative schemes (which involve checking $\|\lambda K\|$ to avoid diverging) are not worthwhile considering. However, 'difficult' problems (or a poor choice of quadrature rule) may lead to much larger values of N, making iteration potentially quite worthwhile. Moreover, in m-dimensional problems the number of nodes becomes raised to the mth power and the situation changes radically. Then, we may have physical or other information which suggests that only a few iterations of the equation (4.1.2c) are necessary; these iterations may prove very much faster than a complete solution of (4.1.9).

[†](4.1.7a) is not a completely standard Jacobi scheme; the standard scheme has the form

$$[\mathbf{I} - \mathbf{diag}(\lambda\mathbf{K})]\mathbf{x}_{n+1} = \mathbf{y} + \lambda[\mathbf{K} - \mathbf{diag}(\mathbf{K})]\mathbf{x}_n$$

where $\mathbf{diag}(\mathbf{K})$ is a diagonal matrix containing the diagonal elements of $\mathbf{K}$.

If the Neumann series diverges, as it usually will for large $\| \lambda K \|$, the iterative scheme (4.1.2c) will usually diverge also; we can however then consider alternative iterative schemes for the solution of (4.1.9). We return to this topic in Section 4.6.2.

4.2.2   Accuracy of the Nystrom method: qualitative considerations

For a given quadrature rule R, generally applicable bounds for the Nystrom method are given by equation (4.1.14a). This equation contains two factors:

(i) A factor $(1 + \| \lambda H \|)$ dependent only on the integral equation and not at all on the quadrature rule used. This factor reflects the sensitivity of the solution to small changes in the integral operator or driving term; that is, the inherent conditioning of the problem.

(ii) A factor $\| E_t \lambda K(s,t) x_R(t) \|$ dependent on both the kernel and the solution. We can hope to make this factor small by a suitable choice of quadrature rule. It represents a measure of the error in integrating the function $\lambda K(s,t) x_R(t)$ over the variable t for fixed s; if this error is small for all $a \leq s \leq b$ then this factor will be small. If, in particular, the rule chosen will integrate $\lambda K(s,t) x_R(t)$ exactly for every fixed s, then the Nystrom method will yield the exact solution to the integral equation.

The same conclusions follow from the alternative bound (4.1.18), although in this case the factor $\| (I - \lambda K)^{-1} \|$ depends, rather weakly, on the quadrature rule R.

We can attempt to get some feeling for the accuracy we might achieve in practice by looking at some particular examples.

Example 4.2.1.

$$x(s) = \left(1 - \frac{\alpha}{2} \right) s - \frac{\alpha}{3} + \alpha \int_0^1 (t+s) x(t) \mathrm{d}t. \tag{1}$$

Solution: $x(s) = s$, independent of α.

In this case the integral term $\int_0^1 K(s,t)x(t)\mathrm{d}t$ clearly has an integrand which is quadratic in t. Hence, any rule of degree at least 2 will yield the exact solution to this equation (the error functional term in (4.1.17a) is zero). For example, Simpson's rule will achieve this infinite accuracy.

Example 4.2.2.

$$x(s) = \sin s - \alpha s + \alpha \int_0^{\pi/2} t s x(t) \mathrm{d}t. \tag{2}$$

Solution: $x(s) = \sin s$.

In this case no rule of finite degree will give an exact solution, but since both the kernel and the solution are smooth (indeed analytic) functions of their arguments, we would expect that a rule of high degree will lead to high accuracy.

Example 4.2.3.

Consider the differential equation

$$x''(s) + \cos^2 s \; x(s) = -\sin^3 s,$$

$$x(0) = 0,$$

$$x\left(\frac{\pi}{2}\right) = 1. \tag{3}$$

Solution: $x(s) = \sin s$.

We saw in Chapter 0 that this has the equivalent integral equation representation

$$x(s) = y(s) + \int_0^{\pi/2} K(s, t) x(t) \mathrm{d}t \tag{4}$$

where

$$y(s) = \frac{2s}{\pi} + \int_0^{\pi/2} K_0(s, t) \sin^3 t \, \mathrm{d}t,$$

$$K(s, t) = K_0(s, t) \cos^2 t,$$

$$K_0(s, t) = \begin{cases} \dfrac{2t}{\pi}\left(\dfrac{\pi}{2} - s\right), & 0 \le t \le s, \\[2mm] \dfrac{2s}{\pi}\left(\dfrac{\pi}{2} - t\right), & s \le t \le \dfrac{\pi}{2}. \end{cases} \tag{5}$$

For this equation the solution is smooth but the kernel function $K(s, t)$ is not (it has discontinuous first partial derivatives at $s = t$). The discussion of Sections 2.8 and 2.9 then suggests that any sequence of rule R_N which does not take explicit account of this discontinuity, is likely to lead to errors which decrease only slowly as N is increased.

4.2.3   Convergence and rates of convergence

This last example brings us to an important characteristic of any numerical method for the solution of (4.1.1). In practice, we require not only an approximate solution but also an estimate of its accuracy. The provision of practical error estimates for a given N-point quadrature rule R is discussed below in Section 4.3. However, the simplest and most commonly used procedure for estimating the achieved accuracy is to choose a family of rules R_N and to compute an approximate solution for members of this family with increasing N, and hence (hopefully) decreasing error, until the results appear to have settled down to the required accuracy. It is usually straightforward to guarantee convergence; that is, to ensure that

$$\lim_{N \to \infty} \| e_{R_N} \| = 0. \tag{6}$$

Sufficient conditions for this are given, for example, in the following theorem.

Theorem 4.2.1. Let $K(s, t)$ be continuous in the square $a \leq s, t \leq b$, $y(s)$ be continuous in $a \leq s \leq b$ and $x(s)$ be the unique continuous solution of (4.1.1). Let $\{R_N\}$ be a sequence of quadrature rules for the interval $[a, b]$, convergent for all continuous functions; that is, for all $f \in C[a, b]$:

$$\lim_{N \to \infty} | If - R_N f | = 0.$$

Then the Nystrom method used with this sequence of rules is convergent; that is, equation (6) holds.

Proof. See Baker (1977), Chapter 2, Section 15.

The conditions of this theorem are met (see Section 2.8) by the following families of rules, on any finite interval $[a, b]$:

   (i) The M-panel Newton–Cotes rules $NC(P, M)$ of degree P, for any fixed P and increasing M
   (ii) The Gauss–Legendre and open and closed Gauss–Chebyshev N-point rules for increasing N.

They are *not* met by the sequence $NC(P, M)$ for fixed M and increasing degree P. It is usual to avoid such a sequence; the most common choice of rules in practice seems to include the repeated Trapezoid or Simpson's rules, and a holistic (that is, one-panel) sequence of Gauss (Legendre or Chebyshev) rules. Either choice guarantees convergence but equally important is the *rate* at which convergence is achieved. Referring back to Section 2.7, the error estimates derived there for N-point rules of various types had the form

$$|E_t f| = \left| \int_a^b f(t)\mathrm{d}t - \sum_{i=1}^N w_i f(t_i) \right| \leq C(f) N^{-p} \tag{7}$$

where $C(f)$ is a constant dependent on the function $f(t)$ and the exponent p depends either on the degree of the rule (if $f(t)$ is smooth enough) or on the continuity properties of $f(t)$ (if the degree of the rule is high enough). We refer to p as the *order of convergence*.

Now applying (7) to the bound (4.1.14a) we note that the amplitude $C(f)$ may depend upon the variable s (we assume that p does not); (4.1.14a) takes the form

$$\|e_R\| \leq (1 + \|\lambda H\|)\|C(\lambda K(s,t)x_R(t))\|N^{-p}. \tag{8}$$

If C is uniformly bounded in s:

$$|C(\lambda K(s,t)x_R(t))| \leq C_0, \quad a \leq s \leq b. \tag{9}$$

Then, taking the maximum norm $\|\cdot\|_\infty$:

$$\|e_R\|_\infty \leq (1 + \|\lambda H\|_\infty)C_0 N^{-p}; \tag{10}$$

and the convergence rate of the Nystrom method is the same as that of the underlying family of quadrature rules. Since relation (9) is satisfied, at least for the commonly used rules, under quite mild assumptions on the kernel $K(s,t)$, the same qualitative remarks concerning the most useful choice of rule apply to the Nystrom method as were made in Chapter 2. That is, high-order rules are likely to be much more effective than low-order rules on 'easy' problems and no less effective on 'difficult' problems. However, we must add here the proviso that special classes of 'difficult' integral equations are important enough to warrant the production of special rules to deal with them effectively; see Section 4.4. below.

4.2.4 Numerical examples

We illustrate the discussion of the last two sub-sections by some numerical examples. In each case, we plot the maximum error $\|e_{R_N}\| = \max_{a \leq s \leq b} |x(s) - x_{R_N}(s)|$ obtained using the Nystrom method for the solution of (4.1.1), with a number of alternative families of rules:

(i) Gauss: The N-point Gauss–Legendre rule, scaled to the interval $[a,b]$

(ii) $NC(q,M)$: The repeated (M-panel) q-point closed Newton–Cotes rule (of degree $q-1$ for q even and q for q odd). The total number of points for such a rule is $N = M(q-1) + 1$; the choices $q = 2, 3$ correspond to the repeated Trapezoid and Simpson's rules respectively.

Example **4.2.4.**

$$x(s) = y(s) + \int_0^1 e^{st}x(t)\mathrm{d}t,$$

$$y(s) = e^s - (e^{s+1} - 1)(s+1).$$

Table 4.2.1. *The maximum error* $\|e_{R_N}\|$ *for the Nystrom solution of the integral equation of Example 4.2.4, using alternative N-point quadrature rules.*

Gauss		Trapezoid		Simpson		NC(5, M)		NC(7, M)	
N	$\|e_{R_N}\|$	N	$\|e_{R_N}\|$	N	$\|e_{R_N}\|$	N	$\|e_{R_N}\|$	N	$\|e_{R_N}\|$
2	1.1×10^{-2}	3	3.8, -1						
4	3.2, -7	5	9.8, -2	5	1.2, -3			7	3.7, -7
6	1.5, -10	9	2.4, -2	9	7.7, -5	9	1.5, -6	13	1.4, -9
		17	6.1, -3	17	4.8, -6	17	2.4, -8		
		33	1.5, -3	33	2.8, -7	33	3.6, -10		
		65	3.8, -4	65	1.9, -8				
		129	9.6, -5	129	1.2, -9				

Exact solution: $x(s) = e^s$.

This example typifies equations with smooth kernels, driving term and solution. For such an apparently easy problem a cheap numerical solution should certainly be possible. The results obtained with the Nystrom method are given in Table 4.2.1; the worldlength of the computer used (an ICL 1906S) was about 11 decimal digits. As expected, the higher order rules are strongly favoured by these results, with the Gauss rule a clear overall winner; a fit to these results shows that the behaviour of the error norm $\|e_{R_N}\|$ for the Newton–Cotes rules is well represented by equation (10), with convergence rates

$$p(NC) \approx 2(\text{Trapezoid}),$$
$$\approx 4(\text{Simpson}),$$
$$\approx 6(NC(5, M)).$$

These figures also represent the degree of the rule, in agreement with the theoretical bound (7).

Example 4.2.5.

$$x(s) = y(s) + \int_0^{\pi/2} s^\beta t^\beta x(t) dt.$$

$$y(s) = s^\alpha - \left(\frac{\pi}{2}\right)^{\alpha + \beta + 1} \frac{s^\beta}{(\alpha + \beta + 1)}.$$

Exact solution: $x(s) = s^\alpha$.

This rather contrived example is intended to display the performance of the Nystrom method on problems with analytic 'difficulties' in the kernel and

Table 4.2.2. *Maximum error* $\|e_{R_N}\|$ *for the Nystrom solution of the integral equation of Example 4.2.5a (with $\alpha = \beta = \frac{3}{2}$).*
The error is essentially zero for any rule of polynomial degree ≥ 3.

Gauss		Trapezoid		Simpson	
N	$\|e_{R_N}\|$	N	$\|e_{R_N}\|$	N	$\|e_{R_N}\|$
2	2.9×10^{-11}	2	6.2, -1	3	1.5, -11
		3	3.5, -1		
		5	1.3, -1		
		9	3.6, -2		
		17	9.4, -3		
		33	2.4, -3		
		65	5.9, -4		
		129	1.5, -5		

driving term. We give α, β non-integer values; then none of K, g or the exact solution x are smooth over the regions involved. Each of the rules we consider is therefore expected to yield only slow convergence in general. We consider two particular cases:

***Example* 4.2.5a:** $\alpha = \beta = \frac{3}{2}$.
The results for this choice of parameters are given in Table 4.2.2. They are at first sight unexpected. Those for the Trapezoid rule indeed converge slowly with convergence rate (see (10)):

$$p \text{ (Trapezoid)} \approx 2,$$

as for Example 4.2.4. However, the results for the Gauss 2-point rule and the single panel Simpson's rule are exact to machine accuracy. This illustrates the remark made in Section 4.2.2: the error term $E_t(Kx)$ is zero in this example, since

$$K(s, t)x(t) = s^{\frac{3}{2}}t^{\frac{3}{2}} \cdot t^{\frac{3}{2}} = s^{\frac{3}{2}}t^3$$

and for all s this function is integrated exactly by any rule of degree ≥ 3. This behaviour would be quite difficult to predict without a knowledge of the exact solution, since the exact form of the solution depends on a cancellation between the kernel and driving term singularities.

***Example* 4.2.5b:** $\alpha = \frac{1}{2}$, $\beta = \frac{3}{4}$.
With this choice of parameters, $K(s, t)x(t) = s^{\frac{3}{2}}t^{\frac{1}{2}}$ is badly behaved near $t = 0$. The results are shown in Table 4.2.3 and the expected slow

Table 4.2.3. *Maximum error $\|e_{R_N}\|$ for the Nyström solution of the integral equation of Example 4.2.5b (with $\alpha = \frac{1}{2}, \beta = \frac{3}{4}$). The values of p shown are estimates of the convergence rate (see equation (10)).*

Gauss		Trapezoid		Simpson		NC(4, M)		NC(5, M)		NC(6, M)		NC(7, M)	
N	$\|e_{R_N}\|$	N	$\|e_{R_N}\|$	N	$\|e_{R_N}\|$	N	$\|e_{R_N}\|$	N	$\|e_{R_N}\|$	N	$\|e_{R_N}\|$	N	$\|e_{R_N}\|$
2	1.7×10^{-2}	3	1.9, -1	3	4.0, -2	4	2.3, -2	5	6.7, -3	6	4.8, -3	7	2.2, -3
4	1.1, -3	5	6.5, -2	5	8.8, -3	7	5.0, -3	9	1.4, -3	11	1.0, -3	13	4.6, -4
8	6.0, -5	9	1.9, -2	9	1.9, -3	13	1.1, -3	17	3.0, -4	21	2.1, -4	25	9.8, -5
16	3.0, -6	17	5.1, -3	17	3.9, -4	25	2.2, -4	33	6.2, -5	41	4.4, -5	49	2.1, -5
32	1.4, -7	33	1.3, -3	33	8.3, -5	49	4.7, -5	65	1.3, -5	81	9.4, -6	97	4.3, -6
64	6.4, -9	65	3.5, -4	65	1.7, -5	97	9.9, -6						
$p \approx 4.5$		$p \approx 2$		$p \approx 2.25$		$p \approx 2.25$		$p \approx 2.25$		$p \approx 2.25$		$p \approx 2.25$	

convergence is realised for all the rules, fits to the tabulated errors yielding convergence rates with exponent

$$p \approx 2.25 \text{(Newton–Cotes, degree} > 2),$$

$$\approx 4.5 \text{(Gauss–Legendre)}.$$

For this example, the high-order Newton–Cotes rules show no advantage over the low-order Simpson's rule. On the other hand, they show no disadvantage; these results, together with those for Examples 4.2.4 and 4.2.5a, suggest that the Gauss rule in particular might be recommended for general use. They also suggest however that an explicit treatment of singularities would be well worthwhile and we return to this in Section 4.4.

4.2.5 Sufficient conditions for a smooth solution

The estimates of convergence rate above and the numerical examples show that the quadrature error will be rapidly convergent to zero if the degree of the rule is high and if for every fixed s, $K(s, t)x(t)$ is a smooth function of t with a high number of continuous derivatives. Under what conditions can we guarantee that this is so? We note that if y and K are continuous then so is $x(t)$ (for we then seek the unique continuous solution of the equation). Under this assumption we can discuss the analyticity properties of x:

Theorem 4.2.2. Let $y(s) \in C^{(p)}[a, b]$ and $K(s, t) \in C^{(p)}[a, b] \times C^{(p)}[a, b]$. Then $x(s) \in C^{(p)}[a, b]$.

Proof. We have

$$x(s) = y(s) + \lambda \int_a^b K(s, t)x(t)dt,$$

whence

$$x'(s) = y'(s) + \lambda \int_a^b \frac{\partial}{\partial s} K(s, t)x(t)dt.$$

But, for $p \geq 1$, $[(\partial/\partial s)K(s, t)]$ is continuous by hypothesis, and then the integral $[\int_a^b (\partial/\partial s)K(s, t) \cdot x(t)dt]$ is a continuous function of s. Hence if $y'(s)$ is continuous so is $x'(s)$.

Similarly for any $q \leq p$:

$$x^{(q)}(s) = y^{(q)}(s) + \lambda \int_a^b \frac{\partial^q}{\partial s^q} K(s, t)x(t)dt;$$

whence, proceeding inductively, it follows that $x \in C^{(p)}[a, b]$.

4.3   Practical error estimates for the Nystrom method

The error estimate (4.2.10) is useful in demonstrating the rate of convergence that can be expected if we use a repeated rule of fixed degree with the Nystrom method.

The underlying quadrature error estimates on which it depends however, in general estimate the parameter C_0 in terms of bounds on some high-order derivative of the solution $x(s)$ and these are usually not available numerically (see Section 2.7). Hence, although the exponent p can often be estimated *a priori*, C_0 is much more difficult to estimate, and as a result (4.2.10) does not lead at all easily to computable error estimates; that is, to estimates that can be provided routinely as part of the solution process. For a low-order rule, such as the repeated Trapezoid or Simpson's rule, the required derivatives could be estimated by differencing the solution but this is impractical for the (usually more efficient) high-order Gauss rules. In practice, two ways have been proposed and used to provide error estimates within the Nystrom framework:

4.3.1   Brute force error estimates

Given the result x_N for an N-point rule, we compute also the result x_{MN} for an MN-point rule and use as an error estimate

$$\varepsilon_E = \| x_N - x_{MN} \|. \tag{1}$$

This estimate assumes that x_{MN} is much closer to x than is x_N, since the exact error is

$$\varepsilon = \| x_N - x \| \le \| x_N - x_{MN} \| + \| x_{MN} - x \|. \tag{2}$$

This assumption will be justified if either x_N is converging rapidly, or $MN \gg N$. In practice, values of M in the range $M = 1.5$–2.0 are used; different authors have adopted different approaches to the problem: shall I return x_{MN} as the solution? (in which case the error estimate is likely to be conservative but none better is available) or x_N? (which seems a pity when the presumably more accurate x_{MN} has been computed, but the user probably specified an N-point solution).

4.3.2   Comparisons of two rules using the same quadrature points

The method of the previous section seems the only practical one available for a Nystrom method based on a Gauss rule. If, however, we have decided to use a repeated Newton–Cotes rule with equally spaced points, it is possible to arrange to re-use the kernel evaluations by repeating the calculation using fewer panels of a higher order rule, or of the same order but with a smaller error constant, and again to compare the two solutions to obtain an error estimate. The idea, which is familiar within the context of predictor–corrector methods for ordinary differential equations, is sufficiently illustrated in Figure 4.3.1, which ignores difficulties over 'odd points' and shows two compatible rules: a three-panel Simpson's rule and a two-panel rule based on the four-point Newton–Cotes rule (the 'three-eighths' rule):

$$\int_a^{a+3h} f(s)ds = \frac{3h}{8}[f(a) + 3f(a+h) + 3f(a+2h)$$

$$+ f(a+3h)] - \frac{3h^5}{80} f^{(iv)}(s).$$

These rules use the same seven equally spaced points. The underlying quadrature error for these two rules in an interval $[a,b]$ is

$$E(\text{Simpson}) = -\frac{h^4(b-a)}{180} f^{(iv)}(\xi_1), \tag{3}$$

$$E(\text{three-eighths}) = -\frac{h^4}{80}(b-a)f^{(iv)}(\xi_2),$$

where $a < \xi_1, \xi_2 < b$. Because of the similarity between these two error forms, we refer to the rules as 'companion' rules. Now if we solve (4.1.1) using the repeated Simpson and three-eighths rules, we obtain two solutions:

Simpson: $x_1 = x + e_1$

three-eighths: $x_2 = x + e_2$

where from (4.1.14) and (3) the error functions satisfy the relation

$$e_1(s) = -\frac{h^4(b-a)}{180}[I + \lambda H]\left[\frac{\partial^4}{\partial t^4} \lambda K(s,t)x_1(t)\right]_{t=\xi_1(s)},$$

$$e_2(s) = -\frac{h^4(b-a)}{80}[I + \lambda H]\left[\frac{\partial^4}{\partial t^4} \lambda K(s,t)x_2(t)\right]_{t=\xi_2(s)}.$$

Ignoring the distinction between x_1 and x_2, and between ξ_1 and ξ_2, on the right hand side of these relations yields the approximate relation

$$e_1(s) \approx \tfrac{4}{9}e_2(s) \tag{4a}$$

Figure 4.3.1. Two rules sharing the same points.

3-panel Simpson's rule

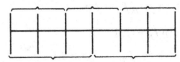

2-panel Newton-Cotes 4-point rule

and finally the estimate

$$x(s) = x_1(s) + \tfrac{4}{5}(x_2(s) - x_1(s)).\tag{4b}$$

Equation (4b) can be used either to correct the solution x_1, or to yield the error estimate:

$$\|e_1\| \approx \tfrac{4}{5}\|x_2 - x_1\|.\tag{5}$$

Both this method and that of Section 4.3.1 are costly to implement, since producing an approximate solution x_N involves the solution of an $N \times N$ full set of linear equations. If we assume the asymptotic relation $\mathrm{cost}(x_N) \sim AN^3$ then the relevant factor for this method is $A = 2$, while that for the method of Section 4.3.1 is $1 + M^3$, that is, $A = 4.4$ for $M = 1.5$ and $A = 9$ for $M = 2$. Clearly, there is pressure to keep M as small as possible and equally clearly, the cost of producing a practical error estimate is high with the Nystrom scheme.

4.3.3   Direct estimation of the error term

An alternative approach, pursued by Thomas (1975), is to estimate the error term directly, again using numerical quadrature. The details of Thomas' approach are rather complex; as for the method of the previous section, the analysis is limited by its nature to the low-order repeated rules; Thomas gave an analysis only for Simpson's rule. Again, it costs as much to produce the error estimate with this approach as the original solution and we shall not pursue this method further here.

4.3.4   The difference correction

The methods of Sections 4.3.2, 3 produce an error function $e(s)$ as well as an error norm $\|e\|$. This function can be used to yield an error estimate or can be added as a correction term to improve the accuracy of the calculated $x(s)$. An alternative approach, which leads to a sequence of such corrections, derives from an alternative form for the error in a repeated (N-panel) Newton–Cotes rule in terms of differences of the integrand. For the repeated Trapezoid rule this takes the form

$$\frac{1}{h} \int_a^{a+Nh} f(s)\,ds = \tfrac{1}{2}f_0 + f_1 + \cdots + f_{N-1} + \tfrac{1}{2}f_N + \Delta \tag{6}$$

where

$$\Delta = (-\tfrac{1}{12}\Delta + \tfrac{1}{24}\Delta^2 - \tfrac{19}{720}\Delta^3 + \cdots)(f_N - f_0)\tag{7a}$$

$$= (-\tfrac{1}{12}\mu\delta + \tfrac{11}{720}\mu\delta^3 + \cdots)(f_N - f_0)\tag{7b}$$

$$= -(\tfrac{1}{12}\nabla + \tfrac{1}{24}\nabla^2 + \tfrac{19}{720}\nabla^3 + \cdots)f_N$$
$$+ (\tfrac{1}{12}\Delta - \tfrac{1}{24}\Delta^2 + \tfrac{19}{720}\Delta^3 + \ldots)f_0.\tag{7c}$$

In these formulae Δ, ∇, δ are the forward, backward and central difference operators and μ here is the averaging operator. The forms (7*a*) and (7*b*) always use points outside the range of integration, while at least the first few terms of (7*c*) (the *Gregory* form) use only interior points. This is therefore the most widely used form; however, we can generate points outside the region $[a, b]$ if needed, directly from the integral equation, provided that the kernel K is defined outside $[a, b] \times [a, b]$, using (4.1.6) as an extrapolation formula.

The use of any of the forms (7) is now straightforward. We first solve the set of uncorrected Nystrom equations

$$\mathbf{A}\mathbf{x}^{(0)} = \mathbf{y} \tag{8}$$

where

$$\mathbf{A} = (\mathbf{I} - \lambda\mathbf{K}).$$

For each fixed $i = 0, 1, \ldots, N$ we may difference the approximate value $K(t_i, t_j)x_j^{(0)}$ (augmented by exterior values via (4.1.6) if necessary) to produce the difference correction to the integral

$$\int_a^b K(t_i, t)x(t)\mathrm{d}t = \Delta_i(x^{(0)}) \text{ say.}$$

In computing Δ_i we retain as many differences as appear significant. Then from the vector $\mathbf{\Delta}(x^{(0)})$ we can solve the correction equations

$$\mathbf{A}\mathbf{x}^{(1)} = h\mathbf{\Delta}(x^{(0)}), \tag{9}$$

$$\mathbf{A}\mathbf{x}^{(2)} = h\mathbf{\Delta}(x^{(1)}),$$

$$\vdots$$

the final solution being given by

$$\mathbf{x} = \mathbf{x}^{(0)} + \mathbf{x}^{(1)} + \mathbf{x}^{(2)} + \cdots.$$

The number of iterations needed depends on the interval h; if we choose h too large the difference corrections may in fact not converge. In practice, the step size h is normally chosen so that only one or two iterations are required for convergence.

We illustrate the method with an example taken from Fox and Goodwin (1953), p. 62.

Example 4.3.1.

$$x(s) = -\frac{2}{\pi}\cos s + \frac{4}{\pi}\int_0^{\pi/2} \cos(s - t)x(t)\mathrm{d}t.$$

Solution: $x(s) = \sin s$.

Using a nine-point Trapezoid rule and solving the equations (8), (9) for $x^{(0)}$ and $x^{(1)}$ we find the results shown in Table 4.3.1.

Table 4.3.1. Solution of the problem of Example 4.3.1

s	0°	10°	20°	30°	40°	50°	60°	70°	80°	90°
$x^{(0)}$	0.00004	0.17546	0.34556	0.50516	0.64941	0.77394	0.87494	0.94937	0.99494	1.01018
$x^{(1)}$	−0.00001	−0.00180	−0.00354	−0.00518	−0.00668	−0.00796	−0.00900	−0.00976	−0.01022	−0.01038
$x^{(0)}+x^{(1)}$	0.0000	0.1737	0.3420	0.5000	0.6427	0.7660	0.8659	0.9396	0.9847	0.9998
$\sin s$	0.0000	0.1736	0.3420	0.5000	0.6428	0.7660	0.8660	0.9397	0.9848	1.0000

It is clear that the correction $x^{(1)}$ is very effective in reducing the error in the basic Nystrom solution $x^{(0)}$. The cost of producing this correction can be estimated as follows. If we produce the difference vectors $\Delta(x^{(i)})$ using P differences in (7c) the cost of producing each component of $\Delta(x^{(i)})$ is approximately $\frac{1}{2}P^2$ multiplications (μ) and additions (α). The total cost of producing the vector $\Delta(x^{(i)})$ is therefore approximately $\frac{1}{2}NP^2(\mu + \alpha)$. The cost of solving (9) for each iterate $x^{(i)}$ is only $\mathcal{O}(N^2)$, since the same matrix is involved as for $x^{(0)}$; hence, the total cost of producing one correction term $x^{(i)}$ is approximately $N(N + \frac{1}{2}P^2)(\mu + \alpha)$. For a crude estimate of the error it is sufficient to take $P = 1$ or 2 (see (7c)) and then the cost of this estimate is only $\mathcal{O}(N^2)$, which is much cheaper than the $\mathcal{O}(N^3)$ cost of those discussed previously. However, this only follows because we have chosen to base the method on the low-order Trapezoid rule. If we attempt to improve this rule by adding further corrections we will need to use higher differences. There are $(N - 1)$ of these available; using all of them, the total cost rises to almost $\frac{1}{2}N^3(\mu + \alpha)$, which is approximately the same as that of producing $x^{(0)}$.

4.3.5 Deferred approach to the limit

An alternative way of including the difference corrections is via a 'deferred approach to the limit'. This technique makes use of the fact that the repeated Trapezoid rule with steplength h has (like many other rules) an error expansion of the form

$$E(h) = A_2 h^2 + A_4 h^4 + A_6 h^6 + \cdots. \qquad (10)$$

If we compute approximate solutions x_r for interval $h_r, r = 1, 2$ then we have two solutions $x_1(s)$, $x_2(s)$ with errors $e_1(s)$, $e_2(s)$ satisfying

$$e_1 = x - x_1 = \sum_{i=1}^{\infty} A_{2i} h_1{}^{2i},$$

$$e_2 = x - x_2 = \sum_{i=1}^{\infty} A_{2i} h_2{}^{2i}. \qquad (11)$$

Taking these two approximations, we can eliminate A_2 to find

$$x = x_2 + \frac{h_2{}^2}{h_1{}^2 - h_2{}^2}(x_2 - x_1) + \sum_{i=2}^{\infty} B_{2i} h^{2i} \qquad (12)$$

and we could, given more basic approximations, x_3, x_4, eliminate more terms in the series. This elimination can be carried out in a systematic manner (cf. Romberg integration for a single integral). The process is neat and effective even though the underlying series (which we have written as infinite series) are usually at best asymptotic; but again the process represents a rather costly way of improving the performance of a low-order rule. For, when the kernel and driving terms are smooth, a Gauss rule will generally yield high accuracy more cheaply, while for badly behaved examples, the form (10) may have only limited validity, or may in fact be invalid; then the 'corrected' result (12) will be no more accurate than the basic approximations x_1, x_2, and the method fails.

4.4   The product Nystrom method

The examples of Section 4.2.4 clearly demonstrate the need for special rules to approximate the integral in

$$x(s) = y(s) + \lambda \int_a^b K(s,t)x(t)\mathrm{d}t \tag{1}$$

when the integrand is *singular* in the sense that one or more of its derivatives is not well defined within the range of integration. For example, if

$$K(s,t) = \ln|s - t|\bar{K}(s,t)$$

where $\bar{K}(s,t)$ is a smooth function of its arguments, then K is singular since it is unbounded when $s = t$. We can often factor out the singularity in K by writing

$$K(s,t) = p(s,t)\bar{K}(s,t) \tag{2}$$

where, in the above example, $p(s,t) = \ln|s - t|$. We assume here that such a factorisation is available. It is not necessary for p to be unbounded to cause trouble; for example, if p takes the form $p(s,t) = (s - t)^{\frac{1}{2}}$, all derivatives with respect to t are unbounded when $s = t$ so that K is again singular in the sense used here.

We therefore rewrite (1) in the form

$$x(s) = y(s) + \lambda \int_a^b p(s,t)\bar{K}(s,t)x(t)\mathrm{d}t \tag{3}$$

where p and $\bar{K}$ are respectively 'badly behaved' and 'well behaved' functions of their arguments. This nomenclature is obvious in the case that $p(s,t)$ is unbounded at some point which is used as an abscissa by the quadrature rule, since then the rule yields an undefined result; however we emphasise (see Section 2.9) that the presence of unbounded low-order derivatives can seriously affect the accuracy and rate of convergence of quadrature rules which do not take explicit cognisance of them, and we assume that in such cases these unbounded derivatives stem from the factor $p(s,t)$.

The obvious way of taking account of the nature of $p(s,t)$ is via the use of *product integration* which treats $p(s,t)$ exactly and approximates only that part of the integrand which is smooth by a suitable Lagrange interpolation polynomial. Thus, when $s = s_i$, we approximate the integral term in (3) by

$$\int_a^b p(s_i,t)\bar{K}(s_i,t)x(t)dt \approx \sum_{j=1}^N w_{ij}\bar{K}(s_i,t_j)x(t_j) \tag{4}$$

where $s_i = t_i$, $i = 1,\ldots,N$. The weights w_{ij} are constructed by insisting that the rule in (4) be exact when $\bar{K}(s_i,t)x(t)$ is a polynomial in t of degree $\leq r$ say. Then we shall require, for each value of i, the existence of (and a knowledge of) the $(r+1)$ integrals or moments

$$m_{ij} = \int_a^b t^j p(s_i,t)dt, \quad j=0,1,\ldots,r. \tag{5}$$

The weights w_{ij} can then be constructed from the moments m_{ij} using the method of undetermined coefficients (see Section 2.2). They are clearly problem-dependent; that is, they obviously depend on the form of $p(s,t)$. As a result, the method is normally restricted to a 'standard' range of singularities (those for which the moments m_{ij} can be calculated). However, this covers a number of cases which arise in practice so that the approach has practical value. The weights also depend on the quadrature points s_i so that actually calculating the weights adds significantly to the cost of the algorithm.

We illustrate the method by a simple example which approximates the integral term in (4) by a product integration form of Simpson's rule; it is obvious that the procedure can be easily extended to give product integration analogues of higher degree rules. We may write

$$\int_a^b p(s_i,t)\bar{K}(s_i,t)x(t)dt = \sum_{j=0}^{(N-2)/2} \int_{t_{2j}}^{t_{2j+2}} p(s_i,t)\bar{K}(s_i,t)x(t)dt$$

where $s_i = t_i = a + ih$ with $h = (b-a)/N$ and N even. Now if we approximate the nonsingular part of the integrand over each interval $[t_{2j},t_{2j+2}]$ by the second degree Lagrange interpolation polynomial which interpolates it at the points $t_{2j}, t_{2j+1}, t_{2j+2}$ we find

$$\int_a^b p(t_i,t)\bar{K}(t_i,t)x(t)dt$$

$$\approx \sum_{j=0}^{(N-2)/2} \int_{t_{2j}}^{t_{2j+2}} p(t_i,t)\left\{ \frac{(t_{2j+1}-t)(t_{2j+2}-t)}{2h^2} \bar{K}(t_i,t_{2j})x(t_{2j}) \right.$$

$$+ \frac{(t-t_{2j})(t_{2j+2}-t)}{h^2} \bar{K}(t_i,t_{2j+1})x(t_{2j+1})$$

$$+ \left. \frac{(t-t_{2j})(t-t_{2j+1})}{2h^2} \bar{K}(t_i,t_{2j+2})x(t_{2j+2}) \right\}dt$$

$$= \sum_{j=0}^N w_{ij}\bar{K}(t_i,t_j)x(t_j) \tag{6}$$

where

$$w_{i0} = \frac{1}{2h^2} \int_{t_0}^{t_2} p(t_i, t)(t_1 - t)(t_2 - t)dt,$$

$$w_{i,2j+1} = \frac{1}{h^2} \int_{t_{2j}}^{t_{2j+2}} p(t_i, t)(t - t_{2j})(t_{2j+2} - t)dt,$$

$$w_{i,2j} = \frac{1}{2h^2} \int_{t_{2j-2}}^{t_{2j}} p(t_i, t)(t - t_{2j-2})(t - t_{2j-1})dt$$

$$+ \frac{1}{2h^2} \int_{t_{2j}}^{t_{2j+2}} p(t_i, t)(t_{2j+1} - t)(t_{2j+2} - t)dt,$$

$$w_{iN} = \frac{1}{2h^2} \int_{t_{N-2}}^{t_N} p(t_i, t)(t - t_{N-2})(t - t_{N-1})dt. \tag{6a}$$

If we define

$$\alpha_j(t_i) = \frac{1}{2h^2} \int_{t_{2j-2}}^{t_{2j}} p(t_i, t)(t - t_{2j-2})(t - t_{2j-1})dt,$$

$$\beta_j(t_i) = \frac{1}{2h^2} \int_{t_{2j-2}}^{t_{2j}} p(t_i, t)(t_{2j-1} - t)(t_{2j} - t)dt,$$

$$\gamma_j(t_i) = \frac{1}{2h^2} \int_{t_{2j-2}}^{t_{2j}} p(t_i, t)(t - t_{2j-2})(t_{2j} - t)dt, \tag{7}$$

it follows that

$$w_{i0} = \beta_1(t_i),$$

$$w_{i,2j+1} = 2\gamma_{j+1}(t_i),$$

$$w_{i,2j} = \alpha_j(t_i) + \beta_{j+1}(t_i),$$

$$w_{iN} = \alpha_{N/2}(t_i).$$

Example 4.4.1. We take $p(s, t) = \ln|s - t|$ and following Atkinson (1976, Part II, Chapter 3.2), who considered the product integration form of the Trapezoid rule for this singular weight function, we introduce the change of variable $t = t_{2j-2} + \mu h$, $0 \le \mu \le 2$. Then

$$\alpha_j(t_i) = \frac{h}{3} \ln h + \frac{h}{2} \int_0^2 \mu(\mu - 1) \ln|i - 2j + 2 - \mu| d\mu,$$

$$\beta_j(t_i) = \frac{h}{3} \ln h + \frac{h}{2} \int_0^2 (1 - \mu)(2 - \mu) \ln|i - 2j + 2 - \mu| d\mu,$$

$$\gamma_j(t_i) = \frac{2h}{3} \ln h + \frac{h}{2} \int_0^2 \mu(2 - \mu) \ln|i - 2j + 2 - \mu| d\mu. \tag{8}$$

Now if we define

$$\psi_0(k) = \int_0^2 \ln|k - \mu| d\mu,$$

$$\psi_1(k) = \int_0^2 \mu \ln|k - \mu| d\mu,$$

$$\psi_2(k) = \int_0^2 \mu^2 \ln|k - \mu| d\mu, \tag{9}$$

we find

$$\alpha_j(t_i) = \frac{h}{3} \ln h + \frac{h}{2} [\psi_2(k) - \psi_1(k)],$$

$$\beta_j(t_i) = \frac{h}{3} \ln h + \frac{h}{2} [2\psi_0(k) - 3\psi_1(k) + \psi_2(k)],$$

$$\gamma_j(t_i) = \frac{2h}{3} \ln h + \frac{h}{2} [2\psi_1(k) - \psi_2(k)], \tag{10}$$

where $k = i - 2j + 2$, and it only remains to determine the $\psi_i(k)$, $i = 0, 1, 2$, in order to completely define the weights of the quadrature formula (6). We use integration by parts to obtain

$$\psi_0(k) = k \ln|k| - (k - 2) \ln|k - 2| - 2,$$
$$\psi_1(k) = \tfrac{1}{2} [(k - 2)^2 \ln|k - 2| - k^2 \ln|k|] + \tfrac{1}{4} [k^2 - (k - 2)^2]$$
$$\quad + k\psi_0(k),$$
$$\psi_2(k) = \tfrac{1}{3} [k^3 \ln|k| - (k - 2)^3 \ln|k - 2|] + \tfrac{1}{9} [(k - 2)^3 - k^3]$$
$$\quad + k[2\psi_1(k) - k\psi_0(k)]. \tag{11}$$

These formulae have the advantage of yielding completely explicit forms for the weights w_{ij}, hence avoiding the need to solve the undetermined equations numerically for each collocation point s_i. We emphasise that this is unusual; for an arbitrary weight function and quadrature rule, no explicit solution is available. We caution also that the formulae (11) may lead to cancellation error and care should be exercised in calculating the required weights from them.

It may not always be possible to factor out the singularity in the kernel function as in (2). Consider for example

$$K(s, t) = \ln|\cos s - \cos t|, \quad 0 \le s, t \le \pi.$$

It is not immediately clear what form to choose for the singular factor $p(s, t)$. In such cases we consider

$$K(s, t) = \sum_{i=1}^r p_i(s, t) \bar{K}_i(s, t)$$

where each p_i is badly behaved and each $\bar{K}_i$ is well behaved. Now we may write (see Atkinson, 1976, Part II, Chapter 3.2)

$$\ln|\cos s - \cos t| = \ln\left\{\frac{\sin((s-t)/2)\sin((s+t)/2)}{((s-t)/2)(s+t)(2\pi-s-t)}\right\} + \ln|s-t|$$
$$+ \ln(2\pi - s - t) + \ln(s+t).$$

That is

$$\ln|\cos s - \cos t| = \sum_{i=1}^{4} p_i(s,t)\bar{K}_i(s,t) \qquad (12)$$

where

$$p_1(s,t) = 1,$$
$$p_2(s,t) = \ln|s-t|,$$
$$p_3(s,t) = \ln(2\pi - s - t),$$
$$p_4(s,t) = \ln(s+t),$$

$$K_1(s,t) = \ln\left\{\frac{\sin\left(\dfrac{s-t}{2}\right)\sin\left(\dfrac{s+t}{2}\right)}{\left(\dfrac{s-t}{2}\right)(s+t)(2\pi-s-t)}\right\},$$

$$\bar{K}_2(s,t) = \bar{K}_3(s,t) = \bar{K}_4(s,t) = 1,$$

and we may treat equation (1) in the form

$$x(s) = y(s) + \lambda\int_a^b p_1(s,t)\bar{K}_1(s,t)dt + \lambda\int_a^b p_2(s,t)\bar{K}_2(s,t)dt$$

$$+ \lambda\int_a^b p_3(s,t)\bar{K}_3(s,t)dt + \lambda\int_a^b p_4(s,t)\bar{K}_4(s,t)dt \qquad (13)$$

by applying a product integration rule to each of the integrals in (13).

***Example* 4.4.2.** The following equation

$$x(s) = 1 + \int_0^\pi \ln|\cos s - \cos t|x(t)dt, \quad 0 \le s \le \pi, \qquad (14)$$

has been solved by Atkinson (1976, Part II, Chapter 3.2) using the product integration form of the repeated (M-panel) Simpson's rule with the above split for the kernel function. Clearly we cannot apply, to any degree of success, the standard repeated Simpson's rule for the solution of (14) since the integral is unbounded when $\cos s = \cos t$. The results $x_N(s)$ (due to Atkinson) obtained at the point $s = 0$, where N specifies the total number of

Table 4.4.1. *Solution of*
$x(s) = 1 + \int_0^\pi \ln|\cos s - \cos t| x(t)dt, \ 0 \le s \le \pi.$

M	N	$x_N(0)$	Error
1	3	0.314 490 489 76	2.14×10^{-4}
2	5	0.314 687 817 70	1.65, -5
4	9	0.314 703 169 78	1.13, -6
8	17	0.314 704 225 50	7.25, -8
16	33	0.314 704 293 45	4.56, -9

These results are taken from Atkinson (1976), p.117.

points for the product rule, are shown in the first column of Table 4.4.1. Equation (14) has the exact solution

$$x(s) = \frac{1}{1 + \pi \ln 2} \approx 0.314\,704\,298\,02$$

and the second column of the table shows the error in the approximation. The above results show that the rule yields rapid convergence with convergence rate $p \approx 4$. Thus we see that, on using a special rule which takes into account the nature of the singularity in the kernel function, we can match the convergence rate which we would expect to obtain for the solution of a nonsingular equation using the standard repeated Simpson's rule.

4.5 The method of El-gendi

In Chapter 7 we discuss methods based, not only on the direct approximation of the integral operator, but also on the provision of an approximating expansion for the solution, of the form

$$x(s) \approx \sum_{i=1}^{N} a_i^{(N)} h_i(s)$$

where $\{h_i(s)\}$ is a known set of functions (the 'expansion set'). Such methods then provide algorithms for computing the expansion coefficients $a_i^{(N)}$, $i = 1,\ldots,N$. They look at first sight very different from the Nystrom method considered here, but in practice there are many points of contact, some of which are discussed in Section 8.6. We discuss here one method, due to El-gendi (1969), in which an expansion of the solution is introduced but for which the points of contact with the Nystrom method are so close that the method is in fact completely equivalent to a particular Nystrom method; the algorithm then yields a halfway house between the two approaches. The derivation which follows differs from that given by El-gendi but makes this link very clear.

We approximate the solution by a finite Chebyshev expansion

$$x(s) \approx x_N(s) = \sum_{i=0}^{N}{}' a_i T_i(s) \tag{1}$$

and consider the solution of the second kind equation

$$x(s) = y(s) + \lambda \int_{-1}^{1} K(s,t)x(t)\mathrm{d}t \tag{2}$$

where the range of integration has been normalised for convenience to $[-1, 1]$. Now if we approximate the integral in (2) by means of the Clenshaw–Curtis quadrature rule of Section 2.4.4 and set $s = s_i$, $i = 0, 1, \ldots, N$, corresponding to the quadrature points of the rule, we have the following set of linear equations to solve:

$$x_N(s_i) = y(s_i) + \lambda \sum_{j=0}^{N}{}'' w_j K(s_i, t_j)x_N(t_j),$$

$$i = 0, 1, \ldots, N, \tag{3}$$

where

$$\left. \begin{aligned} w_j &= \frac{4}{N} \sum_{\substack{r=0 \\ r \text{ even}}}^{N}{}'' \frac{1}{1 - r^2} \cos\left(\frac{rj\pi}{N}\right) \\ s_j &= t_j = \cos\left(\frac{j\pi}{N}\right) \end{aligned} \right\}, j = 0, 1, \ldots, N.$$

Solution of the above system for $x_N(s_i)$, $i = 0, 1, \ldots, N$, enables the coefficients a_i in (1) to be determined via the relation

$$a_i = \frac{2}{\pi} \int_{-1}^{1} \frac{x(s) T_i(s)\mathrm{d}s}{(1 - s^2)^{\frac{1}{2}}} \simeq \frac{2}{N} \sum_{j=0}^{N}{}'' x_N(s_j) \cos\frac{ij\pi}{N}, \quad i = 0, 1, \ldots, N, \tag{4}$$

where we have used the closed Gauss–Chebyshev quadrature rule of Section 2.4.3 to approximate the integral representation for the a_i. Equation (4) yields the El-gendi solution of (2). This derivation views equation (1) as an interpolation formula to generate approximate solutions at points other than the quadrature points, with the underlying solution stemming from the Nystrom method with the Clenshaw–Curtis quadrature rule. The choice of this rule is most easily motivated by the introduction of the Chebyshev expansion (1) but it has advantages even within the framework of a Nystrom calculation, since it provides a sequence of high-order rules with easily calculable weights and pre-assigned points.

4.6   Techniques for the solution of the Nystrom equations
4.6.1   Iterative or direct methods?

We return to the question of efficient solution techniques for the

Nystrom equations

$$\mathbf{x} = \mathbf{y} + \lambda \mathbf{K}\mathbf{x}. \tag{1}$$

In practice, (1) is usually solved by direct Gauss elimination, with associated cost for an N-point rule of about $\frac{1}{3}N^3$ multiplications and additions.

This algorithm is 'safe' in the sense that it cannot lead to unbounded errors as the number of points N is increased, provided that the *condition number* $Q(N)$ of the Nystrom matrix

$$Q(N) = \|\mathbf{I} - \lambda \mathbf{K}\| \|(\mathbf{I} - \lambda \mathbf{K})^{-1}\| \tag{2}$$

does not become too large.

Now, if λ is close to a characteristic value of K, the condition number may indeed be large. But we are interested here in the behaviour for fixed λ and increasing N, and because the matrix $\mathbf{K}$ represents an approximation to the integral operator K, we expect that

$$\lim_{N \to \infty} Q(N) = \|I - \lambda K\| \cdot \|(I - \lambda K)^{-1}\|. \tag{2a}$$

Equation (2a) represents a restriction on the family of N-point rules considered; we use it only to indicate that, in practice, the Nystrom equations are unlikely to be illconditioned expect for λ close to a characteristic value, and that the condition number is likely to be approximately independent of N. Thus, Gauss elimination should cause no problems of accuracy. However, for large N its $\mathcal{O}(N^3)$ cost dependence can lead to surprisingly long computing times, even for the one-dimensional problems considered in this book, and we seek some way of reducing this cost. The obvious approach is to construct an iterative scheme for (4.1.9); one such scheme is in fact given by the Neumann iterative solution (4.1.7a). The cost of one iteration using (4.1.7a) is about $N^2(\mu + \alpha)$; hence if P iterations are carried out the total cost is about $N^2 P(\mu + \alpha)$. We can estimate the number of iterations required as follows. Let $\mathbf{x}_R$ be the exact solution of (1) and as in (4.1.10) set $\mathbf{e}_n = \mathbf{x}_R - \mathbf{x}_n$ where $\mathbf{x}_n$ satisfies the recurrence relation (4.1.7a), that is:

$$\mathbf{x}_{n+1} = \mathbf{y} + \lambda \mathbf{K}\mathbf{x}_n. \tag{3}$$

From (1), (3) we derive the recurrence relation for $\mathbf{e}_n$:

$$\mathbf{e}_{n+1} = \lambda \mathbf{K}\mathbf{e}_n \tag{4}$$

with solution

$$\mathbf{e}_n = \lambda^n \mathbf{K}^n \mathbf{e}_0 \tag{4a}$$

and corresponding error bound

$$\|\mathbf{e}_n\| \leq \|\lambda \mathbf{K}\|^n \|\mathbf{e}_0\|. \tag{4b}$$

Now suppose that we seek a solution of (1) accurate to within absolute error ε; that is, we require to choose P to satisfy the inequality

$$\|e_P\| \le \varepsilon. \tag{5}$$

Then provided $\|\lambda K\| < 1$, (5) is satisfied if

$$P > (\ln(\varepsilon) - \ln\|e_0\|)/\ln\|\lambda K\|. \tag{5a}$$

Now provided, as above, that $\|\lambda K\|$ is approximately independent of N, the number of iterations is also approximately independent of N and the total solution cost is $\mathcal{O}(N^2)$.

4.6.2   Alternative iterative schemes

As $\|\lambda K\| \to 1$ the number of iterations predicted by (5a) tends to infinity, reflecting the fact that the iterative scheme (3) will in general diverge for $\|\lambda K\| > 1$. As a result the Neumann scheme (3) does not yield a practical general algorithm; we seek alternative schemes which have weaker convergence conditions but which retain an $\mathcal{O}(N^2)$ overall cost. Theorem 4.6.1 below provides us with a prescription for choosing such a scheme.

Theorem 4.6.1. Let A be a linear operator in a Hilbert space X. Then the linear equation

$$Ax = y \tag{6}$$

has a unique solution $x \in X$ for every $y \in X$ if and only if there exists a linear operator P in X such that P^{-1} exists and

$$\|I - PA\| < 1. \tag{7a}$$

The solution x of (6) is given by

$$x = \sum_{i=0}^{\infty} (I - PA)^i Py. \tag{7b}$$

Proof. See Rall (1955).

It follows directly from this theorem that if an unique solution $x \in X$ of (6) exists for every $y \in X$, then an operator P in X exists such that condition (7a) is satisfied and the iterative process

$$\begin{cases} x_{n+1} = Py + (I - PA)x_n, \\ x_0 = Py, \end{cases} \tag{8a}$$

has the solution

$$x_n = \sum_{i=0}^{n} (I - PA)^i Py, \tag{8b}$$

which as $n \to \infty$ yields the formal relation given by (7b). Now when A is the linear integral operator $I - \lambda K$ with $P = I$, equation (6) takes the form

$$(I - \lambda K)x = y \tag{9}$$

and Theorem 4.6.1 shows that the Neumann series

$$x = \sum_{i=0}^{\infty} \lambda^i K^i y \tag{9a}$$

converges to the solution of (9) provided that $\| \lambda K \| < 1$. This suggests that with a judicious choice of the linear operator P we can relax this somewhat restrictive condition. Consider the following iterative scheme,

$$\left.\begin{aligned} x_{n+1} &= \mu y + (1 - \mu)x_n + \mu \lambda K x_n, \quad 0 < \mu \le 1, \\ x_0 &= \mu y, \end{aligned}\right\} \tag{10}$$

for the solution of (9), which clearly reduces to the Neumann scheme,

$$\left.\begin{aligned} x_{n+1} &= y + \lambda K x_n, \\ x_0 &= y, \end{aligned}\right\} \tag{10a}$$

when $\mu = 1$. Thus the choice $P = \mu I$ yields an iterative method of solution which is akin to the method of successive overrelaxation applied to the solution of a set of simultaneous linear equations, when the operator A in (6) is a nonsingular matrix. If we set $e_n = x_n - x$ then on subtracting (9) from (10) we find

$$e_{n+1} = [(1 - \mu)I + \mu \lambda K]e_n, \tag{11}$$

whence it follows that

$$(1 - \|(1 - \mu)I + \mu \lambda K\|)\|e_n\| \le \|x_{n+1} - x_n\|. \tag{12}$$

Thus providing that we know or can estimate the quantity $\|(1 - \mu)I + \mu \lambda K\|$ and

$$\|(1 - \mu)I + \mu \lambda K\| < 1 \tag{13}$$

the iterative scheme (10) will converge to the solution of (9) and the following inequality

$$\|e_n\| \le \frac{\|x_{n+1} - x_n\|}{1 - \|(1 - \mu)I + \mu \lambda K\|}, \tag{14}$$

will provide us with a computable bound on the error. Atkinson (1976, Part II, Chapter 1.0) has shown that when $K : X \to X$ is a positive definite self-adjoint compact operator the iteration (10) will converge for any $\lambda < 0$ provided that $0 < \mu < -2\lambda/(\|K\| - \lambda)$.

Thus for a choice of μ inside this range the scheme in (10) will converge to the solution of (9) for $\lambda < 0$ whereas the scheme (10a) may not.

Further iterative schemes have been produced which, subject to restrictions on the parameter μ and the form of the kernel $K(s, t)$, converge to the solution of (9); for details of these schemes the reader is referred to papers by Petryshyn (1963), Rall (1955), Samuelson (1953), Wagner (1951), Bückner (1948).

In practice however the success of these schemes depends on the choice of μ; we give below details of a scheme, presented by Atkinson (1976, Part II, Chapter 4.2) and originated by Brakhage (1960), which is also based on the Nystrom method and which uses an iterative technique to solve the resulting sets of linear equations, but for which the parameters involved are easier to fix satisfactorily. The method is based on the introduction of a 'coarse' solution which is then used to define an approximate inverse of the Nystrom matrix.

We wish to solve the equation

$$x(s) = y(s) + \lambda \int_a^b K(s, t)x(t)\mathrm{d}t \tag{15}$$

which we may approximate, using numerical quadrature, by

$$x_M(s) = y(s) + \lambda \sum_{j=1}^{M} w_{M,j}K(s, t_{M,j})x_M(t_{M,j}) \tag{16}$$

where the subscript M signifies the use of an M-point quadrature rule. Now if we set $s = t_{M,i}, i = 1, \ldots, M$, we have the following set of linear equations to solve:

$$x_M(t_{M,i}) = y(t_{M,i}) + \lambda \sum_{j=1}^{M} w_{M,j}K(t_{M,i}, t_{M,j})x_M(t_{M,j}),$$

$$i = 1, \ldots, M. \tag{17}$$

We assume that for M small we can solve the above problem directly for $x_M(t_{M,i}), i = 1, \ldots, M$, using Gauss elimination and we show how to use this to define an iterative method for solving (17) for $N > M$ abscissae $t_{N,i}$, $i = 1, \ldots, N$. To do this we note that equation (16) defines a solution $x_M(s)$ for all s; the $M \times M$ system of Nystrom equations (17) may be viewed as serving only to define values for the parameters $x_M(t_{M,j})$ which appear on the right hand side of (16). It is important that we can take this viewpoint, since we want to deal simultaneously with discretisations using different numbers of quadrature points. Since (16), (17) are linear, they together define $x_M(s)$ as the solution of a linear operator equation which we write in the standard operator form

$$(I - \lambda K_M)x_M = y \tag{16a}$$

where I, K_M are operators on the space of continuous functions. Now introduce a second quadrature rule with $N > M$ points and corresponding Nystrom solution x_N and discretised operator K_N. Then

$$(I - \lambda K_N)x_N = y$$

and it follows that

$$(I - \lambda K_M)x_N = y + \lambda(K_N - K_M)x_N$$

whence

$$(I - \lambda K_M)x_N = y + \lambda(K_N - K_M)y + \lambda^2(K_N - K_M)K_N x_N. \qquad (18)$$

Thus given $x_{N,0}$ we can define the obvious iterative scheme

$$(I - \lambda K_M)x_{N,n+1} = y + \lambda(K_N - K_M)y + \lambda^2(K_N - K_M)K_N x_{N,n} \qquad (19)$$

for $n \geq 0$. If we write $e_n = x_N - x_{N,n}$ it follows that

$$e_{n+1} = (I - \lambda K_M)^{-1}\lambda^2(K_N - K_M)K_N e_n \qquad (20)$$

and we have

$$\|e_{n+1}\| \leq \|(I - \lambda K_M)^{-1}\lambda^2(K_N - K_M)K_N\| \cdot \|e_n\|. \qquad (21)$$

For M sufficiently large, Atkinson (1976, Part II, Chapter 4.2) asserts that for all $N > M$,

$$\|(I - \lambda K_M)^{-1}\lambda^2(K_N - K_M)K_N\| < 1. \qquad (21a)$$

This assertion is essentially a restriction on the family of quadrature rules considered; however, it is clearly a reasonable restriction since

$$\|(I - \lambda K_M)^{-1}\lambda^2(K_N - K_M)K_N\|$$
$$\leq \|\lambda(K_N - K_M)\| \|(I - \lambda K_M)^{-1}\lambda K_N\|$$

and on the right hand side we expect that $\lim_{N,M \to \infty} \|K_N - K_M\| = 0$. Given (21a), the iterative scheme (19) converges to the solution of (15). Equation (19) may be written equivalently as

$$\left.\begin{aligned} r_{N,n} &= y - (I - \lambda K_N)x_{N,n}, \\ x_{N,n+1} &= x_{N,n} + [I + (I - \lambda K_M)^{-1}\lambda K_N]r_{N,n} \\ &= x_{N,n} + r_{N,n} + \delta, \end{aligned}\right\} \qquad (22)$$

where

$$(I - \lambda K_M)\delta = \lambda K_N r_{N,n}.$$

This has the form of a residual correction and is better suited to computational use than (19) since the latter requires considerably more kernel evaluations.

Equation (22) in fact yields a rather satisfactory iterative scheme for the solution of (15). It has one parameter, M, to be adjusted, but this merely has to be set 'sufficiently large'. It is simple to make an initial guess and to increase M if the iterations diverge. Note however that (22) is still phrased in terms of continuous operators; in practice we shall wish to work wholly with $N \times N$ matrices and N-vectors giving the approximate solution at the quadrature points $\{t_{N,i}\}$. The steps needed to do this are the following:

(i) solve the system (17) for a given value of M to give

$$x_M(t_{M,i}), \quad i = 1, \ldots, M$$

(ii) choose $N > M$ and use the interpolating formula of equation (16) to provide an initial value for $x_{N,0}(t_{N,i})$, $i = 1, \ldots, N$:

$$x_{N,0}(t_{N,i}) = y(t_{N,i}) + \lambda \sum_{j=1}^{M} w_{M,j} K(t_{N,i}, t_{M,j}) x_M(t_{M,j}),$$
$$i = 1, \ldots, N$$

(iii) for $n \geq 0$ form

$$r_{N,n}(t_{N,i}) = y(t_{N,i}) - x_{N,n}(t_{N,i})$$
$$+ \lambda \sum_{j=1}^{N} w_{N,j} K(t_{N,i}, t_{N,j}) x_{N,n}(t_{N,j}), \quad i = 1, \ldots, N$$

(iv) solve the two systems

$$(a) \quad \delta(t_{M,i}) - \lambda \sum_{j=1}^{M} w_{M,j} K(t_{M,i}, t_{M,j}) \delta(t_{M,j})$$
$$= \lambda \sum_{j=1}^{N} w_{N,j} K(t_{M,i}, t_{N,j}) r_{N,n}(t_{N,j}), \quad i = 1, \ldots, M$$

$$(b) \quad \delta(t_{N,i}) - \lambda \sum_{j=1}^{M} w_{M,j} K(t_{N,i}, t_{M,j}) \delta(t_{M,j})$$
$$= \lambda \sum_{j=1}^{N} w_{N,j} K(t_{N,i}, t_{N,j}) r_{N,n}(t_{N,j}), \quad i = 1, \ldots, N$$

(v) form $x_{N,n+1}(t_{N,i}) = x_{N,n}(t_{N,i}) + r_{N,n}(t_{N,i}) + \delta(t_{N,i}), \quad i = 1, \ldots, N$.

Atkinson (1976, Part II, Chapter 5.1) describes a routine, IESIMP, based on these steps together with a mechanism for producing asymptotic error estimates which in turn allow the rate of convergence of the iterative method to be monitored as the solution process proceeds. IESIMP uses Simpson's rule to approximate the integral in (15); M is chosen to be odd (≥ 3) and on successful completion of step (i) the value of $N = 2M - 1$ is set. Some or all of steps (i)–(v) are repeated if and when various convergence criteria are violated; for details of the particular test strategy adopted in IESIMP we refer the reader to Atkinson (1976, Part II, Chapter 5.1). IESIMP is an example of an *automatic* routine (see Section 10.1) and we discuss in Section 10.4 the numerical performance of this routine when used to solve a particular set of test problems.

Exercises

1. How would you expect the Nystrom method for solving

$$x(s) = y(s) + \lambda \int_a^b K(s,t)x(t)dt$$

to behave if λ is close to a characteristic value of K? How would you detect and treat this possibility?

2. The equation

$$x(s) = y(s) + \int_a^b K(s,t)x(t)dt$$

is to be solved using the repeated Trapezoid and mid-point rules, to give solutions

$$x_1(s) = x(s) + e_1(s)$$

and

$$x_2(s) = x(s) + e_2(s)$$

respectively, where, in the notation of Section 4.3.2, the error functions satisfy the relations

$$e_1(s) = -\frac{h^2}{12}(b-a)(I+H)\left[\frac{\partial^2}{\partial t^2}K(s,t)x_1(t)\right]_{t=\xi_1(s)},$$

$$e_2(s) = \frac{h^2}{24}(b-a)(I+H)\left[\frac{\partial^2}{\partial t^2}K(s,t)x_2(t)\right]_{t=\xi_2(s)},$$

with $a \le \xi_1$, $\xi_2 \le b$. Show that the following error estimate obtains:

$$\|e_1\| \approx \tfrac{2}{3}\|x_2 - x_1\|.$$

3. Gregory's formula is given by

$$\int_a^b f(s)ds = h\sum_{r=0}^n{}'' f_r + C_k + \mathcal{O}(h^{k+2}),$$

where

$$h = \frac{b-a}{n}, \quad f_r = f(\xi_r) = f(a+rh),$$

$$C_k = \left(h\sum_{r=1}^k a_r\nabla^r\right)f_n + \left(h\sum_{r=1}^k (-1)^r a_r\Delta^r\right)f_0,$$

$$a_1 = -\tfrac{1}{12}, \quad a_2 = -\tfrac{1}{24}, \quad a_3 = -\tfrac{19}{720}, \dots .$$

Show that when this formula is used, with $k = 2$, to replace the integral in the equation

$$x(s) = y(s) + \int_a^b K(s,t)x(t)\mathrm{d}t$$

the following set of approximate algebraic equations is obtained, on successively setting $s = \xi_i$, $i = 0, 1, \dots, n$:

$$x_i = y_i + h \sum_{j=0}^{n}{}'' K(\xi_i, \xi_j)x_j + \Delta, \quad i = 0, 1, \dots, n,$$

where

$$\Delta = -\tfrac{1}{8}h\{K(\xi_i, \xi_0)x_0 + K(\xi_i, \xi_n)x_n\} + \tfrac{1}{6}h\{K(\xi_i, \xi_1)x_1$$
$$+ K(\xi_i, \xi_{n-1})x_{n-1}\} - \tfrac{1}{24}h\{K(\xi_i, \xi_2)x_2 + K(\xi_i, \xi_{n-2})x_{n-2}\}.$$

4. Suppose the leading term in the truncation error of an integration formula is proportional to h^q where q is independent of h, and suppose that we use this formula to estimate the integral in the equation

$$x(s) = y(s) + \int_a^b K(s,t)x(t)\mathrm{d}t.$$

The corresponding approximate solution $\tilde{x}(s, h)$ may be expressed in the form

$$x(s) = \tilde{x}(s, h) + A(s)h^q.$$

Show that for two choices of $h = h_1, h_2$ with $h_1 = 2h_2$, the h^q error can be eliminated to provide the improved formula:

$$x(s) = \tilde{x}(s, h_2) + \frac{1}{2^q - 1}[\tilde{x}(s, h_2) - \tilde{x}(s, h_1)].$$

5. Show that with the change of variable

$$t = t_{2j-2} + \mu h, \quad 0 \le \mu \le 2,$$

equations (4.4.7) are transformed to equations (4.4.8).

5

Quadrature methods for Volterra equations of the second kind

5.1 Introduction

Most of this book is concerned with the solution of integral equations of Fredholm form. While methods for Fredholm equations can usually be adapted to solve Volterra problems (see Section 5.3 and Chapter 11) it is not particularly sensible to do this; Volterra integral equations are essentially 'initial value' problems (to borrow a differential equation phrase) and efficient numerical procedures make use of this. In this chapter we consider quadrature methods for second kind Volterra equations; first kind Volterra equations are treated in Chapter 12 and Volterra integro-differential equations in Chapter 13.

We consider then the numerical solution of nonsingular Volterra equations of the second kind of the general form

$$x(s) = y(s) + \int_a^s k(s, t, x(t))dt, \quad a \le s \le b \tag{1}$$

and we assume that the solution is required over a finite interval $[a, b]$, that y is continuous in $[a, b]$, k is continuous in $a \le t \le s \le b$ and k satisfies a uniform Lipschitz condition in x; these conditions will ensure that a unique continuous solution to the problem (1) exists. If the kernel is linear in its third argument, that is, there exists a function K such that

$$k(s, t, x(t)) = K(s, t)x(t) + k_0(s, t) \tag{1a}$$

for all $a \le t \le s \le b$, then equation (1) is said to be linear and reduces to

$$x(s) = \bar{y}(s) + \int_a^s K(s, t)x(t)dt, \quad a \le s \le b, \tag{1b}$$

where

$$\bar{y}(s) = y(s) + \int_a^s k_0(s, t)dt. \tag{1c}$$

We shall take (1b) as the canonical form for a linear Volterra equation and we will not distinguish notationally between y and $\bar{y}$.

We saw in Section 0.3 that any initial value problem for ordinary differential equations can be reformulated as a Volterra equation of the form (1) or (1b) and this naturally suggests the use of *marching methods* for the solution of (1, 1b). These methods produce approximations x_i to $x(s_i)$ at successive mesh points (see below) in $[a, b]$.

5.1.1   Quadrature methods for linear equations

An obvious numerical procedure is to approximate the integral term in (1b) via a quadrature rule which integrates over the variable t for a fixed value of s. It is natural to choose a regular mesh in s and t; thus setting $s = s_i \equiv a + ih$, where $h = (b - a)/N$ is the fixed steplength, we would approximate in an obvious notation the integral term in the linear equation (1b) by

$$\int_a^{s_i} K(s_i, t)x(t)\mathrm{d}t \approx h \sum_{j=0}^{i} w_{ij}K(s_i, t_j)x(t_j)$$

$$= h \sum_{j=0}^{i} w_{ij}K_{ij}x(t_j) \tag{2}$$

where $s_i = t_i$, $i = 0, 1, \ldots, N$. This quadrature rule leads to the following set of equations:

$$x(s_0) = y(s_0),$$
$$x(s_1) = y(s_1) + h[w_{10}K_{10}x(t_0) + w_{11}K_{11}x(t_1)]$$
$$+ E_{1,t}(K(s_1, t)x(t)),$$
$$x(s_i) = y(s_i) + h \sum_{j=0}^{i} w_{ij}K_{ij}x(t_j)$$
$$+ E_{i,t}(K(s_i, t)x(t)), \quad i = 1, 2, \ldots, N, \tag{3}$$

where $E_{i,t}(K(s_i, t)x(t))$ represents the error term in the quadrature rule. If the $E_{i,t}$ are assumed negligible and $(1 - hw_{ii}K_{ii}) \neq 0$ for any i we can clearly solve this set of equations for x_i, $i = 1, \ldots, N$, where x_i is an approximation to $x(s_i)$, by direct forward substitution.

This procedure is obviously numerically very straightforward; however, there remains the problem of choosing suitable weights w_{ij}. We note that, for each i, the set $\{w_{ij}, j = 0, 1, \ldots, i\}$ represents the weights for an $(i + 1)$-point quadrature rule of Newton–Cotes type (equally spaced points) for the interval $[0, ih]$. For large i there are many possible choices of rule; for small $i = 1, 2, \ldots$, the choice is rather limited; yet there seems (and is) little point in choosing an accurate rule for large i if we cannot choose an equally accurate rule for small i. Let us start by considering the simplest possible rule, the repeated Trapezoid rule. Then the rule is of degree 1 for each i (which seems at least consistent) and the weights w_{ij} are given by

$$w_{i0} = w_{ii} = \tfrac{1}{2}; \quad w_{ij} = 1, \quad j = 1, 2, \ldots, i-1$$

so that (3) reduces to the simple form

$$x_0 = y(s_0),$$

$$x_i = y(s_i) + h \sum_{j=0}^{i} {}'' K_{ij} x_j, \quad i = 1, 2, \ldots, N. \tag{3a}$$

Equations (3a) can be solved successively for $i = 1, 2, \ldots, N$ (whence the term 'marching method'); the cost of the solution is then only $\mathcal{O}(N^2)$, so that Volterra problems are clearly 'easier' in some sense than Fredholm problems, for which we need to work quite hard to achieve such operation counts.

However the Trapezoid rule is not very accurate so that the use of (3a) will normally involve quite small steps h; the question arises: can we increase the accuracy without taking very small steps? As in the case of a Fredholm equation there are several possibilities:

(i) the 'deferred approach to the limit' is again valid if K is regular
(ii) we can compute and iterate the difference correction $E_{i,t}$ using the so-called Newton–Gregory formula (see equation (4.3.6, 7c))
(iii) we can use a more accurate rule.

We illustrate the straightforward use of fixes (i) and (ii) by means of an example taken from Fox and Goodwin (1953):

Example 5.1.1.

$$x(s) = \frac{1}{(s+2)^2} - 2 \int_0^s \frac{1}{(s-t+2)^2} x(t) dt. \tag{4}$$

The results obtained with a Trapezoid rule using $h = \tfrac{1}{2}$ and $h = \tfrac{1}{4}$, corresponding respectively to $x_1(s)$ and $x_2(s)$, are shown in the second and third columns of Table 5.1.1. In the fourth column we give the 'deferred approach to the limit' or 'h^2 extrapolation' solution $x_E(s)$ obtained straightforwardly from these using the formula

$$x_E(s) = \frac{4x_2(s) - x_1(s)}{3}$$

(see equation (4.3.12)).

When we try to compute the difference corrections we find that for small s we do not have enough function values inside the range $[0, s]$ to do so. In

Table 5.1.1. *Solution of* $x(s) = 1/(s+2)^2 - 2\int_0^s [x(t)/(s-t+2)^2]\,dt$.

s	$x_1(s)$	$x_2(s)$	$x_E(s)$	$x_C(s)$
		Computed solution		
0.0	0.250 00	0.250 00	0.250 00	0.250 00
0.5	0.124 44	0.124 97	0.125 15	0.125 15
1.0	0.068 72	0.069 24	0.069 41	0.069 42
1.5	0.041 43	0.041 84	0.041 98	0.041 97
2.0	0.026 90	0.027 22	0.027 33	0.027 33
2.5	0.018 59	0.018 83	0.018 91	0.018 91
3.0	0.013 52	0.013 72	0.013 79	0.013 78
3.5	0.010 26	0.010 41	0.010 46	0.010 46
4.0	0.008 05	0.008 18	0.008 22	0.008 21
4.5	0.006 49	0.006 59	0.006 62	0.006 62
5.0	0.005 36	0.005 45	0.005 48	0.005 47

These results are taken from Table 9 of Fox and Goodwin (1953).

principle, since we compute these differences *a posteriori*, we can use function values which *are* available, from outside the range. If we try this at an interval $h = \frac{1}{2}$ we find that the differences fail to converge: a reflection of the pole at $t = s + 2$. However, with $h = \frac{1}{4}$, there is no apparent difficulty and the corrected results $x_C(s)$ obtained are shown in the last column of Table 5.1.1. Clearly the last two columns are in good agreement.

We notice a second feature of this table: the computed solution appears to be converging slowly to zero as s increases. As a result we could very soon increase the interval h without significant loss of accuracy. Indeed, should we wish to double the interval it is a simple matter to do so. There is an obvious connection in this regard between Volterra equations and differential equations with single point boundary conditions, and we return to the topic of step size control in Section 5.4. The correspondence extends also to the difficulties involved in using a higher order rule. Consider for example Simpson's rule applied to the solution of (1b). The first defining equations become:

$$x_0 = y(s_0),$$
$$x_1 = ?,$$
$$x_2 = y(s_2) + \tfrac{1}{3}h[K_{20}x_0 + 4K_{21}x_1 + K_{22}x_2].$$

The third of these equations uses Simpson's rule and defines x_2 given x_0 and x_1; however, Simpson's rule cannot be used to define x_1 since not enough points are available for the rule to be applicable. An alternative method (a 'starting method') is required to define x_1; similarly, a general multistep method, given by

$$x_i = y(s_i) + h \sum_{j=0}^{i} w_{ij} K_{ij} x_j, \quad i = k, k+1, \ldots, N, \tag{5}$$

will require k starting values, that is $x_0, x_1, \ldots, x_{k-1}$, of which only one, x_0, is immediately available. Thus, as is the case for differential equations, we shall require special starting procedures to provide these initial values, unless of course $k = 1$ in which case the method is said to be 'self-starting'.

In cases where the kernel and solution are smooth functions of their arguments we would expect high-order rules, incorporated with an appropriate starting procedure, to lead to high accuracy. Moreover we would choose a high-order rule in preference to one of low order. We leave a discussion of starting procedures to Section 5.1.4.

A further difficulty which has to be resolved is that a method such as (5) will need to be supplemented by methods which produce approximations of the same accuracy as those produced at every kth solution value by (5). Suppose $N = km + \mu$ where m, μ are integers with $\mu < k$; then the system in (5) may be solved for x_{ki}, $i = 1, 2, \ldots, m$, using a composite rule which is the m-panel repeated $(k+1)$-point quadrature rule. Additional methods must be used to supply the starting values $x_0, x_1, \ldots, x_{k-1}$ and the missing function values at the intermediate points $x_{ki+1}, x_{ki+2}, \ldots, x_{k(i+1)-1}$, $i = 1, \ldots, m-1$ and $x_{km+1}, \ldots, x_{km+\mu}$. When $k = 2$ a method frequently used to approximate the solution at the point s_i is the repeated Simpson's rule when i is even with a combination of the repeated Simpson's rule and the 1-panel or single panel Trapezoid rule when i is odd. When i is odd it would not at first sight appear to matter at which end of the range of integration the Trapezoid rule is applied; we shall see later however, in Section 5.1.6, that it is imperative that we apply the Trapezoid rule to the *upper end* of integration in order that the method be stable. An alternative method when $k = 2$ is to use the repeated Simpson's rule over the interval $[s_0, s_i]$ when i is even with a combination of the repeated Simpson's rule over $[s_0, s_{i-3}]$ and the 4-point three-eighths rule applied to $[s_{i-3}, s_i]$ when i is odd. This method has the advantage that, given a suitable starting value x_1, all approximate solution values may be calculated to the same order of accuracy.

5.1.2 Quadrature methods for nonlinear equations

Volterra equations arising in practice are usually nonlinear; however, the nonlinearity adds rather little to the difficulty of obtaining a numerical solution. The methods described above remain applicable; as for differential equations, it is useful to distinguish between *explicit* and *implicit* methods.

In the case of a nonlinear Volterra equation of the form (1) an explicit system of equations for each x_i can be obtained provided that $w_{ii} = 0$, $i = 1, 2, \ldots, N$. When this condition holds we obtain the defining system:

$$x_0 = y(s_0),$$

$$x_i = y(s_i) + h \sum_{j=0}^{i-1} w_{ij} k(s_i, t_j, x_j), \quad i = 1, \ldots, N. \tag{6}$$

If however $w_{ii} \neq 0$ for at least one value of i the system is given by:

$$x_0 = y(s_0),$$

$$x_i = y(s_i) + h \sum_{j=0}^{i} w_{ij} k(s_i, t_j, x_j), \quad i = 1, \ldots, N. \tag{7}$$

These equations define the x_i implicitly and the ith equation must be solved by an iterative technique: usually by simple iteration. Using such a technique, a sequence of values $x_i^{(r)}, r = 0, 1, \ldots,$ is obtained which hopefully converges to x_i, and which is defined by the equations

$$x_i^{(r)} = y(s_i) + h \sum_{j=0}^{i-1} w_{ij} k(s_i, t_j, x_j)$$

$$+ w_{ii}(s_i, t_i, x_i^{(r-1)}), \quad r = 1, 2, \ldots . \tag{8}$$

Given a starting value $x_i^{(0)}$ we may iterate (8) until convergence is achieved, that is, until the absolute difference between successive iterates is less than some prescribed accuracy. Such a method is analogous to the use of a predictor–corrector method for the numerical solution of ordinary differential equations, where the predictor is the method used to provide $x_i^{(0)}$ and the corrector is given by equation (8). In practice, iteration to convergence in the corrector imposes a restriction, in addition to that imposed by stability and accuracy requirements, on the maximum steplength h which can be used.

5.1.3   Runge–Kutta methods

Runge–Kutta methods for the solution of (1) are self-starting methods which determine approximations to the solution at the points $s_i = a + ih$, $i = 1, \ldots, N$, by generating approximations at some intermediate points in $[s_i, s_{i+1}]$, $i = 0, \ldots, N-1$:

$$s_i + \theta_r h, \quad i = 0, 1, \ldots, N-1, \quad r = 1, \ldots, p-1,$$

where

$$0 = \theta_0 \leq \theta_1 \leq \cdots \leq \theta_{p-1} \leq 1. \tag{9}$$

The derivation follows closely that for ordinary differential equations. We recall the general p-stage Runge–Kutta method for the initial value problem

$$x'(s) = f(s, x(s)),$$
$$x(a) = x_0, \tag{10}$$

given by

$$x_{i+1} = x_i + h \sum_{l=0}^{p-1} A_{pl} k_l^i \tag{11}$$

where

$$
\left.
\begin{aligned}
k_0^i &= f(a + ih, x_i), \\
k_r^i &= f(a + (i + \theta_r)h, x_i + h \sum_{l=0}^{r-1} A_{rl} k_l^i), \quad r = 1, \ldots, p-1,
\end{aligned}
\right\} \tag{11a}
$$

$$\sum_{l=0}^{r-1} A_{rl} = \begin{cases} \theta_r, & r = 1, 2, \ldots, p-1, \\ 1, & r = p, \end{cases} \tag{11b}$$

with x_r an approximation to the solution at $s = s_r = a + rh$. The second argument of k_r^i may be regarded as an approximation to $x(a + (i + \theta_r)h)$ and we rewrite equation (11) as

$$x_{i+1} = x_i + h \sum_{l=0}^{p-1} A_{pl} f(s_i + \theta_l h, x_{i+\theta_l}). \tag{12}$$

The parameters A_{pl}, θ_l are chosen in practice to yield a final approximation of specified order; that is, with a local truncation error of $\mathcal{O}(h^{q+1})$ for some chosen q which is the order of the method. This requirement yields a set of nonlinear equations for the unknown parameters; for a given pair (p, q), no solutions, one solution or a family of solutions may exist.

Example 5.1.2. Suppose we choose $p = 2$ in (12). Then it follows that

$$x_{i+1} = x_i + hA_{20}f(s_i, x_i) + hA_{21}f(s_i + \theta_1 h, x_i + hA_{10}f(s_i, x_i)).$$

We use Taylor's theorem for a function of two variables to obtain

$$x_{i+1} = x_i + h(A_{20} + A_{21})f + h^2 A_{21}(\theta_1 f_s + A_{10} f f_x) + \mathcal{O}(h^3)$$

where we have introduced the notation

$$f = f(s_i, x_i), \quad f_s = \frac{\partial}{\partial s} f(s_i, x_i), \quad f_x = \frac{\partial}{\partial x} f(s_i, x_i).$$

Now if we compare this expression term by term with

$$x_{i+1} = x_i + hf + \tfrac{1}{2}h^2(f_s + f f_x) + \mathcal{O}(h^3)$$

(where we have used equation (10)) we have the following set of three equations in four unknowns to satisfy for the method to be of order two:

$$A_{20} + A_{21} = 1,$$
$$A_{21}\theta_1 \quad = \tfrac{1}{2},$$
$$A_{21}A_{10} \quad = \tfrac{1}{2}.$$

Clearly there exist an infinite number of solutions of these equations corresponding to an infinite number of two-stage Runge–Kutta methods of order two. We consider two particular solutions which are popular in practice:

(i) when $A_{20} = A_{21} = \tfrac{1}{2}$ the resulting method is

$$x_{i+1} = x_i + \tfrac{1}{2}h[f(s_i, x_i) + f(s_{i+1}, x_i + hf(s_i, x_i))]$$

or

$$x_{i+1} = x_i + \tfrac{1}{2}h[f(s_i, x_i) + f(s_{i+1}, \tilde{x}_{i+1})]$$

where

$$\tilde{x}_{i+1} = x_i + hf(s_i, x_i).$$

This is the *improved Euler method*.

(ii) when $A_{20} = 0$, $A_{21} = 1$ the resulting method is the *modified Euler method* given by

$$x_{i+1} = x_i + hf(s_i + \tfrac{1}{2}h, x_i + \tfrac{1}{2}hf(s_i, x_i)).$$

When $p = q = 4$, we obtain in a similar way the *classical fourth order Runge–Kutta method* given by the following choice of parameters:

$$\theta_0 = 0, \quad \theta_1 = \theta_2 = \tfrac{1}{2}, \quad \theta_3 = 1,$$
$$A_{10} = \tfrac{1}{2}, \quad A_{20} = 0, \quad A_{21} = \tfrac{1}{2},$$
$$A_{30} = 0 = A_{31}, \quad A_{32} = 1,$$
$$A_{40} = \tfrac{1}{6} = A_{43}, \quad A_{41} = \tfrac{1}{3} = A_{42}.$$

The method defined in equation (12) can be extended to give a class of Runge–Kutta methods for the solution of

$$x(s) = y(s) + \int_a^s k(s, t, x(t))dt, \quad a \le s \le b. \tag{13}$$

Setting $s = s_i$ in (13) we have

$$x(s_i) = y(s_i) + \int_a^{a+ih} k(a + ih, t, x(t))dt, \quad i = 1, \ldots, N,$$

$$= y(s_i) + \sum_{j=0}^{i-1} \int_{a+jh}^{a+(j+1)h} k(a + ih, t, x(t))dt, \quad i = 1, \ldots, N,$$

and we can determine an approximation x_i to $x(s_i)$ from the following equation

$$x_i = y(s_i) + h \sum_{j=0}^{i-1} \sum_{l=0}^{p-1} A_{pl} k(a + ih, a + (j + \theta_l)h, x_{j+\theta_l}). \tag{14}$$

Now for $s \in (s_i, s_{i+1})$ we may write equation (13) in the following form

$$x(s) = y(s) + \sum_{j=0}^{i-1} \int_{s_j}^{s_{j+1}} k(s, t, x(t)) dt + \int_{s_i}^{s} k(s, t, x(t)) dt. \tag{15}$$

Then setting $s = s_i + \theta_v h$, $v = 1, \ldots, p-1$, and approximating the final integral term in (15) by

$$\int_{s_i}^{s_i+\theta_v h} k(s_i + \theta_v h, t, x(t)) dt \approx h \sum_{l=0}^{v-1} A_{vl} k(s_i + \theta_v h, s_i + \theta_l h, x_{i+\theta_l})$$

we see that the Runge–Kutta method for Volterra equations of the form (13) may be expressed as

$$x_{i+\theta_v} = y(s_i + \theta_v h) + h \sum_{j=0}^{i-1} \sum_{l=0}^{p-1} A_{pl} k(s_i + \theta_v h, s_j + \theta_l h, x_{j+\theta_l})$$

$$+ h \sum_{l=0}^{v-1} A_{vl} k(s_i + \theta_v h, s_i + \theta_l h, x_{i+\theta_l}),$$

$$i = 0, 1, \ldots, N-1, \quad v = 1, 2, \ldots, p-1, \tag{16}$$

where $x(a) = y(a)$ and the parameters A_{rj}, θ_j, $r = 1, \ldots, p$, $j = 0, 1, \ldots, p-1$, define the particular method.

5.1.4 Starting procedures

We saw in Section 5.1.1 that special starting procedures are required for use with quadrature methods applied to the solution of equation (1). Consider the solution of (1) by the multistep method:

$$x_i = y(s_i) + h \sum_{j=0}^{i} w_{ij} k(s_i, t_j, x_j), \quad i = k, k+1, \ldots, N, \tag{17}$$

and suppose the method is of order p, that is we may write

$$E_{i,t}(k(s_i, t, x(t))) \approx A h^{p+1}$$

where A is a constant and $E_{i,t}(k(s_i, t, x(t))$ is the error associated with the quadrature rule. Now, suppose that we try to achieve an overall (global) accuracy of $\mathcal{O}(h^p)$; then the method in (17) should (we expect) be of local accuracy $\mathcal{O}(h^{p+1})$ (since we shall use $\mathcal{O}(h^{-1})$ steps) while the starting method need only be of accuracy $\mathcal{O}(h^p)$ (since it is used for only $\mathcal{O}(1)$ steps). Clearly there is a need for high-order starting methods.

If the kernel is sufficiently regular it may be possible to find a Taylor series expansion for x in the neighbourhood of $s = a$ from which the necessary starting values may be found. For example, if we differentiate the integral equation (1) n times, we find

$$x^{(n)}(s) = y^{(n)}(s) + \sum_{j=0}^{n-1} \frac{d^{n-1-j}}{ds^{n-1-j}} k^{(j)}(s,s,x(s))$$

$$+ \int_a^s \frac{\partial^n}{\partial s^n} k(s,t,x(t)) dt \tag{18}$$

where

$$k^{(j)}(s,s,x(s)) = \left[\frac{\partial^j}{\partial s^j} k(s,t,x(t)) \right] \Bigg|_{t=s}.$$

This is a recurrence relation from which we can compute $x^{(n)}(s)$, $n = 1, 2, \ldots$. In particular the values at $s = a$ follow on dropping the integral term, and from these we can predict $x(s)$ for the first step $s = a + h$ via the Taylor series. We could also predict $x(a + 2h)$ in this way but it is preferable to start again by replacing a by $a + h$ in (18) to determine a new set of derivatives at $s = a + h$. In practice however this Taylor series method is generally unattractive since it requires the derivatives of k; depending on the form of the kernel these derivatives may be difficult to evaluate and certainly the majority of users of library routines for this method would be reluctant to provide such derivative functions.

Our prime need is for a high-order method, that is, one which is exact whenever $k(s,t,x(t))$ is a polynomial of high degree. We can construct such rules without introducing derivatives and the simplest way conceptually is to carry out a 'deferred approach to the limit' calculation over the interval $[a, a + h]$ using intervals of $h, h/2, h/4, \ldots$. This process can be carried out iteratively until results to the desired order have been obtained or until the desired accuracy appears to have been reached. For example, let us carry out one stage of such a process for a linear Volterra equation using the Trapezoid rule. Using a steplength h we have, setting the lower limit $a = 0$ in (1b):

$$x(0) = y(0),$$
$$x(h) = y(h) + \tfrac{1}{2}h[K(h,0)x(0) + K(h,h)x(h)] + Ah^2 + \mathcal{O}(h^4). \tag{19}$$

Then with a steplength $\tfrac{1}{2}h$ we have:

$$x(0) = y(0),$$
$$x\left(\frac{h}{2}\right) = y\left(\frac{h}{2}\right) + \tfrac{1}{4}h\left[K\left(\frac{h}{2},0\right)x(0) + K\left(\frac{h}{2},\frac{h}{2}\right)x\left(\frac{h}{2}\right) \right], \tag{20a}$$
$$x(h) = y(h) + \tfrac{1}{4}h\left[K(h,0)x(0) + 2K\left(h,\frac{h}{2}\right)x\left(\frac{h}{2}\right) + K(h,h)x(h) \right]$$
$$+ \tfrac{1}{4}Ah^2 + \mathcal{O}(h^4), \tag{20b}$$

whence it follows from $(20a, b)$ that

$$x(h)[1 - \tfrac{1}{4}hK(h,h)]$$

$$= y(h) + \tfrac{1}{4}h\left[K(h,0)x(0) + 2K\left(h, \frac{h}{2}\right) \frac{\{y(h/2) + \tfrac{1}{4}hK(h/2,0)x(0)\}}{1 - \tfrac{1}{4}hK(h/2, h/2)} \right]$$

$$+ \tfrac{1}{4}Ah^2 + \mathcal{O}(h^4). \tag{21}$$

Now eliminating A between (19) and (20) we find

$$x(h)[1 - \tfrac{1}{6}hK(h,h)] = y(h) + \tfrac{1}{6}hK(h,0)x(0) + \tfrac{2}{3}hK\left(h, \frac{h}{2}\right)x\left(\frac{h}{2}\right) \tag{22}$$

and we note that the above rule is just Simpson's rule applied to equation (1*b*) over the interval $[a, a + h] = [0, h]$; thus we obtain a third order approximation to $x(h)$.

Alternatively, Runge–Kutta type rules can be used for a fixed number of initial steps of the quadrature rule thus providing starting formulae of appropriate orders; or indeed, as we saw in the previous section, they may be used to obtain the entire numerical solution.

Finally, a number of special starting procedures have appeared in the literature, one of which, proposed by Day (1968), provides third order approximations for $x(a + h)$, $x(a + 2h)$ and $x(a + 3h)$:

$$x_1^{(0)} = y(a + h) + hk(a + h, a, x_0), \tag{23}$$

$$x_1^{(1)} = y(a + h) + \tfrac{1}{2}h[k(a + h, a, x_0) + k(a + h, a + h, x_1^{(0)})], \tag{24}$$

$$x_{\frac{1}{2}}^{(0)} = \tfrac{1}{2}(x_0 + x_1^{(1)}), \tag{25}$$

$$x_{\frac{1}{2}}^{(1)} = y\left(a + \frac{h}{2}\right) + \tfrac{1}{4}h\left[k\left(a + \frac{h}{2}, a, x_0\right) + k\left(a + \frac{h}{2}, a + \frac{h}{2}, x_{\frac{1}{2}}^{(0)}\right) \right] \tag{26}$$

$$x_1 = y(a + h) + \tfrac{1}{6}h\left[k(a + h, a, x_0) + 4k\left(a + h, a + \frac{h}{2}, x_{\frac{1}{2}}^{(1)}\right) \right.$$

$$\left. + k(a + h, a + h, x_1^{(1)}) \right], \tag{27}$$

$$x_2^{(0)} = y(a + 2h) + 2hk(a + 2h, a + h, x_1), \tag{28}$$

$$x_2 = y(a + 2h) + \tfrac{1}{3}h[k(a + 2h, a, x_0) + 4k(a + 2h, a + h, x_1)$$

$$+ k(a + 2h, a + 2h, x_2^{(0)})], \tag{29}$$

$$x_3^{(0)} = y(a + 3h) + \tfrac{3}{2}h[k(a + 3h, a + h, x_1) + k(a + 3h, a + 2h, x_2)], \tag{30}$$

$$x_3 = y(a + 3h) + \tfrac{3}{8}h[k(a + 3h, a, x_0) + 3k(a + 3h, a + h, x_1)$$

$$+ 3k(a + 3h, a + 2h, x_2) + k(a + 3h, a + 3h, x_3^{(0)})]. \tag{31}$$

The initial stage of this method, given by equations (23)–(27), uses a predictor–corrector method (namely Euler's method as predictor with the Trapezoid rule as corrector) to provide a second order approximation $x_1^{(1)}$ to $x(a + h)$, followed respectively by linear interpolation and the Trapezoid rule to provide first and second order approximations to $x(a + h/2)$. Finally Simpson's rule is used to provide a third order approximation to $x(a + h)$. The second and third stages use the midpoint rule and a 2-point open Newton–Cotes rule respectively to provide initial approximations to the solution, followed by Simpson's rule to provide third order approximations to $x(a + 2h)$ and $x(a + 3h)$. These starting values may be used, for example, to start a Newton–Gregory method or a repeated 5-point Newton–Cotes rule, or we may simply use x_0 and x_1 to start the repeated Simpson's rule. This construction clearly shares something with the spirit of Runge–Kutta, but is rather more *ad hoc*; we trust that the reader is sufficiently put off by this to prefer the rather more wholehearted Runge–Kutta rules (although it is unlikely in practice to matter which is used).

5.1.5 Convergence

The numerical method is said to be *convergent* if the solution of the approximating set of equations converges to the solution of the exact problem as the steplength h tends to zero; that is, if

$$\lim_{h \to 0} |x(a + ih) - x_i| = 0 \quad \text{with} \quad ih = s - a \text{ fixed.}$$

Consider the equation

$$x(s) = y(s) + \int_a^s k(s, t, x(t)) \mathrm{d}t, \quad a \leq s \leq b, \tag{32}$$

and suppose that at $s = s_i = a + ih$ the quadrature formula

$$\int_a^{a+ih} k(s_i, t, x(t)) \mathrm{d}t = h \sum_{j=0}^{i} w_{ij} k(s_i, t_j, x(t_j)) + E_{i,t}(k(s_i, t, x(t))) \tag{33a}$$

is such that

$$E_{i,t}(k(s_i, t, x(t))) \to 0 \quad \text{as} \quad h \to 0 \text{ with } ih \text{ fixed.} \tag{33b}$$

Then the quadrature formula is said to be *consistent* and we show that when this consistency condition is satisfied[†] the resulting quadrature method for the solution of (32) is convergent. At $s = s_i$, $i = k, k + 1, \ldots, N$, we have, on

---

[†]Those readers familiar with multistep methods for the solution of ordinary differential equations should recall that consistency and zero stability are necessary requirements for convergence of the method. Note however the fixed interval as $h \to 0$ in (33b).

substituting (33*a*) in (32),

$$x(s_i) = y(s_i) + h \sum_{j=0}^{i} w_{ij}k(s_i, t_j, x(t_j))$$

$$+ E_{i,t}(k(s_i, t, x(t))), \quad i = k, k+1, \ldots, N \tag{33}$$

and the corresponding approximating equations are

$$x_i = y(s_i) + h \sum_{j=0}^{i} w_{ij}k(s_i, t_j, x_j), \quad i = k, k+1, \ldots, N. \tag{34}$$

Thus we have

$$x(a + ih) - x_i = h \sum_{j=0}^{i} w_{ij}[k(s_i, t_j, x(t_j)) - k(s_i, t_j x_j)]$$

$$+ E_{i,t}(k(s_i, t, x(t))), \quad i = k, k+1, \ldots, N.$$

Now if the kernel function k satisfies a Lipschitz condition in its third argument with Lipschitz constant L and we set $e_i = x(a + ih) - x_i$, it follows that

$$|e_i| \leq hL \sum_{j=0}^{i} |w_{ij}||e_j| |E_{i,t}(k(s_i, t, x(t)))|, \quad i = k, k+1, \ldots, N.$$

Let $w = \max_{i,j}|w_{ij}|$ and $e = \max_{0 \leq i \leq k-1} |e_i|$. Then, for sufficiently small h, (see Delves and Walsh, 1974, Chapter 11)

$$|e_i| \leq \frac{hLw}{1 - hLw} \sum_{j=0}^{i-1} |e_j| + \frac{1}{1 - hLw}|E_{i,t}(k(s_i, t, x(t)))|,$$

$$i = k, k+1, \ldots, N,$$

and it follows that

$$|e_i| \leq \frac{\{|E_{i,t}(k(s_i, t, x(t)))| + hLw(k-1)e\}}{1 - hLw} \exp\left\{\frac{wL}{1 - hLw}ih\right\}. \tag{35}$$

Hence $|e_i| \to 0$ as $h \to 0$ with ih fixed. We may write equivalently

$$|e_i| = \mathcal{O}(h^{p+1}) + \mathcal{O}(h^{q+1})$$

where the error in the quadrature rule is $\mathcal{O}(h^{p+1})$ and the error in the starting values is $\mathcal{O}(h^q)$. If we set $r = \min(p, q)$ then we say the quadrature rule is *convergent of order r*.

5.1.6 Round-off error and stability

We have seen already that there is a strong analogy with ordinary differential equations with single point boundary conditions. As in that case, we should also consider the build-up of rounding errors. The subject is unfortunately more complicated than for differential equations and we give only a simplified treatment here.

First let us consider how an error made at $s = a$ propagates in the course of solution of the integral equation. Suppose that at $s = a$ we make an error ε. We have the exact solution

$$x(s) = y(s) + \int_a^s k(s, t, x(t))dt; \quad x(a) = y(a). \tag{36}$$

Now consider the case where we have evaluated an approximate solution $\tilde{x}(s)$ with $\tilde{x}(a) = y(a) + \varepsilon$ and where we use the integral equation to generate $\tilde{x}$ between a and s. If we set

$$\tilde{x}(s) = \bar{y}(s) + \int_a^s k(s, t, \tilde{x}(t))dt \tag{37}$$

where

$$\bar{y}(s) = y(s) + \varepsilon, \quad \forall s,$$

then with $e(s) = \tilde{x}(s) - x(s)$, we have

$$e(s) = \bar{y}(s) - y(s) + \int_a^s [k(s, t, \tilde{x}(t)) - k(s, t, x(t))]dt. \tag{38}$$

Now using the Mean Value Theorem we obtain

$$e(s) = \varepsilon + \int_a^s \bar{k}(s, t)e(t)dt \tag{39}$$

where

$$\bar{k}(s, t) = \frac{\partial}{\partial z(t)}k(s, t, z(t))|_{z(t) = \tilde{x}(t) + \theta(x(t) - \tilde{x}(t))}, \quad 0 < \theta < 1.$$

Thus $e(s)$ satisfies a Volterra equation of the second kind and we may write equivalently

$$e'(s) = \frac{\partial}{\partial s}\int_a^s \bar{k}(s, t)e(t)dt$$

$$= \bar{k}(s, s)e(s) + \int_a^s \frac{\partial}{\partial s}\bar{k}(s, t)e(t)dt. \tag{40}$$

This is a homogeneous *integro-differential equation*. If we set $s = a$, we obtain

$$e'(a) = \bar{k}(a, a)e(a) = \varepsilon\bar{k}(a, a). \tag{41}$$

Unfortunately the propagation properties of (40) are hard to analyse in general, and we consider a particular example, with kernel $\bar{k} \equiv \lambda$, that is,

$$x(s) = y(s) + \lambda \int_a^s x(t)dt. \tag{42}$$

Then we have

$$e'(s) = \lambda e(s)$$

with corresponding solution

$$e(s) = C \exp(\lambda s).$$

Thus if λ is positive the error is unbounded in s. By analogy with the situation in differential equations, we refer to this form of instability as *inherent*; its effect is largely independent of the numerical quadrature rule used. The occurrence of this form of instability is of course serious only if the error increases faster than the true solution, since we are normally interested in relative accuracy.

More important, and more in our control, is the possibility of instability *induced* by the choice of quadrature rule. Consider first the solution, by linear multistep methods, of the ordinary differential equation

$$x'(s) = f(s, x(s)),$$
$$x(s_0) = x_0. \tag{43}$$

Such methods produce an approximation to the function value at each step from a fixed number of back values, using a formula of the type

$$x_{n+1} = \sum_{i=0}^{p} a_i x_{n-i} + h \sum_{i=-1}^{p} b_i f(s_{n-i}, x_{n-i}) \tag{44}$$

where the a_i, b_i are constants. (See Lambert, 1973; Henrici, 1962). Equation (44) represents a $(p+1)$-step method which is explicit or implicit according as b_{-1} is zero or nonzero and which requires $p+1$ starting conditions.

We introduce, via an example, the concept of *stability* of numerical methods for the solution of such differential equations.

Example 5.1.3.
In accordance with normal practice, we consider the *test equation*

$$x'(s) = \lambda x(s),$$
$$x(s_0) = x_0, \tag{45}$$

where λ is a complex constant.

Suppose we use Milne's method (see, for example, Lambert, 1973, p. 93) with a steplength $h = 0.1$ to solve equation (45) with $\lambda = -1$, subject to the initial condition $x(0) = 1$. The exact solution of this problem is $x(s) = e^{-s}$. Milne's method uses the predictor–corrector pair

$$x_{n+4}^{(p)} = x_n + \frac{4h\lambda}{3}(2x_{n+1} - x_{n+2} + 2x_{n+3}),$$

$$x_{n+4}^{(c)} = x_{n+2} + \frac{h\lambda}{3}(x_{n+2} + 4x_{n+3} + x_{n+4}^{(p)}),$$

Table 5.1.2.*Milne's method applied to* $x'(s) = -x(s)$, $x(0) = 1$.

s	$x^{(c)}$	e^{-s}
0.3	$7.408\,182\,207\,4 \times 10^{-1}$	$7.408\,182\,206\,9, -1$
$\vdots$	$\vdots$	$\vdots$
8.0	$3.351\,759\,215\,0, -4$	$3.354\,626\,281\,5, -4$
8.1	$3.038\,327\,402\,3, -4$	$3.035\,391\,383\,3, -4$
$\vdots$	$\vdots$	$\vdots$
13.2	$2.382\,129\,469\,5, -7$	$1.850\,601\,201\,2, -6$
13.4	$-2.083\,126\,065\,6, -7$	$1.515\,144\,115\,3, -6$

and we show in Table 5.1.2 the true solution and the corrected values obtained after 'iterating to convergence'. We have used the Taylor series method, outlined in Section 5.1.4, to obtain the necessary starting values.

Note that when $s = 8.0$, 8.1 the results differ by only 3 figures in the seventh decimal place, but when s reaches 13.2 the results differ by a factor of 8. More alarming is the fact that at $s = 13.4$ the calculated value is negative whereas the true solution is a strictly positive function! In order to explain these discrepancies we consider Milne's corrector equation since in the mode of iterating to convergence the predictor equation is irrelevant for the purpose of analysing the numerical results. The corrector equation may be written as

$$ax_{n+4} + 2bx_{n+3} + cx_{n+2} = 0 \tag{46}$$

with

$$a = 1 - \tfrac{1}{3}h\lambda, \quad b = -\tfrac{2}{3}h\lambda, \quad c = -(1 + \tfrac{1}{3}h\lambda),$$

and where for convenience we have omitted all superscripts. Equation (46) is a difference equation, and we seek solutions of the form $x_n = p^n$. Any such solution must satisfy the equation

$$ap^2 + 2bp + c = 0. \tag{47}$$

This is the so-called *characteristic equation* for Milne's method; the left hand side of (47) is referred to as the *characteristic* or *stability polynomial* Π of the method. Now the general solution of (46) is

$$x_n = Ap_+^n + Bp_-^n$$

where $p_\pm$ are the roots of (47) with an obvious notation and A, B are constants determined by the starting values. Now, for small values of $h\lambda$, we find that p_+ is given by (on substituting the values of a, b, c):

$$p_+ = \frac{1 + \tfrac{2}{3}h\lambda + \mathcal{O}(h^2\lambda^2)}{1 - \tfrac{1}{3}h\lambda}.$$

Since $s = nh$ and $\lim\limits_{n \to \infty} (1 + z/n)^n = e^z$ we have

$$p^n_+ = e^{\lambda s}\left[1 + \mathcal{O}\left(\frac{1}{n}\right)\right].$$

Similarly we find

$$p^n_- = (-1)^n e^{-\lambda s/3}\left[1 + \mathcal{O}\left(\frac{1}{n}\right)\right]$$

and it follows that

$$x_n = A e^{\lambda s}\left[1 + \mathcal{O}\left(\frac{1}{n}\right)\right] + B(-1)^n e^{-\lambda s/3}\left[1 + \mathcal{O}\left(\frac{1}{n}\right)\right].$$

In particular, when $\lambda = -1$, we obtain

$$x_n = A e^{-s}\left[1 + \mathcal{O}\left(\frac{1}{n}\right)\right] + B(-1)^n e^{s/3}\left[1 + \mathcal{O}\left(\frac{1}{n}\right)\right]. \tag{48}$$

When $A = 1$ and $B = 0$ equation (48) yields the correct solution, since as $n \to \infty$, $x_n \to e^{-s}$. However the second term in (48) is an increasing exponential which oscillates in sign and for large values of s this term will yield the dominant contribution to x_n, thus swamping the true solution; we therefore say that Milne's method is *unstable*.

This problem of instability arises because there is a component in the solution of the difference equation which diverges. We introduce the following definition:

Definition 5.1.1. The difference method for $x' = \lambda x$ is *absolutely stable* for a given $h\lambda$ if all zeros of Π are less than one in modulus.

Thus in Example 5.1.3 we see that for small values of $h\lambda$

$$p_+ = 1 + h\lambda + \mathcal{O}(h^2\lambda^2), \quad p_- = -1 + \tfrac{1}{3}h\lambda + \mathcal{O}(h^2\lambda^2),$$

and hence when $h\lambda < 0, |p_-| > 1$. Thus the method of Example 5.1.3 is not absolutely stable for this range of $h\lambda$; we say it is *absolutely unstable*. (In fact when $h\lambda > 0, |p_+| > 1$ and the method is absolutely unstable for all nonzero values of $h\lambda$). In practice the stability requirement of Definition 5.1.1 imposes a restriction on the size of the steplength h which in turn has the effect of keeping to an acceptable level any errors which propagate from errors in the starting values, as the solution process proceeds. We should note here that the requirement of absolute stability is very stringent. In many examples, the required solution is itself an increasing function; we may then not object to an increasing error component provided that it does not increase faster than the solution. A more appropriate criterion in such a case is that of *relative stability* (see Lambert, 1973, p. 68).

We now discuss the extension of the concept of stability to the solution of Volterra integral equations.

Consider the solution of equation (1) using the m-panel repeated $(k+1)$-point rule

$$x_i = y(s_i) + h \sum_{j=0}^{i} w_{ij} k(s_i, t_j, x_j), \quad i = k, k+1, \ldots, N, \tag{49}$$

with $N = km - 1$. This method uses *all* previously computed function values to determine the function value at the next step, and this fact complicates the discussion. We note that the discussion for differential equations above was restricted to the solution of the model problem (42); for the integral equation (1) we follow Kershaw (1974) (who follows Mayers, 1962) and present a brief stability analysis for the very simple test equation

$$x(s) = 1 + \lambda \int_0^s x(t) dt \tag{50}$$

which is just the integral equation formulation of equation (42) with $y(s) = 1, a = 0$. We consider the case where the quadrature rule used in (49) is (i) the repeated Simpson's rule when i is even, and (ii) the repeated Simpson's rule with the single panel Trapezoid rule applied to (a) the lower end (method STa) and (b) the upper end (method STb) of the range of integration, when i is odd. The defining equations for these rules are then

(i) (STa or b)

$$x_{2i} = 1 + \tfrac{1}{3}h\lambda \sum_{j=0}^{2i} w_{2i,j} x_j; \tag{51}$$

$$w_{2i,0} = w_{2i,2i} = 1; \quad w_{2i,j} = 3 - (-1)^j, \quad 1 \le j \le 2i - 1$$

(ii) (a) (STa)

$$x_{2i+1} = 1 + \tfrac{1}{2}h\lambda(x_0 + x_1) + \tfrac{1}{3}h\lambda \sum_{j=1}^{2i+1} w_{2i+1,j} x_j; \tag{52}$$

$$w_{2i+1,1} = w_{2i+1,2i+1} = 1; \quad w_{2i+1,j} = 3 - (-1)^j, \quad 2 \le j \le 2i$$

(b) (STb)

$$x_{2i+1} = 1 + \tfrac{1}{3}h\lambda \sum_{j=0}^{2i} w_{2i,j} x_j + \tfrac{1}{2}h\lambda(x_{2i} + x_{2i+1}) \tag{53}$$

where the $w_{2i,j}$ are defined as in equation (51).

Consider first the method STa defined by (i), (ii) (a). Then on setting $i = v, v+1$ in (51) we obtain

$$x_{2v+2} - x_{2v} = \tfrac{1}{3}h\lambda(x_{2v} + 4x_{2v+1} + x_{2v+2}), \quad v = 0, 1, 2, \ldots.$$

Similarly if we set $i = v, v+1$ in (52) we have

$$x_{2v+3} - x_{2v+1} = \tfrac{1}{3}h\lambda(x_{2v+1} + 4x_{2v+2} + x_{2v+3}), \quad v = 0, 1, 2, \ldots.$$

Thus in general

$$(1 - \tfrac{1}{3}h\lambda)x_{k+2} - \tfrac{4}{3}h\lambda x_{k+1} - (1 + \tfrac{1}{3}h\lambda)x_k = 0, \quad k = 0, 1, 2, \ldots$$

and we have already seen in Example 5.1.3 that this method is unstable.

Method STb, defined by (i), (ii)(b) yields the following difference equations

$$x_{2\nu+1} - x_{2\nu} = \tfrac{1}{2}h\lambda(x_{2\nu} + x_{2\nu+1}), \quad \nu = 0, 1, 2, \ldots$$

and

$$x_{2\nu+3} - x_{2\nu+1} = \tfrac{1}{3}h\lambda(x_{2\nu} + 4x_{2\nu+1} + x_{2\nu+2})$$
$$+ \tfrac{1}{2}h\lambda(x_{2\nu+2} + x_{2\nu+3} - x_{2\nu} - x_{2\nu+1}), \quad \nu = 0, 1, 2, \ldots,$$

whence it follows that

$$(1 + \tfrac{1}{2}h\lambda)x_{2\nu} = (1 - \tfrac{1}{2}h\lambda)x_{2\nu+1}, \quad \nu = 0, 1, 2, \ldots, \tag{54}$$

and

$$(1 - \tfrac{1}{2}h\lambda)x_{2\nu+3} - \tfrac{5}{6}h\lambda x_{2\nu+2} - (1 + \tfrac{5}{6}h\lambda)x_{2\nu+1}$$
$$+ \tfrac{1}{6}h\lambda x_{2\nu} = 0, \quad \nu = 0, 1, 2, \ldots. \tag{55}$$

Now using equation (54) to substitute for $x_{2\nu}$, $x_{2\nu+2}$ in (55) we obtain

$$(1 - \tfrac{5}{6}h\lambda + \tfrac{1}{6}h^2\lambda^2)x_{2\nu+3} = (1 + \tfrac{7}{6}h\lambda + \tfrac{1}{2}h^2\lambda^2)x_{2\nu+1}, \quad \nu = 0, 1, 2, \ldots,$$

whence it follows (from (54)) that

$$(1 - \tfrac{5}{6}h\lambda + \tfrac{1}{6}h^2\lambda^2)x_{2\nu+2} = (1 + \tfrac{7}{6}h\lambda + \tfrac{1}{2}h^2\lambda^2)x_{2\nu}, \quad \nu = 0, 1, 2, \ldots.$$

Thus we have in general

$$x_{k+2} = \left(\frac{1 + \tfrac{7}{6}h\lambda + \tfrac{1}{2}h^2\lambda^2}{1 - \tfrac{5}{6}h\lambda + \tfrac{1}{6}h^2\lambda^2}\right)x_k, \quad k = 0, 1, 2, \ldots.$$

This recurrence relation yields the characteristic equation

$$p^2 = \left(\frac{1 + \tfrac{7}{6}h\lambda + \tfrac{1}{2}h^2\lambda^2}{1 - \tfrac{5}{6}h\lambda + \tfrac{1}{6}h^2\lambda^2}\right).$$

The condition for absolute stability of the method is therefore precisely

$$\left|\frac{1 + \tfrac{7}{6}h\lambda + \tfrac{1}{2}h^2\lambda^2}{1 - \tfrac{5}{6}h\lambda + \tfrac{1}{6}h^2\lambda^2}\right| < 1$$

and this is satisfied for $-6 < h\lambda < 0$; the restriction $h\lambda < 0$ is seen most clearly by considering small values of $h\lambda$, when the characteristic roots are

$$p_{\pm} = \pm(1 + h\lambda + \mathcal{O}(h^2\lambda^2)).$$

It is clear from the above approximation that for small $h\lambda$, $|p_{\pm}| < 1$ when $h\lambda < 0$.

There is thus a remarkable difference between methods STa and STb: STa is always absolutely unstable, while STb is stable at least for small negative $h\lambda$. This difference between the two methods is not an isolated one.

Noble (1969) has studied the use of the repeated Simpson's rule when the value of i in (49) is even, and the repeated Simpson's rule with the three-eighths rule applied to the upper or lower end of the range of integration when i is odd. He concluded that again the methods are respectively stable and unstable; let us try to understand why this comes about.

An inspection of the weights associated with these methods led Noble to introduce the concept of a *repetition factor r* for a given method. Suppose we solve (1) using the repeated quadrature rule of (49), where it is assumed that $x_0 = y(s_0)$ and $x_1, \ldots, x_{k-1}$ are determined by some starting procedure. Then the method is said to have a repetition factor r if r is the smallest integer such that

$$w_{i+r,j} = w_{ij}, \quad j = \alpha, \alpha + 1, \ldots, i - \beta, \tag{56}$$

where α and β are integers independent of i. For example, if we use over the interval $[a, a + (2s + 1)h]$, the method defined by (i), (ii) (b) then this method has associated weights $w_{2s+1,j}$ given by

$$\text{ST}b : \frac{6}{h} \{w_{2s+1,j}\}_{j=0,1,\ldots,2s+1} = \{2, 8, 4, 8, \ldots, 8, 5, 3\};$$

while the method defined by (i), (ii)(a) has weights

$$\text{ST}a : \frac{6}{h} \{w_{2s+1,j}\}_{j=0,1,\ldots,2s+1} = \{3, 5, 8, 4, 8, \ldots, 8, 2\}.$$

Thus method STa has a repetition factor 1 and method STb a repetition factor of 2.

Linz (1968) conjectured that methods based on repeated rules which have a repetition factor $r = 1$ tend to be numerically stable, whereas those for which r exceeds unity tend to be unstable. Noble (1969) has shown that such methods with $r = 1$ are indeed in a suitable sense stable, and we outline briefly Noble's error analysis for the solution of equation (1) using the method of equation (49). The solution of (49) provides approximate function values x_i to the exact solutions $x(a + ih)$ and we may write

$$x(a + ih) = y(a + ih) + h \sum_{j=0}^{i} w_{ij} k(a + ih, a + jh, x(a + jh))$$
$$+ E_{i,t}(k(a + ih, t, x(t))), \quad i = k, k + 1, \ldots, N,$$

where $E_{i,t}(k(a + ih, t, x(t)))$ is the discretisation error associated with the quadrature rule. We define $e_i = x(a + ih) - x_i$ and it follows that

$$e_i = h \sum_{j=0}^{i} w_{ij} [k(a + ih, a + jh, x(a + jh)) - k(a + ih, a + jh, x_j)]$$
$$+ E_{i,t}(k(a + ih, t, x(t))), \quad i = k, k + 1, \ldots, N,$$

whence

$$e_i = h \sum_{j=0}^{i} w_{ij} \left[e_j \frac{\partial k}{\partial z}(a + ih, a + jh, z)|_{z=x_j} + \mathcal{O}(e_j^2) \right]$$

$$+ E_{i,t}(k(a + ih, t, x(t))), \quad i = k, k+1, \ldots, N.$$

Now introduce quantities ε_i satisfying the following linear difference equation

$$\varepsilon_i = h \sum_{j=0}^{i} w_{ij} \bar{k}(a + ih, a + jh) \varepsilon_j$$

$$+ E_{i,t}(k(a + ih, t, x(t))), \quad i = k, k+1, \ldots, N, \qquad (57)$$

where

$$\bar{k}(a + ih, a + jh) = \frac{\partial}{\partial z} k(a + ih, a + jh, z)|_{z=x_j}.$$

Following Noble we assume the existence of continuous functions $u_l(s)$, $l = 1, 2, \ldots, k$, and $E_{i,t}(s)$, $i = k, k+1, \ldots, N$, such that

$$u_l(a + [(m-1)k + l - 1]h) = \varepsilon_{(m-1)k+l-1}$$

$$+ \mathcal{O}(h), \quad m = 2, 3, \ldots, \frac{(N+1)}{k},$$

$$E_{i,t}(a + ih) = E_{i,t}(k(a + ih, t, x(t))) + \mathcal{O}(h), \qquad (58)$$

whence it can be shown that (see Noble, 1969)

$$e_i = \varepsilon_i + \mathcal{O}(\varepsilon_i)$$

so that the quantities ε_i describe adequately the behaviour of the error function. Now, under assumptions (58), equation (57) gives rise to (see Noble, 1969) the following system of Volterra integral equations of the second kind:

$$\mathbf{u}(s) = \int_a^s \bar{k}(s, t) \mathbf{z}(t) dt + \mathbf{E}_{i,t}(s) \qquad (59)$$

where

$$\mathbf{u}(s) = (u_1(s), u_2(s), \ldots, u_k(s))^T,$$

$$\mathbf{z}(s) = (z_1(s), z_2(s), \ldots, z_k(s))^T,$$

and each $z_i(s)$ is a linear combination of the $u_i(s)$, $i = 1, \ldots, k$, that is,

$$z_i(s) = \sum_{j=1}^{k} \beta_{ij} u_j(s).$$

Thus we may write equation (59) equivalently as

$$\mathbf{u}(s) = \int_a^s \bar{k}(s, t) \mathbf{C} \mathbf{u}(t) dt + \mathbf{E}_{i,t}(s) \qquad (59a)$$

where $\mathbf{C}$ is a $k \times k$ matrix with $\mathbf{C} = (\beta_{ij})$, $i, j = 1, 2, \ldots, k$, and the β_{ij} are related to the quadrature weights w_{ij} in (57). Now if $\mathbf{C}$ has k linearly independent eigenvectors $\mathbf{p}_i$ with corresponding eigenvalues λ_i, and we denote by $\mathbf{P}$ the matrix whose ith row is $\mathbf{p}_i$, then we have

$$\mathbf{PCP}^{-1} = \mathbf{\Lambda}$$

where $\mathbf{\Lambda} = \mathbf{diag}(\lambda_i)$. Thus premultiplying (59a) by $\mathbf{P}$ yields the system of independent Volterra equations of the second kind:

$$\alpha_i(s) = \lambda_i \int_a^s \bar{k}(s,t)\alpha_i(t)\mathrm{d}t + \mu_i(s), \quad i = 1, \ldots, k, \tag{60}$$

where

$$\mathbf{P}u(s) = \boldsymbol{\alpha}(s),$$

$$\mathbf{PE}_{i,t}(s) = \boldsymbol{\mu}(s).$$

Noble proceeds to show that $\lambda = 1$ is an eigenvalue of $\mathbf{C}$; thus the corresponding integral equation has exactly the same form as equation (39), and of the remaining eigenvalues, those which are non-zero contribute to the instability of the method. But, if the method has a repetition factor of one, Noble shows that the rows of $\mathbf{C}$ are all the same; that is, $\mathbf{C}$ has rank 1. It then follows that the remaining $k - 1$ eigenvalues are all zero, so that the method is automatically stable in this sense.

It is not however true that methods having a repetition factor greater than one are always unstable; see McKee and Brunner (1980) for counterexamples.

5.2   Block methods

These form a class of self-starting methods which produce a block of values at a time; they give up any attempt to solve the problem by marching one step at a time and instead introduce a rule over a small region which uses points over a larger region. Consider the solution of (5.1.1) in the range $a \leq s \leq b$ with $b - a = Nph$; that is, we divide the interval $[a, b]$ into N equal intervals, each of which is then divided into p subintervals of length h.

Now assume that approximate solution values have been calculated for the first $(r - 1)$ blocks; then a typical block method produces at the rth stage the following set of approximations

$$x_{p(r-1)+1}, x_{p(r-1)+2}, \ldots, x_{pr}.$$

For $p(r - 1) < i \leq pr$, $r = 1, \ldots, N$, we may rewrite (5.1.1) in the form

$$x(a + ih) = y(a + ih) + \int_a^{a + p(r-1)h} k(a + ih, t, x(t))\mathrm{d}t$$

$$+ \int_{a + p(r-1)h}^{a + ih} k(a + ih, t, x(t))\mathrm{d}t. \tag{1}$$

Now if we use the following quadrature rules to approximate the integral terms in (1)

$$\int_a^{a+p(r-1)h} k(a+ih,t,x(t))dt$$

$$\approx h \sum_{j=0}^{p(r-1)} w_{ij}k(a+ih,a+jh,x(a+jh)), \qquad (2a)$$

$$\int_{a+p(r-1)h}^{a+ih} k(a+ih,t,x(t))dt$$

$$\approx h \sum_{j=p(r-1)}^{pr} \bar{w}_{ij}k(a+ih,a+jh,x(a+jh)), \qquad (2b)$$

we obtain the set of approximating equations

$$x_0 = y(a),$$

$$x_i = y(a+ih) + h \sum_{j=0}^{p(r-1)} w_{ij}k(a+ih,a+jh,x_j)$$

$$+ h \sum_{j=p(r-1)}^{pr} \bar{w}_{ij}k(a+ih,a+jh,x_j),$$

$$i = p(r-1)+1,\ldots,pr, \quad r=1,\ldots,N. \qquad (3)$$

Note that if $\bar{w}_{ij}=0$, $j>i$, the above method reduces simply to a multistep method of the type discussed in Section 5.1.2. Otherwise, given the weights w_{ij} such that the quadrature rule (2a) has associated error of order p say, the accuracy of the calculation can be maintained by an appropriate choice of the weights $\bar{w}_{ij}$.

This rather general description perhaps obscures the underlying simplicity of the basic approach; let us illustrate it by considering a particular example.

Example **5.2.1.** Consider the case where $p=2$ and $r=n+1<N$. Then equations (3) have the form

$$x_i = y(a+ih) + h \sum_{j=0}^{2n} w_{ij}k(a+ih,a+jh,x_j)$$

$$+ h \sum_{j=2n}^{2n+2} \bar{w}_{ij}k(a+ih,a+jh,x_j), \quad i=2n+1, 2n+2, \qquad (4a)$$

where $x_0 = y(a)$ and $x_1,\ldots,x_{2n}$ are assumed already calculated. Let the weights w_{ij} be those of the repeated Simpson's rule; then the quadrature rule (2a) has associated error of order h^4. Clearly, when $i=2n+2$, Simpson's rule can be used to provide approximations to both the integrals in (1), thus yielding a third order approximation for $x(a+(2n+2)h)$, given by (4a) with $i = 2n + 2$. For $i = 2n + 1$, in order to ensure that the accuracy of the calculation is maintained, we must use a three-point quadrature rule of the form (2b) with associated error of order h^4. Such a rule is provided by the following choice of weights:

$$\bar{w}_{2n+1,2n} = \tfrac{5}{12}, \quad \bar{w}_{2n+1,2n+1} = \tfrac{2}{3}, \quad \bar{w}_{2n+1,2n+2} = -\tfrac{1}{12}. \qquad (4b)$$

(The choice (4b) is such that the quadrature rule (2b) is exact for quadratics.)

Equations (4a) now have the form, with $i = 2n + 1, 2n + 2$:

$$x_{2n+1} = y(a + (2n+1)h) + \tfrac{1}{3}h \sum_{j=0}^{2n} w_{2n+1,j} k(a + (2n+1)h, a + jh, x_j)$$

$$+ \tfrac{1}{12}h\{5k(a + (2n+1)h, a + 2nh, x_{2n})$$
$$+ 8k(a + (2n+1)h, a + (2n+1)h, x_{2n+1})$$
$$- k(a + (2n+1)h, a + (2n+2)h, x_{2n+2})\}, \qquad (5a)$$

$$x_{2n+2} = y(a + (2n+2)h)$$

$$+ \tfrac{1}{3}h \sum_{j=0}^{2n+2} w_{2n+2,j} k(a + (2n+2)h, a + jh, x_j), \qquad (5b)$$

where

$$w_{2n+1,0} = w_{2n+1,2n} = 1,$$
$$w_{2n+1,j} = 3 - (-1)^j, \quad 1 \le j \le 2n - 1,$$

and

$$w_{2n+2,0} = w_{2n+2,2n+2} = 1,$$
$$w_{2n+2,j} = 3 - (-1)^j, \quad 1 \le j \le 2n + 1.$$

Thus at each stage we have a pair of nonlinear equations to solve; these must be solved using an iterative technique.

We notice that the above example requires the kernel to be evaluated at the point $(a + (2n+1)h, a + (2n+2)h, x_{2n+2})$; in general this situation corresponds to evaluations of the kernel in equation (3) at the points $(a + ih, a + jh, x_j)$ where $j > i$; that is, at points outside the domain of definition of the kernel. Although this often causes no problems, it is at best uncomfortable; the following alternative method to that outlined in Example 5.2.1 has been suggested by Linz (1969).

Example 5.2.2. Equation (5b) presents no difficulties and we concentrate on equation (5a) which uses an approximation of the form

$$\int_{a+2nh}^{a+(2n+1)h} k(a + (2n+1)h, t, x(t))\,dt$$

$$\approx h \sum_{j=2n}^{2n+2} \bar{w}_{2n+1,j} k(a + (2n+1)h, a + jh, x_j)$$

where the weights $\bar{w}_{2n+1,j}$ are given by equation (4b). If now we replace the above approximation by

$$\int_{a+2nh}^{a+(2n+1)h} k(a+(2n+1)h,t,x(t))dt$$

$$\approx \tfrac{1}{6}h\{k(a+(2n+1)h, a+2nh, x_{2n})$$
$$+ 4k(a+(2n+1)h, a+(2n+\tfrac{1}{2})h, x_{2n+\frac{1}{2}})$$
$$+ k(a+(2n+1)h, a+(2n+1)h, x_{2n+1})\}$$

so that the kernel is evaluated only at points within its domain of definition, we require only a second order approximation to the solution at $s = a + (2n+\tfrac{1}{2})h$. This is provided by

$$x_{2n+\frac{1}{2}} = \tfrac{1}{8}\{3x_{2n} + 6x_{2n+1} - x_{2n+2}\}$$

corresponding to the value of the interpolating polynomial of degree 2 at $s = a + (2n+\tfrac{1}{2})h$ which interpolates $x(s)$ at $s = 2nh, (2n+1)h$ and $(2n+2)h$.

The restriction imposed by using quadrature rules with equally spaced points may be removed to yield a more general block method (see Baker, 1977, p. 870). However the resulting method will again require evaluations of the kernel function at points outside the domain of definition and the method must be modified appropriately if this is unacceptable.

5.3 A Fredholm-type approach for linear Volterra equations

For any finite b the Volterra equation

$$x(s) = y(s) + \int_a^s K(s,t)x(t)dt, \quad a \le s \le b, \tag{1}$$

can be written equivalently as

$$x(s) = y(s) + \int_a^b \bar{K}(s,t)x(t)dt, \quad a \le s \le b, \tag{2}$$

where

$$\bar{K}(s,t) = \begin{cases} K(s,t), & s \ge t, \\ 0, & s < t. \end{cases}$$

Equation (2) is a Fredholm equation of the second kind; thus the methods of Chapter 4 could be applied to provide an approximate solution of equation (1). Replacing the integral in (2) by a quadrature rule of the form

$$\int_a^b \phi(t)dt \approx h \sum_{j=0}^{N} w_j \phi(t_j)$$

where $a \le t_0 < t_1 < \cdots < t_N \le b$ with $s_i = t_i = a + ih$, we obtain the following set of approximating equations:

$$x_i = y(s_i) + h \sum_{j=0}^{i} w_j K_{ij} x_j, \quad i = 0, 1, \ldots, N. \tag{3}$$

In particular, if $K(s,s) = 0$, the kernel $\bar{K}(s,t)$ is continuous in $a \le s, t \le b$ and the above method would appear to be quite sensible. However, in general, the quadrature errors of the Fredholm formulation tend to be larger than those associated with the Volterra formulation; consider the following example which appears in Baker (1977, p. 760) where the Trapezoid rule

$$\int_{a}^{b} \phi(t) dt \approx h \sum_{j=0}^{N} {}'' \phi(a + jh) \tag{4}$$

is used to approximate the integral in the Fredholm formulation of equation (1). This yields, on setting $\phi(s) = 0$ for $a + ih < s \le a + Nh$, equation (3) with

$$w_0 = \tfrac{1}{2}; \quad w_j = 1, \quad j = 1, \ldots, i.$$

Thus for $K(s,s) \ne 0$ it can be seen that the quadrature error of the rule (4) is $\mathcal{O}(h)$ whereas the quadrature error of the following rule,

$$\int_{a}^{a+ih} \phi(t) dt \approx h \sum_{j=0}^{i} {}'' \phi(a + jh), \tag{5}$$

used to replace the integral in (1), assuming sufficient differentiability of K, is $\mathcal{O}(h^3)$. The resulting set of approximating equations is given by (3) where

$$w_0 = \tfrac{1}{2} = w_i; \quad w_j = 1, \quad j = 1, \ldots, i-1.$$

Methods of this type extend formally to nonlinear equations; however, they then yield a set of N simultaneous nonlinear equations with no 'built-in' initial approximation provided, and hence look distinctly unattractive.

Methods of Fredholm type have been developed in Delves, Abd-Elal and Hendry (1979) for particular classes of singular Volterra equations; the simultaneous treatment of the unknowns then allows an efficient treatment of the singularity.

5.4   Variable step methods

The quadrature methods described in Section 5.1 are fixed step methods; that is, they are methods which divide the range of interest $a \le s \le b$ into N equally spaced intervals of length $h = (b-a)/N$ and which proceed to solve the integral equation (5.1.1) at the set of discrete points $a + ih$, $i = 1, \ldots, N$. Thus in order that the method provide approximations to some prescribed accuracy ε say, that is,

$$|x(a+ih) - x_i| < \varepsilon, \quad i = 1, \ldots, N, \tag{1}$$

the problem must in practice be solved successively using smaller step-lengths until two consecutive sets of approximations agree to the required accuracy. The cost of a method is conventionally assessed in terms of the number of kernel evaluations made in the course of the solution process, and the above procedure is clearly costly in this respect (although the number of kernel evaluations can be considerably reduced by successively halving the steplength and making use of previously calculated kernel values). By analogy with the numerical solution of ordinary differential equations, variable step methods for the solution of Volterra equations of the second kind attempt, at each step of the calculation, to choose an optimal steplength which guarantees that the required error criterion (1) is satisfied. Hopefully, such running estimates remove the need for a recalculation to check the accuracy; moreover, the choice of an 'optimal' step at each stage should (if successful) be more economical than the use of a fixed steplength, at least provided that the overheads involved in making the choice are not too great.

A variable step method requires, at each step in the solution of the approximating equations, an estimate of the local error, that is the neglected error term in the quadrature rule which replaces the integral term in equation (5.1.1). We recall here the form of the general nonsingular Volterra equation of the second kind:

$$x(s) = y(s) + \int_a^s k(s, t, x(t)) \mathrm{d}t, \quad a \le s \le b. \tag{2}$$

Let the interval $[a, b]$ be divided into N (not necessarily equally spaced) subintervals $\{h_i\}$ where

$$h_i = s_{i+1} - s_i, \quad i = 0, 1, \ldots, N-1,$$

with

$$a = s_0 < s_1 < \cdots < s_N = b.$$

Variable step methods proceed by approximating the integral term in (2) for $s = s_i$, $i = 1, \ldots, N$ via a quadrature rule of the form

$$\int_a^{s_i} k(s_i, t, x(t)) \mathrm{d}t \approx \sum_{j=0}^{i} w_{ij} k(s_i, t_j, x(t_j)) \tag{3}$$

where

$$t_i = s_i, \qquad\qquad\qquad i = 0, 1, \ldots, N,$$
$$w_{ij} = w_{ij}(h_j, h_{j-1}, \ldots, h_0), \quad i = 1, \ldots, N, j = 0, 1, \ldots, i.$$

Then, in the notation of Section 5.1, this quadrature rule leads to the following set of equations:

$$x(s_0) = y(s_0),$$

$$x(s_i) = y(s_i) + \sum_{j=0}^{i} w_{ij}k(s_i, t_j, x(t_j))$$

$$+ E_{i,t}(k(s_i, t, x(t))), \quad i = 1, 2, \ldots, N \tag{4}$$

which in turn give rise to the set of approximating equations

$$x_0 = y(s_0),$$

$$x_i = y(s_i) + \sum_{j=0}^{i} w_{ij}k(s_i, t_j, x_j), \quad i = 1, 2, \ldots, N. \tag{5}$$

Two immediate problems arise:

(i) Because the steps h_i are unequal, it is more difficult to make a choice of the w_{ij} leading to a high-order rule.

(ii) We need a detailed estimate of the local error from which to produce a choice of the 'best' steplength h_{i-1}.

The first difficulty can be overcome most easily by restricting the number of changes of steplength (that is, by keeping a constant steplength for a group of steps). We avoid this difficulty here by discussing solely, for ease of presentation, a method (due to Phillips, 1977) based on the Trapezoid rule; we restrict the discussion also to the case that the equation is linear. This method proceeds by replacing the integral term in (2) for $s = s_i$ by a quadrature rule of the form (3) where

$$w_{i0} = \tfrac{1}{2}h_0,$$
$$w_{ij} = \tfrac{1}{2}(h_{j-1} + h_j), \quad j = 1, 2, \ldots, i-1,$$
$$w_{ii} = \tfrac{1}{2}h_{i-1}.$$

This provides an estimate x_i of $x(s_i)$ given by

$$(1 - \tfrac{1}{2}h_{i-1}K(s_i, s_i))x_i = y(s_i) + \sum_{j=0}^{i-1} \tfrac{1}{2}(h_{j-1} + h_j)K(s_i, t_j)x_j \tag{6}$$

where $h_{-1} \equiv 0$. In order to generate an error estimate, a second estimate $\tilde{x}_i$ of $x(s_i)$ is obtained by integrating first over $[a, s_{i-1}]$ and then over $[s_{i-1}, s_i]$ with a steplength of $\tfrac{1}{2}h_{i-1}$, using in both cases the Trapezoid rule. Thus an approximation to $x(s_{i-\frac{1}{2}})$ is required and this is provided by

$$(1 - \tfrac{1}{4}h_{i-1}K(s_{i-\frac{1}{2}}, s_{i-\frac{1}{2}}))x_{i-\frac{1}{2}}$$

$$= y(s_{i-\frac{1}{2}}) + \sum_{j=0}^{i-2} \tfrac{1}{2}(h_{j-1} + h_j)K(s_{i-\frac{1}{2}}, t_j)x_j$$

$$+ \tfrac{1}{2}(h_{i-2} + \tfrac{1}{2}h_{i-1})K(s_{i-\frac{1}{2}}, t_{i-1})x_{i-1}. \tag{7}$$

Then $\tilde{x}_i$ is given by

$$(1 - \tfrac{1}{4}h_{i-1}K(s_i,s_i))\tilde{x}_i = y(s_i) + \sum_{j=0}^{i-2} \tfrac{1}{2}(h_{j-1} + h_j)K(s_i,t_j)x_j$$

$$+ \tfrac{1}{2}(h_{i-2} + \tfrac{1}{2}h_{i-1})K(s_i,t_{i-1})x_{i-1}$$

$$+ \tfrac{1}{2}h_{i-1}K(s_i,t_{i-\frac{1}{2}})x_{i-\frac{1}{2}}. \tag{8}$$

Now using a number of 'localising' assumptions (see Phillips (1977), Section 3.2) the following equations are obtained:

$$(1 - \tfrac{1}{2}h_{i-1}K(s_i,s_i))(x(s_i) - x_i) = \sum_{j=0}^{i-1} A_j^{(i)}h_j^3, \tag{9}$$

$$(1 - \tfrac{1}{4}h_{i-1}K(s_{i-\frac{1}{2}},s_{i-\frac{1}{2}}))(x(s_{i-\frac{1}{2}}) - x_{i-\frac{1}{2}})$$

$$= \sum_{j=0}^{i-2} A_j^{(i)}h_j^3 + \tfrac{1}{8}A_{i-1}^{(i)}h_{i-1}^3, \tag{10}$$

$$(1 - \tfrac{1}{4}h_{i-1}K(s_i,s_i))(x(s_i) - \tilde{x}_i) = \tfrac{1}{2}h_{i-1}K(s_i,t_{i-\frac{1}{2}})x_{i-\frac{1}{2}}$$

$$+ \sum_{j=0}^{i-2} A_j^{(i)}h_j^3 + \tfrac{1}{4}A_{i-1}^{(i)}h_{i-1}^3, \tag{11}$$

where the $A_j^{(i)}, j = 0, 1, \dots, i-1$, correspond to the coefficients of the error terms in the variable step Trapezoid rule and are given by

$$A_j^{(i)} = \frac{-1}{12}\frac{\partial^2}{\partial t^2}K(s_i,t)x(t)\bigg|_{t=\xi_j}, \quad j = 0, 1, \dots, i-1,$$

where $s_j \leq \xi_j \leq s_{j+1}$. The variable step Trapezoid rule applied to the interval $[s_j, s_{j+2}] = [s_j, s_{j+1}] \cup [s_{j+1}, s_{j+2}]$ may be considered as a repeated rule applied to panels of width h_j and h_{j+1}, or as a 2-point rule applied to a single panel of width $(h_j + h_{j+1})$. Phillips provides an algorithm for determining the $A_j^{(i)}, j = 0, 1, \dots, t-1$, as a linear combination of two such quadrature approximations, thus allowing the error term in equation (9) to be determined. From this an 'optimal' value for h_{i-1} can be obtained which will ensure that the error in x_i is acceptable.

Suppose that for any $\alpha \in (0,1)$ h_{i-1} is chosen such that the following conditions hold:

$$h_{i-1} \leq \left| \frac{2\alpha}{K(s_i,s_i)} \right|, \tag{12a}$$

$$|A_j^{(i)}h_j^3| \leq \frac{\varepsilon h_j(1-\alpha)}{b-a}, \quad j = 0, 1, \dots, i-1. \tag{12b}$$

Then it follows that

$$|1 - \tfrac{1}{2}h_{i-1}K(s_i,s_i)| \geq 1 - \alpha$$

and the error term in equation (9) is bounded by

$$\left| \sum_{j=0}^{i-1} A_j^{(i)} h_j^3 / (1 - \tfrac{1}{2} h_{i-1} K(s_i, s_i)) \right| \le \varepsilon.$$

Thus the new steplength h_{i-1} should be chosen to satisfy

$$h_{i-1} < \min \left(\frac{2\alpha}{|K(s_i, s_i)|}, \ \left(\frac{\varepsilon(1-\alpha)}{(b-a)A_{i-1}^{(i)}} \right)^{\frac{1}{4}} \right) \tag{13}$$

subject to condition (12*b*) being satisfied for each value of *j*.

In practice, condition (12*b*) may not be satisfied for a particular back-value of *j*. This will happen, for example, if the kernel $K(s, t)$ is smooth for small s, t (leading to a choice of large steps early in the calculation) but becomes ill-behaved for large s and small t (when the above algorithm requires the use of these same large steps in a now inappropriate situation). This problem is often ignored by algorithm designers; it is solved by Phillips by monitoring the condition (12*b*); if necessary the interval $[s_j, s_{j+1}]$ is bisected before proceeding further; that is, a previously optimal steplength is halved. Note that the approximate solution which is now required at the point $s_{j+\frac{1}{2}}$ will have already been calculated provided that only a single halving is required. The need to bisect early intervals, in order that the requested accuracy ε in all approximate function values be maintained, is referred to as the problem of 'fill-in', since additional approximate function values must be filled in and returned in the solution. The problem of fill-in arises because during the solution process the kernel is effectively treated as a function of one variable.

In the case where the kernel function is nonlinear, error estimation proves more difficult; the reader is referred to Phillips (1977) where this problem is considered in detail along with higher order variable step methods for both linear and nonlinear Volterra equations of the second kind.

5.5   Singular equations: product integration

We now consider the numerical solution of equation (5.1.1) when the kernel function $k(s, t, x(t))$ is singular. Such problems were considered in Section 4.4 in the context of linear Fredholm equations; the discussion given here follows that section closely. We assume that the kernel function or one of its low-order derivatives is badly behaved, and suppose that $k(s, t, x(t)) = p(s, t)\bar{k}(s, t, x(t))$ where p and $\bar{k}$ are respectively singular and well behaved functions of their arguments. Then the method of product integration may be used to solve problems of the form

$$x(s) = y(s) + \int_a^s p(s, t)\bar{k}(s, t, x(t))\mathrm{d}t, \quad a \le s \le b. \tag{1}$$

As in the previous section let the interval $[a, b]$ be divided into N subintervals $\{h_i\}$ where

$$h_i = s_{i+1} - s_i, \quad i = 0, 1, \ldots, N - 1,$$

with

$$a = s_0 < s_1 < \cdots < s_N = b.$$

Then the method proceeds by approximating the integral term in (1), for $s = s_i$, $i = 1, \ldots, N$, via a quadrature rule of the form

$$\int_a^{s_i} p(s_i, t) \bar{k}(s_i, t, x(t)) \mathrm{d}t \approx \sum_{j=0}^{i} w_{ij} \bar{k}(s_i, t_j, x(t_j)) \tag{2}$$

where $t_i = s_i$, $i = 0, 1, \ldots, N$. The weights are constructed by insisting that the rule in (2) be exact when $\bar{k}(s_i, t, x(t))$ is a polynomial in t of degree $\leq r$ say; for each value of i, this requires the existence of (and a knowledge of) the $(r + 1)$ moments

$$\mu_{ij} = \int_a^{s_i} t^j p(s_i, t) \mathrm{d}t, \quad j = 0, 1, \ldots, r. \tag{3}$$

To illustrate the procedure we present here the product integration analogue of the variable step Trapezoid rule applied to equation (1), although we emphasise again that we are not advocating the use of such low-degree rules. We note that the standard Trapezoid rule approximates the integrand over each interval $[s_j, s_{j+1}]$ by a first degree interpolation polynomial interpolating at the points s_j, s_{j+1}. Thus the product integration analogue of this rule proceeds by approximating the non-singular part of the integrand, namely $\bar{k}(s, t, x(t))$, at $s = s_i$ by

$$\bar{k}(s_i, t, x(t)) \approx \frac{(t_{j+1} - t)}{h_j} \bar{k}(s_i, t_j, x(t_j))$$

$$+ \frac{(t - t_j)}{h_j} \bar{k}(s_i, t_{j+1}, x(t_{j+1})).$$

Hence it follows that

$$\int_a^{s_i} p(s_i, t) \bar{k}(s_i, t, x(t)) \mathrm{d}t = \sum_{j=0}^{i-1} \int_{t_j}^{t_{j+1}} p(s_i, t) \bar{k}(s_i, t, x(t)) \mathrm{d}t$$

$$\approx \sum_{j=0}^{i-1} \int_{t_j}^{t_{j+1}} p(s_i, t) \left\{ \frac{(t_{j+1} - t)}{h_j} \bar{k}(s_i, t_j, x(t_j)) \right.$$

$$+ \frac{(t - t_j)}{h_j} \bar{k}(s_i, t_{j+1}, x(t_{j+1})) \right\} \mathrm{d}t$$

$$= \sum_{j=0}^{i} w_{ij} \bar{k}(s_i, t_j, x(t_j)) \tag{4a}$$

Table 5.5.1. *Solution of*
$$x(s) = 1/(1+s)^{\frac{1}{2}} + \pi/8 - \tfrac{1}{4}\sin^{-1}((1-s)/(1+s))$$
$$-\tfrac{1}{4}\int_0^s [x(t)/(s-t)^{\frac{1}{2}}]dt,\ 0 \le s \le 1$$

h	ε
0.2	3.27×10^{-4}
0.1	8.28, -5
0.05	2.11, -5
0.025	5.36, -6
0.0125	1.35, -6

where the weights are calculated from the expressions

$$w_{i0} = \frac{1}{h_0}\int_{t_0}^{t_1} p(s_i,t)(t_1-t)dt,$$

$$w_{ij} = \frac{1}{h_j}\int_{t_j}^{t_{j+1}} p(s_i,t)(t_{j+1}-t)dt$$

$$+ \frac{1}{h_{j-1}}\int_{t_{j-1}}^{t_j} p(s_i,t)(t-t_{j-1})dt,\quad j=1,2,\ldots,i-1,$$

$$w_{ii} = \frac{1}{h_{i-1}}\int_{t_{i-1}}^{t_i} p(s_i,t)(t-t_{i-1})dt. \tag{4b}$$

We can similarly obtain the product integration analogue of Simpson's and higher order rules. These rules will require starting procedures to provide approximate function values at the first few points; product integration analogues of those procedures outlined in Section 5.1.4 may be readily obtained for particular forms of $p(s,t)$.

Example 5.5.1. We solve the equation (taken from Linz, 1969a)

$$x(s) = \frac{1}{(1+s)^{\frac{1}{2}}} + \frac{\pi}{8} - \frac{1}{4}\sin^{-1}\left(\frac{1-s}{1+s}\right) - \frac{1}{4}\int_0^s \frac{x(t)dt}{(s-t)^{\frac{1}{2}}}$$

whose exact solution is $x(s) = 1/(1+s)^{\frac{1}{2}}$ using the product integration analogue of the Trapezoid rule with a constant steplength h. Any attempt to apply the standard Trapezoid rule to this problem will clearly be unsatisfactory as the kernel is unbounded when $s=t$. The example is solved over the range $0 \le s \le 1$ for a number of values of h. The associated weights of the rule are given by (see equation (4b) above and Exercise 4 below)

$$w_{i0} = \tfrac{2}{3}h^{\frac{1}{2}}\{3i^{\frac{1}{2}} + 2[(i-1)^{\frac{3}{2}} - i^{\frac{3}{2}}]\},$$

$$w_{ij} = \tfrac{4}{3}h^{\frac{1}{2}}[(i-j-1)^{\frac{3}{2}} + (i-j+1)^{\frac{3}{2}} - 2(i-j)^{\frac{3}{2}}], \quad j = 1, 2, \ldots, i-1,$$

$$w_{ii} = \frac{4h^{\frac{1}{2}}}{3},$$

and we show in Table 5.5.1 the maximum error ε obtained over $[0, 1]$ using a range of values of h.

The results confirm that the error is decreasing as h^2 in accordance with the expected behaviour of the standard Trapezoid rule applied to the solution of nonsingular equations. We see therefore that the singular nature of the kernel in this example has been satisfactorily accounted for.

5.6   Systems of equations

Coupled systems of equations introduce no significant new features or difficulties. Consider a system of equations of the form

$$x_i(s) = y_i(s) + \sum_{j=1}^{n} \int_a^s k_{ij}(s, t, x_j(t)) dt, \quad a \leq s \leq b,$$

$$i = 1, \ldots, n, \tag{1}$$

where the y_i, k_{ij}, $i, j = 1, \ldots, n$ are known functions. In order to simplify the discussion, let us assume that each k_{ij} is a linear function; we shall show how to solve the simpler system

$$x_i(s) = y_i(s) + \sum_{j=1}^{n} \int_a^s K_{ij}(s, t) x_j(t) dt, \quad a \leq s \leq b, \quad i = 1, \ldots, n, \tag{2}$$

when $n = 2$, using the repeated Trapezoid rule. Suppose the interval $[a, b]$ is divided into N equal subintervals of length $h = (b - a)/N$ such that

$$a = s_0 < s_1 < \ldots < s_N = b$$

with $s_i = a + ih$, $i = 0, 1, \ldots, N$. It is clear from equation (2) that

$$x_1(a) = y_1(a), \quad x_2(a) = y_2(a),$$

and we now determine $x_i(s_j)$ for $i = 1, 2$ and $j = 1, \ldots, N$. Equation (2) reduces, on setting $n = 2$, to

$$x_1(s) = y_1(s) + \int_a^s K_{11}(s, t) x_1(t) dt + \int_a^s K_{12}(s, t) x_2(t) dt,$$

$$x_2(s) = y_2(s) + \int_a^s K_{21}(s, t) x_1(t) dt + \int_a^s K_{22}(s, t) x_2(t) dt. \tag{3}$$

Thus at the first stage when $s = s_1$ we obtain, on approximating each integral term by the Trapezoid rule, the following set of approximating equations

$$x_{11} = y_1(s_1) + h \sum_{j=0}^{1}{}'' K_{11}(s_1,t_j)x_{1j} + h \sum_{j=0}^{1}{}'' K_{12}(s_1,t_j)x_{2j},$$

$$x_{21} = y_2(s_1) + h \sum_{j=0}^{1}{}'' K_{21}(s_1,t_j)x_{1j} + h \sum_{j=0}^{1}{}'' K_{22}(s_1,t_j)x_{2j}, \tag{4}$$

where $x_{ij} \approx x_i(s_j)$. On simplifying the above we obtain

$$\{1 - \tfrac{1}{2}hK_{11}(s_1,t_1)\}x_{11} - \tfrac{1}{2}hK_{12}(s_1,t_1)x_{21}$$
$$= y_1(s_1) + \tfrac{1}{2}hK_{11}(s_1,t_0)x_{10} + \tfrac{1}{2}hK_{12}(s_1,t_0)x_{20}, \tag{5}$$
$$\{1 - \tfrac{1}{2}hK_{22}(s_1,t_1)\}x_{21} - \tfrac{1}{2}hK_{21}(s_1,t_1)x_{11}$$
$$= y_2(s_1) + \tfrac{1}{2}hK_{21}(s_1,t_0)x_{10} + \tfrac{1}{2}hK_{22}(s_1,t_0)x_{20}.$$

Thus x_{11}, x_{21} may be obtained on solution of the following set of simultaneous equations:

$$\begin{pmatrix} 1 - \tfrac{1}{2}hK_{11}(s_1,t_1) & -\tfrac{1}{2}hK_{12}(s_1,t_1) \\ -\tfrac{1}{2}hK_{21}(s_1,t_1) & 1 - \tfrac{1}{2}hK_{22}(s_1,t_1) \end{pmatrix} \begin{pmatrix} x_{11} \\ x_{21} \end{pmatrix}$$
$$= \begin{pmatrix} y_1(s_1) + \tfrac{1}{2}hK_{11}(s_1,t_0)x_{10} + \tfrac{1}{2}hK_{12}(s_1,t_0)x_{20} \\ y_2(s_1) + \tfrac{1}{2}hK_{21}(s_1,t_0)x_{10} + \tfrac{1}{2}hK_{22}(s_1,t_0)x_{20} \end{pmatrix} \tag{6}$$

provided that the coefficient matrix is of course nonsingular. It follows that at the rth stage x_{1r} and x_{2r} are obtained by solving the system:

$$\begin{pmatrix} 1 - \tfrac{1}{2}hK_{11}(s_r,t_r) & -\tfrac{1}{2}hK_{12}(s_r,t_r) \\ -\tfrac{1}{2}hK_{21}(s_r,t_r) & 1 - \tfrac{1}{2}hK_{22}(s_r,t_r) \end{pmatrix} \begin{pmatrix} x_{1r} \\ x_{2r} \end{pmatrix}$$
$$= \begin{pmatrix} y_1(s_r) + h \sum_{j=0}^{r-1}{}' K_{11}(s_r,t_j)x_{1j} + h \sum_{j=0}^{r-1}{}' K_{12}(s_r,t_j)x_{2j} \\ y_2(s_r) + h \sum_{j=0}^{r-1}{}' K_{21}(s_r,t_j)x_{1j} + h \sum_{j=0}^{r-1}{}' K_{22}(s_r,t_j)x_{2j} \end{pmatrix}. \tag{7}$$

Thus the complete solution of equation (2), by the Trapezoid rule when $n = 2$, involves the solution of N sets of simultaneous equations of order two. In the case where the kernel is a nonlinear function we are required to solve at the rth stage $(r = 1, \dots, N)$ a set of nonlinear equations of the form

$$\mathbf{x}_r = \mathbf{g}(\mathbf{x}_r) \tag{8}$$

where $\mathbf{x}_r = (x_{1r}, \dots, x_{nr})^T$. Given a suitable estimate $\mathbf{x}_r^{(0)}$ we can solve this system using the iterative scheme

$$\mathbf{x}_r^{(i+1)} = \mathbf{g}(\mathbf{x}_r^{(i)}), \quad i = 0, 1, 2, \dots, \tag{9}$$

whereby successive iterates $\mathbf{x}_r^{(1)}, \mathbf{x}_r^{(2)}, \dots$ are obtained by solving a sequence of linear problems. The extension to higher order rules, or to more than two coupled equations, is clearly straightforward; software for handling coupled problems has been developed by, for example, Rumyantsev (1965), Logan (1976), Nerinckx (1980).

Exercises

1. The quadrature method

$$x_i = y(s_i) + h \sum_{j=0}^{i} w_{ij} k(s_i, t_j, x_j), \quad i = 1, \dots, N$$

where $h = (b-a)/N$ and $a = s_0 < s_1 < \dots < s_N = b$, may be used to solve the equation

$$x(s) = y(s) + \int_a^s k(s, t, x(t)) dt, \quad a \le s \le b,$$

at the point $s_{2r+1} = a + (2r+1)h$ by employing either of the two methods:

(i) the repeated Simpson's rule over $[s_0, s_{2r-2}]$ with the three-eighths rule over $[s_{2r-2}, s_{2r+1}]$

(ii) the three-eighths rule over $[s_0, s_3]$ and the repeated Simpson's rule over $[s_3, s_{2r+1}]$. Write down the weights associated with methods (i) and (ii) and deduce that method (i) is stable.

2. Show that the linear equation

$$x(s) = y(s) + \int_0^s K(s, t) x(t) dt, \quad s > 0,$$

where $K(s, t) = \sum_{i=1}^{\infty} u_i(s) v_i(t)$ may be written equivalently as

$$x(s) = y(s) + \sum_{i=1}^{\infty} u_i(s) z_i(s)$$

where

$$z_i'(s) = v_i(s) x(s), \quad z_i(0) = 0. \tag{1}$$

Show further how the Runge–Kutta method

$$\tilde{z}_i(\theta_p h) = h \sum_{q=0}^{p-1} A_{pq} v_i(\theta_q h) x(\theta_q h), \quad p = 1, \dots, r,$$

with

$$0 = \theta_0 \le \dots \le \theta_r = 1,$$

when applied to the system of equations (1), may be used to provide an approximation $\tilde{x}(h)$ to $x(h)$.

3. The second kind nonlinear Volterra equation

$$x(s) = y(s) + \int_a^s k(s, t, x(t)) dt, \quad a \le s \le b,$$

150 Quadrature methods for Volterra equations

is to be solved by the variable step analogue of the composite Simpson's rule. Show that the integral term

$$\int_{s_{2j}}^{s_{2j+2}} k(s_{2i}, t, x(t))\,dt$$

may be approximated by

$$\sum_{r=2j}^{2j+2} w_{2i,r} k(s_{2i}, s_r, x_r)$$

where the weights are defined by

$$w_{2i,2j} = \frac{(2h_{2j} - h_{2j+1})(h_{2j} + h_{2j+1})}{6h_{2j}},$$

$$w_{2i,2j+1} = \frac{(h_{2j} + h_{2j+1})^3}{6h_{2j}h_{2j+1}},$$

$$w_{2i,2j+2} = \frac{(2h_{2j+1} - h_{2j})(h_{2j} + h_{2j+1})}{6h_{2j+1}},$$

with

$$h_{2j} = s_{2j+1} - s_{2j}, \quad h_{2j+1} = s_{2j+2} - s_{2j+1}.$$

(*HINT.* Approximate $k(s_{2i}, t, x(t))$ over the interval $[s_{2j}, s_{2j+2}]$ by a second degree Lagrange polynomial interpolating k at the points $s_{2j}, s_{2j+1}, s_{2j+2}.$)

4. The equation

$$x(s) = y(s) + \int_a^s (s-t)^{-\frac{1}{2}} \bar{k}(s, t, x(t))\,dt, \quad a \leq s \leq b,$$

is to be solved by the product integration analogue of the Trapezoid rule using a *constant* steplength h where $h = (b-a)/N$. Show that the method which solves the approximating equations

$$x_i = y(s_i) + \sum_{j=0}^i w_{ij} \bar{k}(s_i, t_j, x_j), \quad i = 1, 2, \dots, N,$$

with $s_i = t_i = a + ih$, $i = 0, 1, \dots, N$, has weights defined by

$$w_{i0} = \frac{2h^{\frac{1}{2}}}{3}\{3i^{\frac{1}{2}} + 2[(i-1)^{\frac{1}{2}} - i^{\frac{1}{2}}]\},$$

$$w_{ij} = \frac{4h^{\frac{1}{2}}}{3}[(i-j-1)^{\frac{1}{2}} + (i-j+1)^{\frac{1}{2}} - 2(i-j)^{\frac{1}{2}}], \quad j = 1, 2, \dots, i-1,$$

$$w_{ii} = \frac{4h^{\frac{1}{2}}}{3}.$$

5. Show that the set of Volterra integral equations

$$x_1(s) + \int_0^s e^{s-t} x_1(t)dt + \int_0^s \cos(s-t) x_2(t)dt = \cosh s + s \sin s,$$

$$x_2(s) + \int_0^s e^{s+t} x_1(t)dt + \int_0^s s \cos t x_2(t)dt = 2 \sin s + s(\sin^2 s + e^s),$$

is satisfied by

$$x_1(s) = e^{-s}, \quad x_2(s) = 2 \sin s.$$

Use the Trapezoid rule with a constant steplength of $h = 0.1$ to find approximations to $x_1(s)$ and $x_2(s)$ at $s = 0.3$.

6

Eigenvalue problems and the Fredholm alternative

6.1   Formal properties of the eigenvalue problem

The homogeneous Fredholm equation of the second kind has the form

$$x(s) = \lambda \int_a^b K(s,t)x(t)\mathrm{d}t. \tag{1}$$

Equation (1) always has the trivial solution $x(s) = 0$. For a given λ it may or may not have a nontrivial solution; see Section 3.3. We therefore have the interesting theoretical question: for what values of λ *does* (1) have nontrivial solutions? We shall see below that the characteristic values are *isolated*; that is, the spectrum of λ is a point spectrum containing a denumerable number of entries with no finite limit point (alternative (a) of Figure 6.1.1), a spectrum such as that shown in alternative (b) being ruled out by the assumptions ($\mathscr{L}^2$ kernel and solutions) we have made. The characteristic values may however be complex rather than real; see Section 6.1.1 for conditions under which it can be guaranteed that they are all real.

The numerical solution of (1) therefore involves finding both an approximate characteristic value (or equivalently eigenvalue – see Section 3.3) and an approximate characteristic function or eigenfunction $x(s)$. Since the product $\lambda x(s)$ enters into (1), this problem is formally nonlinear but the nonlinearity is of a very special type and the solution of (1) is closely related to the solution of the $N \times N$ algebraic eigenvalue problem $[\mathbf{A} - \gamma\mathbf{I}]\mathbf{a} = \mathbf{0}$. Indeed, we can formally reduce (1) (approximately) to this form by applying the Nystrom technique used for the inhomogeneous equation in Chapter 4. Introducing an N-point quadrature rule $Q(w_i, \xi_i, i = 1, \ldots, N)$ we obtain the equivalent equation

$$x(s) = \lambda \sum_{i=1}^N w_i K(s, \xi_i)x(\xi_i) + E(Kx)$$

where E is the error operator for the rule. Ignoring E and successively setting $s = \xi_j, j = 1, \ldots, N$ we obtain the Nystrom equations

$$(\mathbf{K} - \gamma^{(N)}\mathbf{I})\mathbf{x} = \mathbf{0}, \quad \gamma^{(N)} = (\lambda^{(N)})^{-1}, \tag{2}$$

where $\mathbf{x} = \{x(\xi_i)\}$ is an N-vector, $\mathbf{I}$ is the $N \times N$ unit matrix and $\mathbf{K}$ is the $N \times N$ matrix with elements

$$K_{ij} = K(\xi_i, \xi_j)w_j.$$

Under suitable conditions the eigenvalues of (2) will approximate those of (1) (that is, $\gamma^{(N)} \approx \lambda^{-1}$). It is not immediately clear what these conditions might be. Equation (2) will always have N eigenvalues, in general distinct, while we will see later that (1) may have no non-zero eigenvalues, a finite number, or a denumerably infinite number. For example, we have seen already that a Volterra kernel has no non-zero eigenvalues.

If we ignore possible difficulties of this type, the numerical solution of an eigenvalue problem is very similar to that of the second kind equation. In particular, the discussion of Chapter 4 remains valid insofar as it relates to the accuracy with which the matrix $\mathbf{K}$ represents the integral operator. However, before discussing the accuracy of the approximate eigenvalues we look at some of the relations which the exact eigenvalues obey.

6.1.1 Hermitian kernels

We rewrite (1) in its equivalent operator form

$$[K - \gamma I]x = 0, \gamma = \frac{1}{\lambda}.$$

Then if K is Hermitian and x, y are any two $\mathcal{L}^2$ functions

$$(y, Kx) = (Ky, x) = (x, Ky)^* \tag{3}$$

and we have the following simple properties of the eigenfunction, which mirror similar properties for finite matrices:

(*a*) The eigenvalues γ are real.

(*b*) Eigenfunctions belonging to different eigenvalues are orthogonal.

Figure 6.1.1. Possible spectrum for the characteristic values of (1). (*a*) A point spectrum containing a denumerable set of characteristic values. (*b*) A spectrum containing a continuum; alternative (*b*) cannot occur.

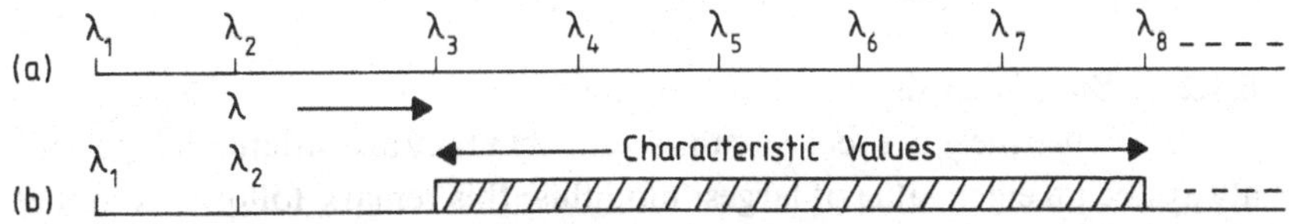

Proof. Let (γ_1, x_1), (γ_2, x_2) be any two eigenvalue–eigenfunction pairs. Then

$$Kx_1 = \gamma_1 x_1,$$
$$Kx_2 = \gamma_2 x_2,$$

and hence

$$(x_2, Kx_1) = \gamma_1(x_2, x_1), \tag{4a}$$

$$(x_1, Kx_2) = \gamma_2(x_1, x_2). \tag{4b}$$

Taking the complex conjugate of (4b) we have

$$(\mathbf{x}_1, Kx_2)^* = \gamma_2^*(x_1, x_2)^*,$$

that is,

$$(Kx_2, x_1) = \gamma_2^*(x_2, x_1) = (x_2, Kx_1) \quad \text{from} \quad (3). \tag{5}$$

Subtracting (5) from (4a) we obtain the identity

$$(\gamma_1 - \gamma_2^*)(x_2, x_1) = 0.$$

Now if we set $\gamma_1 = \gamma_2$, $x_1 = x_2$ we obtain

$$\gamma_1 = \gamma_1^*; \text{ that is, } \gamma_1 \text{ is real;}$$

while if $\gamma_1 \neq \gamma_2$ we see at once that $(x_2, x_1) = 0$.

In addition we prove later that

(c) Every Hermitian kernel has at least one eigenvalue.

It would be nice if the approximate eigenvalues also satisfied these properties, which are of course shared by the eigenvalues of finite Hermitian matrices. The matrix $\mathbf{K}$ of (2) is not symmetric unless all the quadrature weights w_j are equal. However, if the w_j are all positive we can rewrite (2) in the form

$$(\mathbf{D}^{\frac{1}{2}}\mathbf{K}_0\mathbf{D}^{\frac{1}{2}} - \gamma^{(N)}\mathbf{I})\mathbf{y} = \mathbf{0}, \tag{2a}$$

$$\mathbf{y} = \mathbf{D}^{\frac{1}{2}}\mathbf{x}; \mathbf{D} = \mathrm{diag}(w_i) : (\mathbf{K}_0)_{ij} = K(\xi_i, \xi_j)$$

where the matrix $\mathbf{D}^{\frac{1}{2}}\mathbf{K}_0\mathbf{D}^{\frac{1}{2}}$ is Hermitian if the operator K is Hermitian. Thus if all the weights are positive the eigenvalues of (2) are real and (2a) forms a more convenient starting point from which to compute the γ_i (where we have dropped the superscript N) since we may use techniques developed for Hermitian matrices.

6.1.2 Bounds on the γ_i

Since the γ_i are real we can assume that they are ordered. There will always be an eigenvalue of largest modulus; this remark follows from the result, proved earlier in Chapter 3, that all sufficiently small values of $\lambda = 1/\gamma$ are regular values of K. Alternatively we may note directly that for any eigenvalue γ

$$\gamma x = Kx$$

and hence taking norms:

$$|\gamma| \, \|x\| \leq \|K\| \cdot \|x\| \; ; \text{ that is, } |\gamma| \leq \|K\| < \infty.$$

We also show below, in Section 6.3, that if there are an infinite number of eigenvalues, they have a limit point of zero. For a general kernel there may be eigenvalues of both signs.

6.1.3 Non-Hermitian kernels

The discrete equation (2) remains a suitable approximation for non-Hermitian systems and in this respect they need not be treated specially at all. However, the resultant matrix problem is more difficult. A non-Hermitian kernel may have complex eigenvalues or it may have none at all. For example, it was shown in Chapter 3 that a Volterra kernel had no non-zero eigenvalues; a second example given there was that of a kernel of the form

$$K(s,t) = g(s)h(t), \quad (g,h) = 0.$$

It was shown that every finite λ is a regular value of this kernel. Since eigenvalues of K (or of the adjoint kernel $K^\dagger$) may not exist, it proves convenient to introduce the related *singular values* μ and *singular functions* $\{u, v\}$ of K. These are defined by

$$u = \mu K v; v = \mu K^\dagger u \tag{6}$$

and satisfy the following relations:

(a) (i) μ is real
(ii) $(I - \mu^2 K K^\dagger)u = 0$ $\tag{7}$
$(I - \mu^2 K^\dagger K)v = 0$

Proof.

$$v = \mu K^\dagger u = \mu K^\dagger (\mu K v) = \mu^2 K^\dagger K v$$

and similarly for u.

But $K^\dagger K$ is Hermitian positive definite by construction and therefore μ^2 is real and positive, or μ is real.

(b) If $\{u, v\}$ is a pair of singular functions of K belonging to μ, then $\{u, -v\}$ is a pair of singular functions belonging to $-\mu$.

Proof. Obvious.

(c) If $\{u, v\}$ belong to μ and $\{u', v'\}$ belong to μ' and $\mu^2 \neq (\mu')^2$ then $(u, u') = (v, v') = 0$.

Proof. This follows from (a) and the observation that $K^\dagger K, KK^\dagger$ are Hermitian.

In many cases it is the singular functions of K which are required rather than its eigenfunctions. Note that u, v are displayed by (7) as eigenfunctions of the Hermitian operators $KK^\dagger$ and $K^\dagger K$ respectively (so that if $K = K^\dagger$, $u = v = x$). But if the eigenvalues are of interest, clearly we should look at the conditions under which these might exist. The discussions of the next sections are related to this question; at the same time, they lead on to suggest alternative ways of solving the equations.

6.2   Kernels of finite rank (degenerate or separable kernels)

Let $\{a_\nu(t)\}$, $\{b_\nu(t)\}$ be two sequences of linearly independent $\mathscr{L}^2$ functions. A kernel function $K(s, t)$ which can be expressed as the finite sum

$$K(s, t) = \sum_{\nu=1}^{n} a_\nu(s) b_\nu^*(t) \tag{1}$$

is said to be of *finite rank* n; it is often referred to as a *degenerate* or *separable* kernel. We use the notation

$$K = \sum_{\nu=1}^{n} a_\nu \otimes b_\nu. \tag{1a}$$

Then

$$\alpha K = \sum_{\nu=1}^{n} a_\nu \otimes (\alpha^* b_\nu) = \sum_{\nu=1}^{n} (\alpha a_\nu) \otimes b_\nu, \tag{2}$$

$$K^\dagger = \sum_{\nu=1}^{n} b_\nu \otimes a_\nu,$$

$$Kx = \sum_{\nu=1}^{n} (b_\nu, x) a_\nu,$$

$$HK = \sum_{\nu=1}^{n} (Ha_\nu) \otimes b_\nu; \quad KH = \sum_{\nu=1}^{n} a_\nu \otimes (H^\dagger b_\nu). \tag{3}$$

6.2.1   Algebraic solution of finite rank integral equations

For kernels of finite rank we can find an algebraic solution of the inhomogeneous integral equation:

Theorem 6.2.1. Let

$$K = \sum_{v=1}^{n} a_v \otimes b_v$$

and let $(b_\mu, y) = y_\mu$; $(b_\mu, a_v) = k_{\mu v}$; $1 \leq \mu, v \leq n$.
Then if x is an $\mathscr{L}^2$ solution of the equation

$$x = y + \lambda K x \qquad (4)$$

and

$$x_\mu = (b_\mu, x)$$

then

$$x_\mu = y_\mu + \lambda \sum_{v=1}^{n} k_{\mu v} x_v. \qquad (5)$$

Conversely, if x_μ satisfies (5), then

$$x = y + \lambda \sum_{v=1}^{n} x_v a_v \qquad (6)$$

is an $\mathscr{L}^2$ solution of (4) and $x_\mu = (b_\mu, x)$.

Proof. If x satisfies (4), we have

$$x = y + \lambda \sum_{v=1}^{n} (b_v, x) a_v = y + \lambda \sum_{v=1}^{n} x_v a_v$$

whence

$$(b_\mu, x) = (b_\mu, y) + \lambda \sum_{v=1}^{n} k_{\mu v} x_v.$$

Conversely, if x is defined by (6) we have

$$(b_\mu, x) = (b_\mu, y) + \lambda \sum_{v=1}^{n} (b_\mu, a_v) x_v$$

$$= y_\mu + \lambda \sum_{v=1}^{n} k_{\mu v} x_v = x_\mu \text{ from (5)}.$$

Now, substituting in (6), we obtain

$$x = y + \lambda \sum_{v=1}^{n} (b_v, x) a_v = y + \lambda K x.$$

Theorem 6.2.1 reduces the solution of an integral equation with finite rank kernel to that of the set of algebraic equations (5).

Example 6.2.1.

$$u(s) = e^s - \frac{e}{2} + \frac{1}{2} + \frac{1}{2} \int_0^1 u(t) dt.$$

Here, $K = 1$, $\lambda = \frac{1}{2}$; $n = 1$, $a_1(t) = b_1(t) = 1$,

$$k_{11} = \int_0^1 a_1 b_1 \, dt = 1,$$

$$y_1 = \int_0^1 \left(e^s - \frac{e}{2} + \frac{1}{2} \right) ds = e - 1 - \frac{e}{2} + \frac{1}{2} = \frac{e}{2} - \frac{1}{2},$$

$$x_1 = y_1 + \tfrac{1}{2} \cdot 1 \cdot x_1, \quad \text{that is,} \quad x_1 = 2y_1 = e - 1.$$

Hence

$$x = \left(e^s - \frac{e}{2} + \frac{1}{2} \right) + \frac{1}{2} \cdot 1 \cdot (e - 1) = e^s$$

(which you may care to check by direct substitution).

It is obviously a small step from the explicit solution for x to an explicit representation for the resolvent kernel H_λ.

Theorem 6.2.2. With the notation of the previous theorem, let **K** be the matrix with elements $k_{\mu\nu}$.
   Then if $\det (\mathbf{I} - \lambda\mathbf{K}) \neq 0$,

$$H_\lambda = \sum_{\mu,\nu=1}^n D_{\mu\nu}(a_\mu \otimes b_\nu) \tag{7}$$

where the $n \times n$ matrix $\mathbf{D} = (\mathbf{I} - \lambda\mathbf{K})^{-1}$. $\tag{7a}$

Proof. We verify directly that H_λ satisfies the resolvent equations:

$$\lambda K H_\lambda = \lambda \sum_{\mu,\nu=1}^n D_{\mu\nu} K a_\mu \otimes b_\nu$$

$$= \lambda \sum_{\rho,\mu,\nu=1}^n k_{\rho\mu} D_{\mu\nu}(a_\rho \otimes b_\nu), \quad \text{from} \quad (3).$$

But from (7a)

$$(\mathbf{I} - \lambda\mathbf{K})\mathbf{D} = \mathbf{I},$$

that is,

$$\lambda\mathbf{K}\mathbf{D} = \mathbf{D} - \mathbf{I},$$

and hence

$$\lambda \sum_{\mu=1}^n k_{\rho\mu} D_{\mu\nu} = D_{\rho\nu} - \delta_{\rho\nu}.$$

Thus

$$\lambda K H_\lambda = \sum_{\rho,\nu=1}^n (D_{\rho\nu} - \delta_{\rho\nu})(a_\rho \otimes b_\nu)$$

$$= H_\lambda - K.$$

Similarly, we can show that $\lambda H_\lambda K = H_\lambda - K$.

6.2.2 Eigenvalues of a kernel of finite rank

Finally, since the resolvent H_λ can be constructed explicitly, we can discuss the spectrum of a degenerate kernel rather easily:

(a) If, for a fixed value of λ, $(\mathbf{I} - \lambda \mathbf{K})^{-1}$ exists, then so does the resolvent; hence λ is a regular value of the kernel.

(b) If $\mathbf{I} - \lambda \mathbf{K}$ is singular, then there is a nontrivial solution of the equation

$$(\mathbf{I} - \lambda \mathbf{K})\mathbf{x} = \mathbf{0}, \quad \mathbf{x} = (x_\mu), \tag{8}$$

and in that case the non-zero function $x(s)$ defined by

$$x(s) = \lambda \sum_{v=1}^{n} x_v a_v(s)$$

satisfies the homogeneous integral equation. Hence λ is a characteristic value of K.

From (a) and (b) we make two important deductions:

(i) For a degenerate kernel of rank n, every value of λ is either a regular value or a characteristic value.

(ii) Viewed as an eigenvalue equation for λ, equation (8) has r non-zero eigenvalues where r is the rank of $\mathbf{K}$. Hence a kernel of rank r has in general r non-zero eigenvalues, that is, r finite characteristic values. The rank r of $\mathbf{K}$ may be less than n, the rank of K. For example, we consider again $n = 1$ with $k_{11} = (a_1, b_1) = 0$. The rank of $\mathbf{K}$ is then zero and there are no (finite) characteristic values.

There are, however, infinitely many zero eigenvalues (infinite characteristic values) of a finite rank kernel; that is, there are always infinitely many linearly independent functions ξ_i such that

$$K\xi_i = 0, \quad \xi_i \neq 0, \tag{9}$$

since for any ξ,

$$K\xi = \sum_{v=1}^{n} (b_v, \xi)a_v,$$

so that for (9) to hold, we require only that

$$(b_v, \xi) = 0, \quad v = 1, 2, \ldots, n, \tag{10}$$

and this finite set of conditions can be satisfied by infinitely many functions.

6.3 Non-degenerate kernels

We now discuss the application of these results to kernels not of finite rank, that is, 'non-degenerate kernels'. We start by recalling that (see Smithies (1958), Section 3.3), for an $\mathscr{L}^2$ kernel, and any $\varepsilon > 0$, there exists an

$\mathscr{L}^2$ kernel K_0 of finite rank such that

$$\|K - K_0\| < \varepsilon.$$

This follows for a *continuous* kernel from the two-dimensional form of the Weierstrass theorem; but any $\mathscr{L}^2$ kernel may be approximated arbitrarily closely by a continuous kernel.

For any kernel K we may therefore write for any $\varepsilon > 0$

$$K = P + Q \qquad (1a)$$

where P is of finite rank n:

$$P = \sum_{r=1}^{n} a_r \otimes b_r \qquad (1b)$$

and Q is arbitrarily small:

$$\|Q\| < \varepsilon.$$

In what follows we shall be interested in the characteristic values λ of K. Suppose that we wish to discuss the spectrum for $|\lambda| < \omega$, where ω is some large but finite number. Then we choose $\varepsilon = 1/\omega$, so that for all λ, $-\omega < \lambda < \omega$ and

$$\|\lambda Q\| = |\lambda| \, \|Q\| < \omega \varepsilon = 1.$$

Hence for $|\lambda| < \omega$, Q has a resolvent G_λ (see Section 3.6) given by

$$G_\lambda = Q \sum_{i=0}^{\infty} \lambda^i Q^i$$

and which satisfies the resolvent equations

$$G_\lambda - Q = \lambda G_\lambda Q = \lambda Q G_\lambda. \qquad (2)$$

We can now prove the following theorem.

Theorem 6.3.1. With the notation above, let $|\lambda| < \omega$ and

$$x = y + \lambda K x. \qquad (3)$$

In addition let

$$z = y + \lambda G_\lambda y; \quad (x, b_v) = x_v; \quad (z, b_v) = z_v;$$
$$(a_\mu + \lambda G_\lambda a_\mu, b_v) = f_{\mu v}.$$

Then

$$x_\mu = z_\mu + \lambda \sum_{v=1}^{n} f_{\mu v} x_v, \quad 1 \leq \mu \leq n. \qquad (4)$$

Moreover, if x_μ satisfies (4), then

$$x = y + \lambda G_\lambda y + \sum_{v=1}^{n} x_v (a_v + \lambda G_\lambda a_v)$$

is a solution of (3) and $x_v = (x, b_v)$.

Proof. If x satisfies (3), then

$$x = y + \lambda Kx = (y + \lambda Px) + \lambda Qx. \tag{3a}$$

But if we multiply this equation by G_λ we obtain

$$G_\lambda x = G_\lambda(y + \lambda Px) + \lambda G_\lambda Qx$$
$$= G_\lambda(y + \lambda Px) + (G_\lambda - Q)x, \quad \text{from} \quad (2).$$

That is,

$$Qx = \lambda G_\lambda(y + \lambda Px).$$

Hence, substituting in (3a),

$$x = y + \lambda Px + \lambda G_\lambda(y + \lambda Px)$$
$$= y + \lambda G_\lambda y + \lambda(Px + \lambda G_\lambda Px)$$
$$= y + \lambda G_\lambda y + \lambda \sum_{\nu=1}^{n} (x, b_\nu)(a_\nu + \lambda G_\lambda a_\nu), \quad \text{from} \quad (1b)$$
$$= z + \lambda Fx \tag{5}$$

where the operator

$$F = \sum_{\nu=1}^{n} (a_\nu + \lambda G_\lambda a_\nu) \otimes b_\nu \equiv P + \lambda G_\lambda P$$

has a kernel of finite rank n.

Conversely, if $x = z + \lambda Fx = y + \lambda Px + \lambda G_\lambda(y + \lambda Px)$, then since G is the resolvent of Q we have as before

$$G(y + \lambda Px) = Qx.$$

That is,

$$x = y + \lambda Px + \lambda Qx$$
$$= y + \lambda Kx.$$

Comment

This theorem reduces the solution of an arbitrary integral equation (formally) to that of the finite matrix equation (4). We can now use this reduction to characterise the spectrum of K.

We observe that, from the previous theorem, we can obviously construct the resolvent of K. We write (5) in the form

$$(I - \lambda F)x = z$$

and note that the vector $\mathbf{x} = (x_\mu)$ satisfies the set of algebraic equations

$$(\mathbf{I} - \lambda \mathbf{F})\mathbf{x} = \mathbf{z} = (z_\mu); \mathbf{F} = (f_{\mu\nu}).$$

We introduce the inverse matrix $\mathbf{D} = (\mathbf{I} - \lambda \mathbf{F})^{-1} = (d_{\mu\nu})$.

Then we find that the resolvent H_λ of K is given explicitly by

$$H_\lambda = G + \sum_{\mu,\nu=1}^{n} d_{\mu\nu}(a_\mu + \lambda G_\lambda a_\mu) \otimes (b_\nu + \lambda^* G_\lambda^* b_\nu)$$

as can be seen directly by showing that H_λ satisfies the resolvent equations. We then have the following results:

(a) For $|\lambda| < \omega$, λ is a regular value of K unless $\mathbf{I} - \lambda \mathbf{F}$ is singular. The singular values of this matrix equation cannot exceed n in number. Moreover, if det $(\mathbf{I} - \lambda \mathbf{F}) = 0$, there exists a nontrivial solution of the homogeneous integral equation (show this!). Hence, for $|\lambda| < \omega$, every value of λ is either a regular value or a characteristic value. Finally, the resolvent is a meromorphic function of λ for $|\lambda| < \omega$ (why?).

(b) We now let $\omega \to \infty$, and find

(i) Every value of λ is either a regular value or a characteristic value. This important result is the 'Fredholm alternative'.

(ii) The characteristic values are at most enumerable, and have no finite limit point (hence the eigenvalues have a limit point of zero) and the resolvent kernel H_λ is meromorphic in λ for all λ.

(iii) When λ is a characteristic value of K, the homogeneous equation has at most a finite number of linearly independent $\mathscr{L}^2$ solutions.

6.4   Numerical examples

We investigate the use of the Nystrom method to solve the eigenvalue problem

$$\gamma x(s) = \int_a^b K(s,t)x(t)\,dt, \quad a \le s, t \le b. \tag{1}$$

Using the N-point Trapezoid, Simpson and Gauss–Legendre rules for a range of values of N, we tabulate some eigenvalues of interest obtained on solution of the Nystrom equations (6.1.2).

Example 6.4.1.

$$\gamma x(s) = \int_0^1 \left(st - \frac{s^3 t^3}{6}\right)x(t)\,dt, \quad 0 \le s, t \le 1.$$

The kernel is Hermitian and we therefore solve the modified Nystrom equations (6.1.2a). In addition K is degenerate of rank 2 and its non-zero eigenvalues are given on solution of the following characteristic equation

$$\gamma^2 - \tfrac{13}{42}\gamma - \tfrac{2}{1575} = 0,$$

that is, $\gamma_1 = 0.313\,57$, $\gamma_2 = -0.004\,049\,5$ to five significant figures. Computed values obtained for the three eigenvalues of largest modulus are

Table 6.4.1. *Three eigenvalues of largest modulus for the Nystrom solution of the integral equation of Example 6.4.1*

Trapezoid			Simpson			Gauss–Legendre		
N	i	$\gamma_i^{(N)}$	N	i	$\gamma_i^{(N)}$	N	i	$\gamma_i^{(N)}$
3	1	0.34063	3	1	0.31216	2	1	0.31450
	2	-0.86007×10^{-2}		2	$-0.83424, -2$		2	$-0.12267, -2$
	3	0.00000		3	0.00000			
5	1	0.32052	5	1	0.31349	4	1	0.31357
	2	$-0.56771, -2$		2	$-0.43807, -2$		2	$-0.40496, -2$
	3	$-0.15035, -11$		3	$-0.14211, -11$		3	$-0.29084, -12$
7	1	0.31668	7	1	0.31356	6	1	0.31357
	2	$-0.48169, -2$		2	$-0.41172, -2$		2	$-0.40496, -2$
	3	$-0.24856, -11$		3	$-0.25408, -11$		3	$-0.15466, -11$
9	1	0.31532	9	1	0.31357			
	2	$-0.44900, -2$		2	$-0.40712, -2$			
	3	$0.79787, -12$		3	$-0.37824, -11$			
11	1	0.31469	11	1	0.31357			
	2	$-0.43341, -2$		2	$-0.40585, -2$			
	3	$0.30636, -11$		3	$-0.27595, -11$			
15	1	0.31415	15	1	0.31357			
	2	$-0.41959, -2$		2	$-0.40519, -2$			
	3	$0.12004, -11$		3	$-0.13226, -11$			
25	1	0.31377	25	1	0.31357			
	2	$-0.40996, -2$		2	$-0.40498, -2$			
	3	$0.15089, -11$		3	$-0.18504, -11$			
41	1	0.31364	41	1	0.31357			
	2	$-0.40676, -2$		2	$-0.040496, -2$			
	3	$0.18079, -11$		3	$-0.20005, -11$			
49	1	0.31362	49	1	0.31357			
	2	$-0.40621, -2$		2	$-0.40496, -2$			
	3	$-0.38328, -11$		3	$-0.26294, -11$			

given in Table 6.4.1 and we see that the Nystrom method essentially recognises the coefficient matrix as being of rank 2. Note the slow convergence of the repeated Trapezoid rule; Simpson's rule gives five-figure accuracy for γ_1 using only nine points while four points are sufficient for the Gauss–Legendre rule. Thus the higher order Gauss–Legendre rule provides a cheap numerical solution for this problem.

Example 6.4.2.

$$\gamma x(s) = \int_0^1 e^{st} x(t) dt, \quad 0 \le s, t \le 1.$$

Table 6.4.2. *The eigenvalues of largest, fourth largest and seventh largest modulus for the Nystrom solution of the integral equation of Example 6.4.2*

Trapezoid			Simpson			Gauss–Legendre		
N	i	$\gamma_i^{(N)}$	N	i	$\gamma_i^{(N)}$	N	i	$\gamma_i^{(N)}$
3	1	1.4019	3	1	1.3545	2	1	1.3521
5	1	1.3644	5	1	1.3531	4	1	1.3530
	4	1.5310×10^{-4}		4	1.4980, -4		4	7.3823, -5
7	1	1.3580	7	1	1.3530	6	1	1.3530
	4	1.2365, -4		4	9.6279, -5		4	7.6380, -5
	7	7.8973, -11		7	6.4501, -11			
9	1	1.3558	9	1	1.3530	8	1	1.3530
	4	1.0601, -4		4	8.3309, -5		4	7.6380, -5
	7	1.8558, -10		7	1.4590, -10		7	1.6145, -10
11	1	1.3548	11	1	1.3530			
	4	9.6315, -5		4	7.9337, -5			
	7	2.3036, -10		7	2.5879, -10			
13	1	1.3543	13	1	1.3530			
	4	9.0604, -5		4	7.7837, -5			
	7	2.4206, -10		7	2.5075, -10			
25	1	1.3533	25	1	1.3530			
	4	8.0104, -5		4	7.6474, -5			
	7	2.0402, -10		7	1.7632, -10			
41	1	1.3531	41	1	1.3530			
	4	7.7734, -5		4	7.6392, -5			
	7	1.7466, -10		7	1.6071, -10			
49	1	1.3531	49	1	1.3530			
	4	7.7322, -5		4	7.6386, -5			
	7	1.7382, -10		7	1.5808, -10			

This kernel is Hermitian but non-degenerate and Brakhage (1960) demonstrates that

$$1.3527 < \gamma_1 < 1.3534$$

where γ_1 is the maximum eigenvalue of $K(s,t) = e^{st}$.

In Table 6.4.2 we give, to five significant figures, those computed values corresponding to the largest, fourth largest and seventh largest (in modulus) eigenvalue. Here, as in the previous example, the rapid convergence of the Gauss–Legendre rule renders it an obvious 'best choice'.

Example 6.4.3.

$$\gamma x(s) = \int_{-1}^{1} st^2 x(t)\mathrm{d}t, \quad -1 \leq s, t \leq 1.$$

The kernel is non-Hermitian and we solve the Nystrom equations (6.1.2) in their original form. In addition K is degenerate of rank 1 and it is clear that it has no non-zero eigenvalues; for $\gamma x(s) = cs$ where $c = \int_{-1}^{1} t^2 x(t)dt$. Thus $x(s) = s$ (or a multiple of s) is the only possible eigenfunction, with corresponding eigenvalue $\gamma = \int_{-1}^{1} t^2 \cdot t \, dt = 0$. The N-point Trapezoid, Simpson and Gauss–Legendre rules used to solve this problem produced in each case N eigenvalues which were zero to machine accuracy.

Exercises

1. Show that if $x_0(s)$ satisfies the equation

$$x(s) = y(s) + \lambda \int_a^b K(s,t)x(t)dt \tag{1}$$

and the kernel K has characteristic value λ of rank p, that is, corresponding to λ, there are p linearly independent solutions $x_1, \ldots, x_p$ of

$(I - \lambda K)x = 0,$

then the general solution of (1) is given by

$$x(s) = x_0(s) + \sum_{i=1}^{p} \alpha_i x_i(s).$$

2. Show that if $\gamma = 1/\lambda$ is an eigenvalue of the equation

$$x(s) = \lambda \int_a^b K(s,t)x(t)dt$$

or

$(I - \lambda K)x = 0$

in operator form, and x a corresponding eigenfunction, then

$$\gamma = \frac{(x, Kx)}{(x, x)}$$

where

$$(x, y) = \int_a^b x(s)y(s)ds.$$

Show further that if $u = x + \varepsilon$ and K is Hermitian, then

$$\frac{(u, Ku)}{(u, u)} = \gamma + \frac{(\varepsilon, (K - \gamma I)\varepsilon)}{(u, u)}.$$

3. The following equation has kernel of finite rank:

$$x(s) = 3s + 2 + \int_0^1 (1 + s + st)x(t)dt.$$

What is the rank ? Use the results of Theorem 6.2.1 to find the exact solution of this equation and find the eigenvalues of the kernel.

4. Consider the equation

$$x = y + \lambda K x$$

where K is a non-degenerate kernel. In the notation of Theorem 6.3.1 show that the resolvent H_λ of K is given by

$$H_\lambda = G + \sum_{\mu,\nu=1}^{n} d_{\mu\nu}(a_\mu + \lambda G_\lambda a_\mu) \otimes (b_\nu + \lambda^* G_\lambda^* b_\nu).$$

7

Expansion methods for Fredholm equations of the second kind

7.1 Introduction

The theorems of Chapter 6 are pretty in that they characterise the spectrum of K; because of the constructive nature of the proofs they can, in principle at least, be useful numerically. Theorem 6.3.1 in particular looks promising, since it reduces the solution of an *arbitrary* (linear) Fredholm equation of the second kind to that of a finite set of algebraic equations. Of course, there is a snag: to *set up* these equations we must first find an approximating degenerate kernel, and second we must find the resolvent of the 'remainder' kernel Q. In practice we will not know the resolvent G of Q but we could generate it from the Neumann series to any required accuracy. There are two possible numerical approaches, both of which might be useful: (we set $\lambda = 1$ here and omit the suffix λ on G and H)

(i) We approximate K by K_0 sufficiently well that Q is negligible.

(ii) We approximate K only sufficiently well that $\|Q\| < 1$, and then approximate G (by G_0) well enough that $G - G_0$ is negligible.

In both cases we can quantify what we mean by 'sufficiently well' rather simply. Ignoring Q, we compute the solution x_0 to

$$(I - K_0)x_0 = y \tag{1}$$

while the exact solution $x = x_0 + \eta$ satisfies the equation

$$(I - (K_0 + Q))(x_0 + \eta) = y. \tag{2}$$

Introducing the (known) resolvent H_0 of K_0, we have

$$(H_0 - H_0 K_0)\eta = H_0 Q x_0 + H_0 Q \eta.$$

But from the resolvent equation and (1) it follows that

$$(H_0 - H_0 K_0)\eta = K_0 \eta = \eta - Q x_0 - Q \eta.$$

Hence we have

$$\eta = (Q + H_0 Q)x_0 + (Q + H_0 Q)\eta. \tag{3}$$

From this equation we can proceed in two ways. First, we have the immediate bound:

$$\|\eta\| \le \frac{\|Q\|(1 + \|H_0\|)\|x_0\|}{1 - \|Q\|(1 + \|H_0\|)} \tag{4}$$

if $\|Q\|(1 + \|H_0\|) < 1$. This result is appropriate if we wish to neglect Q. Alternatively, if we have merely attempted to reduce Q to a 'reasonable' size, we can observe that (3) has the form

$$\eta = y_0 + \bar{Q}\eta \tag{3a}$$

where $y_0 = \bar{Q}x_0$, $\bar{Q} = Q + H_0 Q$. Moreover, since K_0 approximates K, H_0 approximates H and hence as $\|Q\| \to 0$, $\|H_0\|$ remains bounded (unless λ is a characteristic value of K).

We assume that

$$\|H_0\| < h$$

and then

$$\|\bar{Q}\| \le \|Q\|(1 + h).$$

Then, provided that $\|\bar{Q}\| < 1$, we may iterate (3) for the correction η and this may well be computationally more convenient than iterating the Neumann series for Q. It has the additional advantage that we may at any time estimate the error in the approximate solution. The discussion of Section 4.1 applies directly to (3a). If we iterate with the simpler scheme

$$\eta_{n+1} = y_0 + \bar{Q}\eta_n \tag{5}$$

then from equation (4.1.4) the error $e_n = \eta_n - \eta$ is bounded by

$$\|e_n\| \le \|\eta_{n+1} - \eta_n\|/(1 - \|\bar{Q}\|). \tag{6}$$

7.2   Numerical methods based on approximating the kernel by a separable kernel

We now look at numerical methods based on these ideas. We start by discussing methods based directly on an approximation to the kernel. These methods are mainly of historical interest, but we give a brief discussion because the methods of Section 7.3, although they do not produce an approximation for the kernel directly, do so incidentally, and hence are closely related.

There are several methods by which we can approximate a kernel K by a separable kernel K_0.

(i) Suppose that the sequence $\{h_n(s)\}$ is orthonormal and complete in $\mathscr{L}^2(a,b)$. Then the kernel has the following double (generalised) Fourier expansion:

$$K(s,t) = \sum_{n,m=1}^{\infty} A_{nm} h_n(s) h_m(t). \tag{1}$$

This series is convergent in the mean (see Section 1.5) and the coefficients A_{nm} are given by

$$A_{nm} = \int_a^b \int_a^b K(s,t) h_n^*(s) h_m^*(t) \, ds \, dt. \tag{2}$$

An obvious choice of approximating kernel is the truncated expansion

$$K_0(s,t) = \sum_{n,m=1}^{p} A_{nm} h_n(s) h_m(t).$$

Then

$$\|Q\|_E^2 = \|K - K_0\|_E^2 = \int_a^b \int_a^b |(K - K_0)|^2 \, ds \, dt$$

$$= \int_a^b \int_a^b \left| \sum_{n=1}^{p} \sum_{m=p+1}^{\infty} A_{nm} h_n(s) h_m(t) \right.$$

$$\left. + \sum_{n=p+1}^{\infty} \sum_{m=1}^{\infty} A_{nm} h_n(s) h_m(t) \right|^2 \, ds \, dt$$

$$= \sum_{n=1}^{p} \sum_{m=p+1}^{\infty} |A_{nm}|^2 + \sum_{n=p+1}^{\infty} \sum_{m=1}^{\infty} |A_{nm}|^2 \tag{3}$$

where we have used the orthonormality of the set $\{h_i\}$:

$$\int_a^b h_i^*(s) h_j(s) \, ds = \delta_{ij}. \tag{4}$$

The sum (3) is not directly computable, but can be made as small as desired by appropriate choice of the parameter p; because the series (1) is convergent, the coefficients A_{nm} are guaranteed to converge to zero as n, m increase and clearly we should try to choose the set h_i so that this convergence is as rapid as possible.

(ii) Rather more generally, we can choose two expansion sets $\{U_n\}, \{V_n\}$ and introduce the expansion

$$K(s,t) = \sum_{n,m=1}^{\infty} A_{nm} U_n(s) V_m(t) \tag{5}$$

leading to

$$A_{nm} = \int_a^b \int_a^b K(s,t) U_n^*(s) V_m^*(t) \, ds \, dt \tag{6}$$

where we assume that $\{U_n\}, \{V_n\}$ are both orthonormal sequences complete in $\mathscr{L}^2(a,b)$.

If K is Hermitian the choice $V = U$ ensures that K_0 retains this symmetry; if K is not Hermitian it may be that an alternative choice is more appropriate.

(iii) Suppose that the sequences $\{U_n(s)\}, \{V_n(s)\}$ are complete but not orthonormal in $\mathscr{L}^2(a, b)$. Then the sequence $\{U_m(s)V_n(t)\}$ is complete but not orthonormal in $\mathscr{L}^2(a, b) \times (a, b)$, and we can set

$$K_0(s, t) = \sum_{n,m=1}^{p} A_{nm} U_n(s) V_m(t) \qquad (7)$$

and choose the coefficients to minimise the error $\| K - K_0 \|$, where $\| \cdot \|$ is any operator norm. For example, with the L_2 norm we seek the values of the coefficients A_{nm} which minimise

$$\int_a^b \int_a^b |K(s, t) - K_0(s, t)|^2 ds\, dt.$$

This criterion leads in a standard way to the system of least squares equations for A_{nm}:

$$\sum_{n,m=1}^{p} X_{n'm',nm} A_{nm} = Y_{n'm'}, \quad n', m' = 1, \ldots, p,$$

where

$$X_{n'm',nm} = \int_a^b U_n(s) U_{n'}{}^*(s) ds \int_a^b V_m(t) V_{m'}{}^*(t) dt,$$

$$Y_{n'm'} = \int_a^b \int_a^b K(s, t) U_{n'}{}^*(s) V_{m'}{}^*(t) ds\, dt. \qquad (8)$$

Note that alternative error norms can also be used; for example, the L_1 or L_∞ norm might in some cases be appropriate (see below).

If the interval (a, b) is finite we might take $U_n(s) = V_n(s) = s^{n-1}$ giving a multinomial approximation to K which will converge uniformly if K is continuous. We may determine the coefficients by a least squares or L_∞ procedure; or, if the kernel K has a known Taylor series expansion, we might make use of this.

Example 7.2.1.

$$x(s) = y(s) + \int_0^1 \sin st\, x(t) dt.$$

Approximating $\sin st$ from its Taylor series expansion, we set $K_0(s, t) = st - s^3 t^3/6$, which of course is separable of rank 2, and solve the approximate equation

$$\bar{x}(s) = y(s) + \int_0^1 \left(st - \frac{s^3 t^3}{6} \right) \bar{x}(t) dt.$$

We can estimate the error in the approximation using the bounds derived previously. We have, in the notation of Section 7.1, recalling the remainder term in the Taylor series expansion:

$$|Q(s,t)| = |K(s,t) - K_0(s,t)| \le \frac{s^5 t^5}{120}.$$

Hence we find

$$\|Q\| \le \frac{1}{120} \left[\int_0^1 \int_0^1 s^{10} t^{10} \, ds \, dt \right]^{\frac{1}{2}} = \frac{1}{1320}.$$

Similarly

$$\|K_0\| = \left\{ \int_0^1 \int_0^1 \left(st - \frac{s^3 t^3}{6} \right)^2 ds \, dt \right\}^{\frac{1}{2}} = (\tfrac{1}{9} - \tfrac{1}{75} + \tfrac{1}{1764})^{\frac{1}{2}} < \tfrac{1}{3}.$$

Hence we can use the bound (since $H_0 - K_0 = H_0 K_0$)

$$\|H_0\| \le \frac{\|K_0\|}{1 - \|K_0\|} < \frac{\frac{1}{3}}{\frac{2}{3}} = \tfrac{1}{2}$$

and we find

$$\|\eta\| \le \tfrac{1}{1320} \times \frac{\frac{3}{2}}{1 - \frac{1}{1320} \times \frac{3}{2}} \|\bar{x}\|$$

$$\approx \tfrac{1}{880} \|\bar{x}\|.$$

7.3   Methods based on an expansion of the solution

7.3.1   Introduction

We now look at methods which start by approximating, not the kernel, but the solution x. If we write

$$x \approx x_N = \sum_{i=1}^{N} a_i^{(N)} h_i \tag{1}$$

where the set $\{h_i\}$ is complete in $\mathscr{L}^2(a,b)$, then for some choice of the $a_i^{(N)}$, and for sufficiently large N, we may approximate x as closely as we please by x_N.

An *expansion method* is then an algorithm for determining the $a_i^{(N)}$ for either an arbitrary or a specified choice of the set $\{h_i\}$. There are many possible algorithms and we consider only the most common.

7.3.2   Residual minimisation methods

The simplest method conceptually again appeals to approximation theory. We write the integral equation in the form (again we set $\lambda = 1$)

$$Lx = y, \quad L = I - K, \tag{2}$$

and introduce the residual function r_N and error function ε_N

$$\varepsilon_N = x - x_N,$$ (3)

$$r_N = y - Lx_N.$$

To compute r_N requires no knowledge of x but, since $y - Lx = 0$, we have the identity

$$r_N = (y - Lx_N) - (y - Lx) = L(x - x_N) = L\varepsilon_N.$$ (3a)

We now choose the vector $\mathbf{a}^{(N)} = (a_i^{(N)})$ from the minimisation criterion

$$\mathbf{a}^{(N)} : \min_{\mathbf{a}^{(N)}} \| r_N \|.$$ (4)

There are many norms available in which to minimise; before discussing those methods used in practice, let us go through them by the now familiar process of generating an error bound.

From (2) and (3a) we have at once

$$\| r_N \| \le (1 + \| K \|) \| \varepsilon_N \|.$$

That is,

$$\| \varepsilon_N \| \ge \frac{\| r_N \|}{1 + \| K \|}.$$ (5)

Thus a small residual is a *necessary* condition for a small error. We would rather have an upper bound on ε_N of course; this is harder to provide in general and we content ourselves for now with the following. We rewrite (3a) as

$$\varepsilon_N = r_N + K\varepsilon_N$$

whence

$$\| \varepsilon_N \| \le \| r_N \| + \| K \| \cdot \| \varepsilon_N \| \text{ and hence if } \| K \| < 1$$

$$\| \varepsilon_N \| \le \frac{\| r_N \|}{1 - \| K \|}.$$ (6)

7.3.3   Chebyshev norm

The above results hold true in any norm; the choice of norm is influenced in practice by the analyticity properties of the kernel K. For continuous kernels a possible choice is the Chebyshev (L_∞) norm:

$$\| x \| = \max_{a \le s \le b} | x(s) |$$

which yields a pointwise bound on the error. In this norm it can be shown that

$$\| K \| = \max_{a \le s \le b} \int_a^b | K(s,t) | \mathrm{d}t$$

and in practice this can be evaluated numerically by

    (i) performing the integral for several fixed values of s
    (ii) performing a linear search in s for the maximum.

Neither of these operations need be carried out very accurately, since one or two significant figures is all that we need to attain in $\|K\|$.

Perhaps more important is the method of computing x_N. Again, since the bounds above are valid for any x_N, we need not in principle compute the minimum very precisely. We seek to compute

$$\|r_N\| = \min_{a_i^{(N)}} \max_{a \le s \le b} |y(s) - x_N(s) + \int_a^b K(s,t)x_N(t)dt|.$$

With $x_N(s) = \Sigma_{i=1}^N a_i^{(N)} h_i(s)$ we have, setting $k_i(s) = \int_a^b K(s,t)h_i(t)dt$,

$$\|r_N\| = \min_{a_i^{(N)}} \max_s |y(s) - \sum_{i=1}^N a_i^{(N)}(h_i(s) - k_i(s))|. \qquad (7)$$

The functions $k_i(s)$ can be estimated by numerical quadrature for any fixed value of s; the usual approach for finding the min max is to search, not over the interval $[a,b]$, but over a discrete point set. If we set

$$h_i(s) - k_i(s) = l_i(s) \qquad (8)$$

then we see that the process is completely equivalent to finding a minimax approximation y_N to $y(s)$ with an expansion of the form

$$y_N(s) = \sum_{i=1}^N a_i^{(N)} l_i(s). \qquad (9)$$

Over a discrete point set $\{\xi_i, i=1,\ldots,q\}$, (7) is approximated by

$$\|r_N\| \approx \min_{\mathbf{a}^{(N)}} \max_{k=1}^q |y(\xi_k) - \sum_{i=1}^N a_i^{(N)} l_i(\xi_k)|. \qquad (10)$$

Numerical techniques for solving (10) are considered briefly in the next chapter.

Example 7.3.1. As an example of the technique, we consider Love's equation (see Barrodale and Young, 1970):

$$x(s) = 1 + \int_{-1}^1 \frac{1}{\pi[1 + (s-t)^2]} x(t)dt.$$

If we set

$$x_N(s) = \sum_{i=1}^N a_i^{(N)} s^{2i-2}$$

then with 101 equally spaced points on $[-1, 1]$ Barrodale and Young find for the residual norm the results shown in Table 7.3.1.

Table 7.3.1

N	1	2	3	4	5
$\|r_N\|_\infty$	1.3×10^{-1}	$4.9, -3$	$3.8, -4$	$4.4, -5$	$1.8, -6$

For this equation we have

$$\| K \| = \max_{-1 \le s \le 1} \int_{-1}^{1} \left| \frac{1}{\pi[1 + (s-t)^2]} \right| dt = \tfrac{1}{2} \qquad (11)$$

whence for $N = 5$ we find from (5), (6):

$$1.2 \times 10^{-6} \le \max_{-1 \le s \le 1} |x(s) - \sum_{i=1}^{N} a_i^{(N)} s^{2i-2}| \le 3.6 \times 10^{-6}. \qquad (12)$$

Thus, where applicable, the error bounds are sufficiently tight to be useful.

7.3.4 L_2 norm: least squares approximation

Alternatively we may choose to minimise $\| y - Lx \|_2$, the L_2 norm; we then have the defining equations

$$x_N : \min_{a_i^{(N)}} I(\mathbf{a}^{(N)}) = \int_a^b \left[x_N(s) - y(s) - \int_a^b K(s,t) x_N(t) dt \right]^2 ds \qquad (13)$$

whence on inserting the form of x_N and setting $\partial I / \partial a_i^{(N)} = 0$, $i = 1, \dots, N$, we find the set of linear equations for the $a_i^{(N)}$:

$$\mathbf{L}_{LS}^{(N)} \mathbf{a}^{(N)} = \mathbf{y}_{LS}^{(N)}, \qquad (14)$$

where

$$(\mathbf{L}_{LS}^{(N)})_{ij} = \int_a^b (Lh_i)^*(s) Lh_j(s) ds, \quad i, j = 1, \dots, N,$$

$$(\mathbf{y}_{LS}^{(N)})_i = \int_a^b y(s) Lh_i(s) ds, \quad i = 1, \dots, N,$$

$$Lh_i(s) = h_i(s) - \int_a^b K(s,t) h_i(t) dt.$$

Equations (14) are *linear* but to generate the matrix $\mathbf{L}_{LS}^{(N)}$ we require to perform multiple integrals. In particular, $\mathbf{L}_{LS}^{(N)}$ contains the term

$$\int_a^b \left[\int_a^b K^*(s,t) h_i^*(t) dt \right] \left[\int_a^b K(s,t') h_j(t') dt' \right] ds \qquad (15)$$

and we have N^2 of these apparently triple integrals to perform.

The cost of a least squares method depends strongly on how these integrals are carried out and we consider appropriate numerical techniques in the next chapter, noting here that these N^2 integrals are related closely to each other so that the cost in practice can be reduced to manageable proportions.

We illustrate the use of the least squares technique with a simple example for which the integrals can be performed analytically.

Example 7.3.2.

$$x(s) = \sin s - \tfrac{1}{4}s + \tfrac{1}{4}\int_0^{\pi/2} stx(t)dt.$$

Exact solution: $x(s) = \sin(s)$

Taking $h_1(s) = s, h_2(s) = s^3$ we have

$$x(s) \approx x_2(s) = a_{s1}s + a_{s2}s^3$$

and (see equation (14))

$$Lh_i(s) = h_i(s) - \tfrac{1}{4}\int_0^{\pi/2} sth_i(t)dt$$

whence we find

$$Lh_1(s) = s - \tfrac{1}{4}\int_0^{\pi/2} st^2dt = s\left(1 - \frac{\pi^3}{96}\right),$$

$$Lh_2(s) = s^3 - \tfrac{1}{4}\int_0^{\pi/2} st^4dt = s^3 - \frac{\pi^5 s}{5 \times 128},$$

and hence

$$(\mathbf{L}_{LS})_{11} = \int_0^{\pi/2} (Lh_1)^2 ds \approx 0.592\,159\,55,$$

$$(\mathbf{L}_{LS})_{21} = (\mathbf{L}_{LS})_{12} = \int_0^{\pi/2} (Lh_1)(Lh_2)ds \approx 0.876\,657\,09,$$

$$(\mathbf{L}_{LS})_{22} = \int_0^{\pi/2} (Lh_2)^2 ds \approx 1.837\,176\,89,$$

$$(\mathbf{y}_{LS})_1 = \int_0^{\pi/2} (Lh_1)(\sin s - \tfrac{1}{4}s)ds \approx 0.458\,353\,30,$$

$$(\mathbf{y}_{LS})_2 = \int_0^{\pi/2} (Lh_2)(\sin s - \tfrac{1}{4}s)ds \approx 0.600\,327\,51.$$

Inserting these into (14) and solving for a_{s1}, a_{s2} we find

$$a_{s1} \approx 0.988\,792\,2, \quad a_{s2} \approx -0.145\,061\,8,$$

compared with the exact first two terms in the expansion

$$\sin s = 1.0 - \frac{s^3}{6}$$

$$= 1.0 - 0.166\,667s^3.$$

Note that to produce the 2×2 symmetric matrix $\mathbf{L}_{LS}$ we have in fact performed seven one-dimensional integrations rather than the three three-dimensional integrations which (15) at first glance suggests are necessary.

7.3.5   Method of moments or Ritz–Galerkin

An alternative approach, which also leads to equations of the linear form (14), can be introduced as follows. We note that our aim is to make the residual vector $r(s)$ zero. Now if the set $\{h_i\}$ is complete and orthonormal in $\mathscr{L}^2(a, b)$, the statement $r(s) = 0$ is equivalent to the statement

$$r(s) \text{ is orthogonal to each of the set } \{h_i(s), \quad i = 1, 2, \dots, \infty\}. \quad (16)$$

Now with only N parameters $a_i^{(N)}$ at our disposal the best we can do is to make the residual $r_N(s)$ orthogonal to the first N functions $h_1, \dots, h_N$; that is, to set

$$\int_a^b h_i^*(s)[y(s) - (Lx_N)(s)]\,ds = 0, \quad i = 1, \dots, N. \quad (17)$$

This leads to the defining equations

$$\mathbf{L}_G^{(N)}\mathbf{a}^{(N)} = \mathbf{y}_G^{(N)} \quad (18)$$

where now

$$(\mathbf{L}_G^{(N)})_{ij} = \int_a^b h_i^*(s)(Lh_j(s))\,ds$$

$$= \int_a^b h_i^*(s)h_j(s)\,ds - \int_a^b \int_a^b h_i^*(s)K(s, t)h_j(t)\,ds\,dt,$$

$$(\mathbf{y}_G)_i = \int_a^b h_i^*(s)y(s)\,ds, \quad i, j = 1, \dots, N. \quad (19)$$

These equations are similar to, but simpler in structure than, those given by the least squares method.

In general, the functions h_i may be complete but not orthogonal; then the reasoning behind the approximation as presented is no longer valid. But we may proceed as follows:

We assume that the functions h_i are linearly independent. Then the set $\{h_i\}$ may be orthogonalised by a Gram–Schmidt process to yield the orthonormal set $\{\bar{h}_i\}$:

$$\bar{h}_i = \sum_{j=1}^i T_{ij}h_j, \quad (20)$$

where $\mathbf{T}$ is a (triangular) orthogonalising matrix.

The sets $\{\bar{h}_i, i = 1, \ldots, N\}$ and $\{h_i, i = 1, \ldots, N\}$ span the same space and we may set

$$x_N = \sum_{i=1}^{N} a_i h_i = \sum_{i=1}^{N} \bar{a}_i \bar{h}_i = \sum_{i=1}^{N} \bar{a}_i \sum_{j=1}^{i} T_{ij} h_j$$

$$= \sum_{j=1}^{N} \left(\sum_{i=j}^{N} \bar{a}_i T_{ij} \right) h_j \tag{21}$$

so that

$$a_j = \sum_{i=j}^{N} T_{ij} \bar{a}_i,$$

that is,

$$\mathbf{a} = \mathbf{T}^T \bar{\mathbf{a}}.$$

We now make the residual r orthogonal to each of the set $\{\bar{h}_i\}$, as above, and obtain the set of equations

$$\bar{\mathbf{L}}_G^{(N)} \bar{\mathbf{a}}^{(N)} = \bar{\mathbf{y}}_G^{(N)}$$

where

$$(\bar{\mathbf{L}}_G^{(N)})_{ij} = \int_a^b \bar{h}_i^*(s) L\bar{h}_j(s) ds = \sum_{k=1}^{i} \sum_{l=1}^{j} T_{ik} T_{jl} \int_a^b h_k^*(Lh_l) ds$$

$$= \sum_{k=1}^{i} \sum_{l=1}^{j} T_{ik} (\mathbf{L}_G^{(N)})_{kl}.$$

That is,

$$\bar{\mathbf{L}}_G^{(N)} = \mathbf{T} \mathbf{L}_G^{(N)} \mathbf{T}^T,$$

$$\bar{\mathbf{y}}_G^{(N)} = \mathbf{T} \mathbf{y}_G^{(N)},$$

whence we find

$$\mathbf{T} \mathbf{L}_G^{(N)} \mathbf{a}^{(N)} = \mathbf{T} \mathbf{y}_G^{(N)}; \quad \mathbf{L}_G^{(N)} \mathbf{a} = \mathbf{y}_G^{(N)}. \tag{22}$$

This is the same set of equations as if we had imposed the orthogonality conditions directly on the non-orthonormal set. Again, we illustrate the technique with a simple example.

Example 7.3.3. We solve the same problem as in Example 7.3.2, using the same form for the approximate solution: $x_G(s) = a_{G1} s + a_{G2} s^3$ but solving for the coefficients a_{G1}, a_{G2} from the Galerkin equations (18). The functions $Lh_j(s)$, $j = 1, 2$, are as given in Example 7.3.2; substituting these in (19) we find

$$(\mathbf{L}_G)_{11} \approx 0.874\,658\,6; (\mathbf{L}_G)_{12} = (\mathbf{L}_G)_{21} \approx 1.294\,880\,1;$$

$$(\mathbf{L}_G)_{22} \approx 2.456\,331\,3; y_{G1} \approx 0.677\,018\,0; \quad y_{G2} \approx 0.924\,047\,5$$

and the solution of (18) yields

$$a_{G1} \approx 0.988\,792\,0; \quad a_{G2} \approx -0.145\,062\,0$$

which is extremely close to that obtained by the least squares method. We return to the close connection between these methods in Chapter 8.

7.3.6   The variational derivation of the Galerkin equations

If K is a Hermitian operator, the Galerkin equations can be derived in an alternative manner which makes the result (22) obvious.

Consider the functional $F(x)$:

$$F(x) = (x, Lx) - (x, y) - (y, x). \tag{23}$$

A *stationary point* of $F(x)$ is defined by the condition

$$F(x + \varepsilon g) = F(x) + \mathcal{O}(\varepsilon^2), \quad \forall g, \tag{24}$$

or equivalently by the relations

$$\left. \frac{\partial F(x + \varepsilon g)}{\partial \varepsilon} \right|_{\varepsilon = 0} = 0, \quad \forall g.$$

But

$$\left. \frac{\partial F}{\partial \varepsilon} \right|_{\varepsilon = 0} = (g, Lx) + (x, Lg) - (g, y) - (y, g).$$

Now, if L is Hermitian, $(x, Lg) = (Lx, g)$ and we have

$$\left. \frac{\partial F}{\partial \varepsilon} \right|_{\varepsilon = 0} = (g, Lx - y) + (Lx - y, g). \tag{25}$$

(25) is zero for all g iff $Lx = y$; that is, the functional F is stationary at the solution point of the equation. Now suppose we look for an approximate stationary point in the space spanned by $\{h_1, \ldots, h_N\}$. That is, we make the expansion (1) and set

$$\frac{\partial F}{\partial a_i^{(N)}} = 0, \quad i = 1, 2, \ldots, N.$$

These conditions yield directly equation (22) without requiring the orthogonality of the $\{h_i\}$; the orthogonalisation argument yielding (22) via (20) can be summed up by noting that the orthogonalised set $\{\bar{h}_i\}$ spans the same subspace (of $\mathscr{L}^2$) as the set $\{h_i\}$, and hence defines the same stationary point of F.

7.3.7   A relation between the method of moments and the methods of Section 7.2

We now show that (a particular case of) the method of moments is equivalent to replacing the kernel K by a degenerate kernel. As above, we assume without loss of generality that the set $\{h_i\}$ is orthonormal. Then let us construct a degenerate kernel K_N as follows:

$$K_N(s,t) = \sum_{i=1}^{N} k_i(t)h_i(s) \qquad (26)$$

where

$$k_i(t) = \int_a^b K(s,t)h_i^*(s)\mathrm{d}s.$$

The corresponding integral equation for $x(s)$ is

$$L_N(x) = x(s) - \int_a^b K_N(s,t)x(t)\mathrm{d}t = y(s). \qquad (27)$$

If we solve this equation by the method of moments with the set $\{h_i\}$ we shall obtain the defining equations (18) with $\mathbf{y}_G$ given by (19) and matrix $\mathbf{L}_N^{(N)}$ given by

$$(\mathbf{L}_N^{(N)})_{ij} = \int_a^b h_i^*(s)(L_N h_j(s))\mathrm{d}s = \int_a^b h_i^*(s)h_j(s)\mathrm{d}s$$

$$- \int_a^b h_i^*(s)\left[\int_a^b \sum_{l=1}^{n} k_l(t)h_l(s)h_j(t)\mathrm{d}t\right]\mathrm{d}s$$

$$= \int_a^b h_i^*(s)h_j(s)\mathrm{d}s$$

$$- \sum_{l=1}^{n}\left[\int_a^b h_i^*(u)h_l(u)\mathrm{d}u\right]\left[\int_a^b h_j(t)\int_a^b h_l^*(s)K(s,t)\mathrm{d}s\,\mathrm{d}t\right]$$

$$= \delta_{ij} - \int_a^b\int_a^b h_i^*(s)K(s,t)h_j(t)\mathrm{d}s\,\mathrm{d}t = (\mathbf{L}_G^{(N)})_{ij}$$

where we have used the orthonormality of the $\{h_i\}$. Thus, the solution of the modified system (27) by the method of moments yields the same result as the solution of the original equation. But the modified equation, which has a kernel that is degenerate of rank N, has an exact solution of the form

$$x(s) = y(s) + \sum_{i=1}^{N}\left\{\int_a^b x(t)h_i^*(t)\mathrm{d}t\right\}h_i(s).$$

Thus, if (and only if) $y(s)$ has an exact expansion of the form

$$y(s) = \sum_{i=1}^{N} \alpha_i h_i(s) \qquad (28)$$

we find the exact solution of the degenerate kernel equation, and the method of moments is identical to solving exactly the corresponding degenerate kernel equation.

Condition (28) is certainly satisfied if we choose the first expansion function $h_1(s) \equiv y(s)$; for then $\alpha_1 = 1$, $\alpha_i = 0$, $i > 1$. This choice is sometimes made in practice.

7.4   Expansion methods for eigenvalue problems
7.4.1   The Ritz–Galerkin method

The expansion methods of the previous section may be used with only rather obvious modifications to solve the homogeneous Fredholm equation of the second kind

$$x(s) = \lambda \int_a^b K(s,t)x(t)dt. \tag{1}$$

Equation (1) may be reformulated as

$$\gamma x(s) = \int_a^b K(s,t)x(t)dt \tag{2}$$

where $\gamma = \lambda^{-1}$. An expansion method for (2) then proceeds by approximating the eigenfunctions in the form

$$x(s) \simeq x^{(N)}(s) = \sum_{i=1}^N a_i^{(N)} h_i(s) \tag{3}$$

and the eigenvalues γ by $\gamma^{(N)}$. Defining the residual function $r_N(s)$ by

$$r_N(s) = \int_a^b K(s,t)x^{(N)}(t)dt - \gamma^{(N)}x^{(N)}(s) \tag{4}$$

the particular expansion method is determined by the restrictions imposed on the residual function; we recall that the aim of an expansion method is to determine the coefficients $a_i^{(N)}$, $i = 1,\dots,N$ in (3) in such a way that some measure of $r_N(s)$ is small.

For example if, as in Section 7.3.5, we require that $r_N(s)$ be orthogonal to each of the functions $h_i(s)$, $i = 1,\dots,N$ we generate the Ritz–Galerkin method which for equation (2) is

$$\int_a^b h_i^*(s)\left[\int_a^b K(s,t)x^{(N)}(t)dt - \gamma^{(N)}x^{(N)}(s)\right]ds = 0, \quad i = 1,\dots,N.$$

This leads to the defining equations

$$\mathbf{L}\mathbf{a}^{(N)} = \gamma^{(N)}\mathbf{M}\mathbf{a}^{(N)} \tag{5}$$

where

$$L_{ij} = \int_a^b \int_a^b h_i^*(s)K(s,t)h_j(t)ds\,dt,$$

$$M_{ij} = \int_a^b h_i^*(s)h_j(s)ds, \quad i,j = 1,\dots,N.$$

Equation (5) is a generalised eigenvalue problem; in the case where the set $\{h_i(s)\}$ is orthonormal $\mathbf{M}$ is the identity matrix and the problem reduces to a standard eigenvalue problem.

Equation (5) represents by far the most common expansion method for eigenvalue problems; its popularity stems from the variational character of the eigenvalues which it yields in the (rather common) case that the kernel K is real and symmetric (or complex and self-adjoint). This character is displayed in the following two theorems:

Theorem 7.4.1. Let K be a real symmetric kernel and $\tilde{x}(s) = x(s) + e(s)$ be any approximation to an eigensolution $x(s)$ corresponding to eigenvalue γ, and let

$$\tilde{\gamma} = \frac{\int_a^b \tilde{x}(s) \int_a^b K(s,t)\tilde{x}(t)dt\,ds}{\int_a^b |\tilde{x}(s)|^2\,ds} = F(\tilde{x}). \tag{6}$$

Then

$$\tilde{\gamma} = \gamma + \mathcal{O}(\|e\|^2). \tag{7}$$

Hence $\tilde{\gamma}(\tilde{x})$ is *stationary* at the solution point.

Theorem 7.4.2. Let $(\mathbf{a}_i^{(N)}, \gamma_i^{(N)})$ be the ith eigensolution of (5), ordered so that

$$\gamma_1^{(N)} \ge \gamma_2^{(N)} \ge \cdots \ge \gamma_i^{(N)} \ge \cdots \ge \gamma_N^{(N)} \tag{8}$$

and let the exact eigenvalues γ_i be ordered similarly with $x_i^{(N)}(s) = \sum_{j=1}^N (\mathbf{a}_i^{(N)})_j h_j(s)$. Then under the conditions of Theorem 7.4.1:
(1) $\gamma_i^{(N)}$ is a stationary point of the functional (6); that is

$$\gamma_i^{(N)} = \left(F\left(\sum_{j=1}^N \alpha_j h_j \right) \right) \atop \substack{\text{stat} \\ \alpha} \tag{9}$$

and the functional is stationary at $\alpha_j = (\mathbf{a}_i^{(N)})_j, j = 1, \ldots, N$.
(2) If K is in addition positive definite the stationary point is a maximum point for each i (see equation (18) below) and the approximate eigenvalues satisfy the 'interlacing' condition:

$$\gamma_{i+1}^{(N+1)} \ge \gamma_i^{(N)} \ge \gamma_i^{(N+1)}, \quad i = 1, 2, \ldots, N. \tag{10}$$

(3) Hence in addition

$$\gamma_i \ge \gamma_i^{(N)}, \quad i = 1, 2, \ldots, N, \quad \forall N \tag{11}$$

Comments
These results are of great practical importance. They imply that (for Hermitian problems):

(i) Equations (5) lead to eigenvalue estimates which are much more accurate than the approximate eigenfunctions. Often only the eigenvalues are of interest; then the second order accuracy for $\tilde{\gamma}$ implied by (7) is extremely useful.

(ii) If K is positive definite the computed eigenvalues are lower bounds on the correspondingly numbered exact eigenvalues. This in itself is sometimes (but not always) important; it implies that convergence of each eigenvalue is monotonic in N and this is *certainly* useful numerically.

(iii) Positive definite Hermitian kernels which have finite rank M have only M non-zero eigenvalues; it follows from (10), (11) that for $N > M$, the approximate solution will include $N - M$ *exactly zero* eigenvalues, so that zero eigenvalues are reproduced exactly by equations (5).

All of these remarks assume, with the theorems, that the integrals in (5) are carried out exactly; however, they remain numerically interesting even in the presence of quadrature errors and their importance is such that it seems worthwhile giving a detailed proof of the theorems. For brevity we write (2) in operator form

$$Kx = \gamma x \tag{2a}$$

and the functional F defined in (6) in inner product notation:

$$F(u) = \frac{(u, Ku)}{(u, u)}. \tag{6a}$$

This functional is usually referred to as the *Rayleigh quotient*. It is obviously independent of the normalisation of u and satisfies the equation

$$F(x) = \gamma \tag{12}$$

where x is any eigenfunction.

Moreover, if $\tilde{x} = x + e$,

$$F(\tilde{x}) = \frac{[(\tilde{x}, Kx) + (\tilde{x}, Ke)]}{(\tilde{x}, \tilde{x})}.$$

But

$$(\tilde{x}, Kx) = \gamma(\tilde{x}, x)$$
$$= \gamma(\tilde{x}, \tilde{x} - e)$$
$$= \gamma[(\tilde{x}, \tilde{x}) - (x, e) - (e, e)]$$

and

$$(\tilde{x}, Ke) = (x, Ke) + (e, Ke)$$
$$= (Kx, e) + (e, Ke), \quad (K \text{ Hermitian})$$
$$= \gamma(x, e) + (e, Ke).$$

Whence we find

$$F(\tilde{x}) = \gamma + \frac{(e, (K - \gamma I)e)}{(\tilde{x}, \tilde{x})}. \tag{13}$$

Equation (13) displays the Rayleigh quotient as a variational estimate for γ; the error term is explicitly of order $\|e\|^2$.

Moreover, if we consider the largest eigenvalue γ_1 with eigenfunction x_1, we have for all e

$$(e, (K - \gamma_1 I)e) \leq 0$$

and hence for all $\tilde{x}$

$$F(\tilde{x}) \leq \gamma_1. \tag{14}$$

Thus the Rayleigh quotient yields directly a lower bound on the largest eigenvalue.

To derive the Galerkin equations (5) we now seek a stationary point of $F(\tilde{x})$ for $\tilde{x}$ of the form $x^{(N)}$ (equation (3)). We have at once

$$F(x^{(N)}) = \frac{\mathbf{a}^{(N)\mathrm{T}}\mathbf{L}\mathbf{a}^{(N)}}{\mathbf{a}^{(N)\mathrm{T}}\mathbf{M}\mathbf{a}^{(N)}} \tag{15}$$

where $\mathbf{L}, \mathbf{M}$ are the matrices of (5). Hence setting $\partial F/\partial a_i^{(N)} = 0$, $i = 1, \ldots, N$ and omitting superscripts on $\mathbf{a}^{(N)}$ we find

$$\frac{(\mathbf{a}^{\mathrm{T}}\mathbf{M}\mathbf{a})\mathbf{L}\mathbf{a} - (\mathbf{a}^{\mathrm{T}}\mathbf{L}\mathbf{a})\mathbf{M}\mathbf{a}}{(\mathbf{a}^{\mathrm{T}}\mathbf{M}\mathbf{a})^2} = 0,$$

that is,

$$\mathbf{L}\mathbf{a} - \mu^{(N)}\mathbf{M}\mathbf{a} = 0, \quad \mu^{(N)} = \frac{\mathbf{a}^{\mathrm{T}}\mathbf{L}\mathbf{a}}{\mathbf{a}^{\mathrm{T}}\mathbf{M}\mathbf{a}}. \tag{16}$$

Equation (16) is identical to (5) if we identify $\mu^{(N)} = \gamma^{(N)}$; indeed any solution of (5) clearly satisfies

$$\gamma^{(N)} = \frac{\mathbf{a}^{T}\mathbf{L}\mathbf{a}}{\mathbf{a}^{T}\mathbf{M}\mathbf{a}} \equiv \mu^{(N)} = F(x^{(N)}). \tag{17}$$

The form of $\mu^{(N)}$ shows explicitly that the calculated eigenvalue is indeed a stationary point of $F(x^{(N)})$; if K is positive definite we therefore have from (14)

$$\gamma_1^{(N)} \leq \gamma_1. \tag{18}$$

Thus the stationary point of $F(x^{(N)})$ is a maximum point. It then follows that convergence is monotonic:

$$\gamma_1^{(N+1)} \geq \gamma_1^{(N)} \tag{19}$$

since the maximum with $(N + 1)$ parameters is at least as large as the maximum with N (rewrite equation (3) with N replaced by $N+1$ and set $a_{N+1}^{(N+1)} = 0$).

These results constitute proofs of Theorem 7.4.1 and of part (1) of Theorem 7.4.2. Equation (19) is a special case of (11); however it is less simple to provide a compact proof of parts (2) and (3) of Theorem 7.4.2, which follow most simply from the Courant–Fischer minimax characterisation of the eigenvalues, and we refer the reader to Wilkinson (1965) Chapter 2, Sections 42–7 for the relevant proof.

7.4.2 Numerical examples

Numerical algorithms based on expansion methods are considered in more detail in the next chapter; however we pause here to illustrate the rather nice properties of the Ritz–Galerkin scheme by means of three simple examples based on the use of equations (5), for which the integrals involved can be evaluated exactly. The equations of Examples 7.4.1–3 were previously solved using the Nystrom method in Section 6.4; here we approximate the eigenfunctions by $x(s) \approx \sum_{i=1}^{N} a_i^{(N)} h_i(s)$ with $h_i(s) = s^{i-1}$.

Example 7.4.1.

$$\gamma x(s) = \int_0^1 (st - \tfrac{1}{6}s^3 t^3) x(t) \mathrm{d}t, \quad 0 \leq s, t \leq 1. \tag{20}$$

The Ritz–Galerkin method applied to the solution of equation (20) yields the following set of equations

$$\mathbf{L a}^{(N)} = \gamma^{(N)} \mathbf{M a}^{(N)} \tag{21}$$

where (see equation (5))

$$\left.\begin{array}{l} L_{ij} = \dfrac{1}{(i+1)(j+1)} - \dfrac{1}{6(i+3)(j+3)}, \\[3mm] M_{ij} = \dfrac{1}{i+j-1}, \end{array}\right\} \quad i,j = 1,\ldots,N.$$

Equation (21) has been solved using a generalised eigenvalue routine and we list in Table 7.4.1 the three computed eigenvalues of largest modulus obtained for a range of values of N. Note that the matrix $\mathbf{M}$ is in fact the Hilbert matrix of order N which is notoriously illconditioned. We would therefore expect to produce a meaningful solution for only small values of $N \leq N_0$ where N_0 is a threshold value in the sense that when N exceeds N_0 the effects of illconditioning seriously impair the accuracy, and indeed the credibility, of the computed solution. This illconditioning stems from the use of the monomials s^{i-1} as basis functions. The kernel in equation (20) is Hermitian and therefore has N real eigenvalues of which only two are non-

Table 7.4.1. *Three eigenvalues of largest modulus for the solution of equation (20) by an expansion method.*

N	i	$\gamma_i^{(N)}$
2	1	$0.313\,44$
	2	$-0.177\,24 \times 10^{-2}$
3	1	$0.313\,57$
	2	$-0.398\,64,\ -2$
	3	$0.244\,14,\ -9$
4	1	$0.313\,57$
	2	$-0.404\,96,\ -2$
	3	$-0.222\,36,\ -8$
5	1	$0.313\,57$
	2	$-0.404\,97,\ -2$
	3	$-0.169\,93,\ -8 \pm 0.133\,09,\ -8\,i$
6	1	$0.313\,57$
	2	$-0.404\,95,\ -2$
	3	$-0.115\,75,\ -5$
8	1	$0.313\,59$
	2	$-0.195\,95,\ -2 \pm 0.165\,47,\ -2\,i$
	3	$-0.817\,86,\ -3$

zero and given by

$$\gamma = 0.313\,57, \quad \gamma = -0.004\,049\,5,$$

to five significant figures. We see from Table 7.4.1 that $N_0 = 4$ and for values of $N \leq 4$, $\gamma_i^{(N)}$ converges to γ_i, $i = 1, 2$, with $\gamma_3^{(N)}$ effectively zero. For $N > 4$ however $\gamma_3^{(5)}$ and $\gamma_2^{(8)}$ are complex and $\gamma_3^{(N)}$ increases in modulus as N increases.

Thus we see how an illconditioned basis can lead to poor determination of the solution; we expand on this point further in the next chapter where we also consider the use of orthogonal basis functions.

Example 7.4.2.

$$\gamma x(s) = \int_0^1 e^{st} x(t) \mathrm{d}t, \quad 0 \leq s, t \leq 1. \tag{22}$$

The kernel is Hermitian and the matrix **M** is identical to that of the previous example. The elements of **L** may be calculated using the following set of recurrence formulae:

$$J_1 = e - 1,$$
$$J_i = e - (i - 1)J_{i-1}, \quad i = 2, \ldots, N,$$

Table 7.4.2. *The eigenvalues of*
largest, second largest and fourth
largest modulus of equation (22)
by an expansion method.

N	i	$\gamma_i^{(N)}$
2	1	1.3527
	2	1.0408×10^{-1}
3	1	1.3530
	2	1.0597, -1
4	1	1.3530
	2	1.0598, -1
	4	7.5032, -5
5	1	1.3530
	2	1.0598, -1
	4	7.7408, -5
6	1	1.3530
	2	1.0600, -1
	4	7.8257, -4
8	1	1.5763, $+1$
	2	1.3536
	4	-3.2421, -1

$$L_{11} = \int_0^1 \frac{e^s - 1}{s}\,ds,$$

$$L_{i1} = L_{1i} = J_{i-1} - \frac{1}{i-1}, \quad i = 2, \ldots, N,$$

$$L_{ij} = J_{i-1} - (j-1)L_{i-1,j-1}, \quad i, j = 2, \ldots, N.$$

We give in Table 7.4.2 those computed eigenvalues of largest, second largest and fourth largest modulus. Again we note two features of the calculation:

(i) Extremely rapid convergence of the eigenvalues as N is increased.
(ii) The dominance for 'large' N ($N > 5$) of round-off errors stemming from the illconditioned basis used.

Clearly, we should try to retain the former feature while removing the latter, and we refer forward again to Chapter 8.

Example 7.4.3.

$$\gamma x(s) = \int_{-1}^1 st^2 x(t)\,dt, \quad -1 \le s, t \le 1. \tag{23}$$

We saw in Section 6.4 that this kernel has rank 1 with no non-zero eigenvalues. The matrices **L** and **M** have elements given by

$$L_{ij} = \frac{1+(-1)^i}{i+1} \times \frac{1+(-1)^{j+1}}{j+2},$$

$$M_{ij} = \frac{1+(-1)^{i+j}}{i+j-1}, \quad i,j = 1, \ldots, N,$$

and on inspection **L** is clearly singular of rank 1. This example therefore yields eigenvalues which are exactly zero, apart from round-off noise, thus illustrating the result of Theorem 7.4.2, Comment (iii).

Exercises

1. The integral equation

$$x(s) = y(s) + \int_0^1 (1 - \cos \tfrac{1}{2} st) x(t) dt$$

may be solved by approximating the kernel $K(s,t) = (1 - \cos \tfrac{1}{2} st)$ by a rank 1 kernel $K_1(s,t)$, using a truncated Taylor series expansion for $\cos \tfrac{1}{2} st$:

$$K(s,t) = 1 - \left\{ 1 - \frac{1}{2!} \left(\frac{1}{2} st \right)^2 + \frac{1}{4!} \left(\frac{1}{2} st \right)^4 + \cdots \right\}$$

$$\approx \frac{1}{2!} \left(\frac{1}{2} st \right)^2 = K_1(s,t),$$

and by finding the approximate solution $x_1(s)$ of

$$x_1(s) = y(s) + \int_0^1 K_1(s,t) x_1(t) dt. \qquad (A)$$

Show that the error in the solution may be bounded by

$$\| x - x_1 \| < \tfrac{1}{3360} \| x_1 \|.$$

2. An expansion method for the solution of $x = y + Kx$ seeks an approximation x_N for x of the form

$$x_N(s) = \sum_{i=1}^{N} a_i^{(N)} h_i(s).$$

Derive the set of linear equations (7.3.14) which define the method of least squares approximation and use these to show that the two term approximation $(a_0 + a_1 s)$ provides the exact solution, $x_1(s) = \alpha s$, of equation (A) above, when $y(s) = \alpha(s - s^2/32)$, $K_1(s,t) = \tfrac{1}{8} s^2 t^2$.

3. Show that the exact solution of

$$x(s) = s + \int_0^1 K(s, t)x(t)dt, \quad 0 \le s \le 1,$$

where

$$K(s, t) = \begin{cases} s, & s \le t, \\ t, & s \ge t, \end{cases}$$

is $x(s) = \sec 1 \sin s$, and use Galerkin's method to find a two term approximation of the form $(a_0 + a_1 s)$ to the solution of the integral equation.

4. The method of *collocation* solves the integral equation

$$x(s) = y(s) + \int_a^b K(s, t)x(t)dt$$

using the approximation

$$x(s) \approx x_N(s) = \sum_{i=1}^N a_i^{(N)} h_i(s)$$

and the prescription

$$r_N(s_i) = x_N(s_i) - \int_a^b K(s_i, t)x_N(t)dt - y(s_i) = 0$$

for $\{s_i\} \in [a, b]$, $i = 1, \ldots, N$. Use this method to find an approximation of the from $(a_0 + a_1 s)$ to $x(s)$ when $x(s)$ is given by

$$x(s) = y(s) + \int_0^1 e^{st}x(t)dt,$$

$$y(s) = e^s - \frac{(e^{s+1} - 1)}{s + 1}.$$

Use the end points of the interval as collocation points and compare your answer with the exact solution $x(s) = e^s$.

8

Numerical techniques for expansion methods

8.1 Introduction

We now consider in some detail the numerical implementation of the expansion methods introduced in the last chapter. There, we gave no consideration to the evaluation of the integrals involved or to the choice of expansion sets $\{h_i\}$, or to the solution of the defining equations. However, these questions are crucial in determining both the accuracy and the efficiency (speed) of the methods and we address them here. The equally important question of the provision of error estimates is pursued in Chapter 10.

We assume throughout that the equation to be solved is

$$(Lx)(s) = (I - K)x(s) = y(s), \quad a \le s \le b, \tag{1}$$

and we consider methods based on an approximation x_N for x of the form

$$x_N = \sum_{i=1}^{N} a_i^{(N)} h_i. \tag{2}$$

For any given choice of the $\{a_i^{(N)}\} = \mathbf{a}^{(N)}$, this defines a residual r_N:

$$r_N = y - Lx_N. \tag{3}$$

Following Chapter 7 we define the set $\{l_i(s)\}$:

$$l_i = (I - K)h_i = h_i(s) - \int_a^b K(s, t)h_i(t)\mathrm{d}t. \tag{4a}$$

We frequently need to evaluate $l_i(s)$ numerically; we introduce a quadrature rule $Q_p : \{w_k, \xi_k, k = 1, 2, \ldots, p\}$ and approximate

$$l_i(s) \approx \bar{l}_i(s) = h_i(s) - \sum_{k=1}^{p} w_k K(s, \xi_k)h_i(\xi_k). \tag{4b}$$

The quadrature errors involved in (4b) may contribute significantly to the total error in x_N and if the expansion (2) converges rapidly (which we hope it does) they may even dominate. We discuss their estimation in Chapter 9 and ignore them in this chapter.

We shall not require $\bar{l}_i(s)$ for all s, but only on a discrete point set $\{\eta_k, k = 1, \ldots, q\}$ which may or may not coincide with the set $\{\xi_k\}$.

We introduce the $q \times N$ matrix $\mathbf{A}$ and q-vectors $\mathbf{r}, \mathbf{y}$, with elements

$$A_{ij} = \bar{l}_j(\eta_i); \quad i = 1, \ldots, q; \quad j = 1, \ldots, N; \tag{5}$$

$$r_i = r_N(\eta_i); \quad y_i = y(\eta_i);$$

$$r_i = y_i - \sum_{j=1}^{N} A_{ij} a_j^{(N)}; \tag{6}$$

that is,

$$\mathbf{r} = \mathbf{y} - \mathbf{A}\mathbf{a}^{(N)}.$$

8.2 L_∞ residual minimisation

This method was outlined in Section 7.3.3 and chooses $\mathbf{a}^{(N)}$ to minimise $\| r_N \|_\infty$; that is, we require a solution of the problem

$$\min_{\mathbf{a}^{(N)}} \max_{a \le s \le b} |r_N(s)|. \tag{1}$$

This minimisation problem is *not* simple to solve numerically but because the maximum need not be located very accurately it is possible to replace it by the much simpler problem (7.3.10):

$$\min_{\mathbf{a}^{(N)}} \max_{i=1}^{q} |r_N(\eta_i)| = \min_{\mathbf{a}^{(N)}} \max_{i=1}^{q} |r_i| \tag{2}$$

where $\{\eta_i\}$ is the point set introduced above and r_i is given by (8.1.6); for simplicity we shall omit the superfix on the coefficients $a_i^{(N)}$. Problem (2) is formally that of providing a linear discrete minimax approximation to the function $r_N(s)$ using the basis $\{\bar{l}_i(s)\}$; viewed in this way, its integral equation origin is completely masked, the problem reducing to a standard one in approximation theory. The classical algorithm for finding such approximations is the Remes exchange algorithm (see, for example, Rice, 1969, p. 176). However, this algorithm can be guaranteed to converge only if the set $\{\bar{l}_i\}$ satisfies the Haar condition (see Definition 2.2.1) on the interval $[a, b]$. The monomials $\{s^{i-1}\}$ satisfy this condition and hence so does any orthogonal polynomial basis $\{h_i\}$; however, the relation between the sets $\{h_i\}, \{\bar{l}_i\}$ depends upon the kernel $K(s, t)$ in (8.1.1) so that the Haar condition may not be satisfied. Even if the Haar condition *is* satisfied by the set $\{l_i\}$, an illconditioned set of basis functions can cause trouble. A scheme which does not require satisfaction of the Haar condition is obtained by noting that (2) may be phrased as a problem in linear programming; for the condition

$$|r_i| \leq M, \quad i = 1,\ldots,q,$$

is equivalent to the two linear inequalities

$$r_i - M \leq 0, \quad i = 1,\ldots,q, \tag{3a}$$
$$-r_i - M \leq 0, \quad i = 1,\ldots,q,$$

or equivalently

$$-\sum_{j=1}^{N} A_{ij} a_j - M \leq -y_i, \quad i = 1,\ldots,q, \tag{3b}$$

$$\sum_{j=1}^{N} A_{ij} a_j - M \leq y_i, \quad i = 1,\ldots,q. \tag{3c}$$

We can therefore phrase (2) as:

minimise M subject to the constraints (3b), (3c) (4)

which is a standard linear programming (L.P.) problem with $(N+1)$ variables $(a_i, i = 1,\ldots,N,$ and $M)$ and $2q$ constraints. As such, it can be solved using the standard simplex method of linear programming. However, in practice, tailored L.P. algorithms can and have been written to take advantage of the special features of this problem (see Barrodale and Phillips, 1974). Because the number of constraints is usually much greater than the number of unknowns, it is better to solve the dual problem to (4) and it is possible to save on storage space by noting that, for each i, only one of the pairs of constraints (3b), (3c) will be active for any choice of the vector **a** (unless $r_i \equiv 0$, when the two are equivalent).

Example 8.2.1. We illustrate this L_∞ algorithm by setting up the equations for the problem of Example 7.3.2:

$$x(s) = \sin s - \tfrac{1}{4}s + \tfrac{1}{4}\int_0^{\pi/2} stx(t)\mathrm{d}t \tag{5}$$

with the simple approximating function

$$x(s) \approx x_\infty(s) = a_{\infty 1}s.$$

We choose to minimise the residual norm on the (rather spartan) point set:

$$\eta_1 = 0, \quad \eta_2 = \frac{\pi}{4}, \quad \eta_3 = \frac{\pi}{2}. \tag{6}$$

For illustrative purposes we use the exact form of the functions $l_i(s)$ given in Example 7.3.2:

$$l_1(s) = \left(1 - \frac{\pi^3}{96}\right)s = 0.677\,018s. \tag{7}$$

Then equations (8.1.6) take the form

$$r_1 = r(0) = 0 - l_1(0) \, a_{\infty 1} = 0,$$

$$r_2 = r\left(\frac{\pi}{4}\right) = \frac{1}{\sqrt{2}} - \frac{\pi}{16} - l_1\left(\frac{\pi}{4}\right) a_{\infty 1} = 0.5108 - 0.5317 a_{\infty 1},$$

$$r_3 = r\left(\frac{\pi}{2}\right) = 1 - \frac{\pi}{8} - l_1\left(\frac{\pi}{2}\right) a_{\infty 1} = 0.6073 - 1.0635 a_{\infty 1}. \tag{8}$$

There is no value of $a_{\infty 1}$ for which all three residuals are zero, although due to the choice of trial functions, r_1 is zero for all $a_{\infty 1}$. The minimisation problem (2) defining the 'best' value for the coefficient $a_{\infty 1}$ is for this problem:

$$\underset{a_{\infty 1}}{\text{minimise}} \max \left(|r_1|, |r_2|, |r_3| \right), \tag{9}$$

while, introducing an additional parameter M, the linear programming formulation of (9) is given by

minimise M subject to the constraints

From r_1:

$$0 - M \leq 0,$$

$$0 - M \leq 0; \tag{10a}$$

Figure 8.2.1. Graphical solution of the optimisation problem (9). Residual r_1 is identically zero; $|r_2|$ and $|r_3|$ are shown on the graph. The optimal choice of $a_{\infty 1}$ is marked.

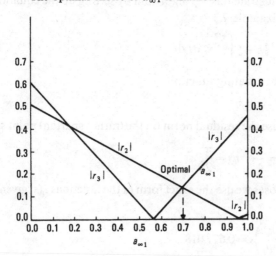

From r_2:

$$-0.5317a_{\infty 1} - M \leq -0.5108,$$
$$0.5317a_{\infty 1} - M \leq 0.5108; \tag{10b}$$

From r_3:

$$-1.0635a_{\infty 1} - M \leq -0.6073,$$
$$1.0635a_{\infty 1} - M \leq 0.6073. \tag{10c}$$

These two formulations are completely equivalent. Clearly, equations (10a) contribute nothing in this example and we can rather easily solve (9) graphically. Figure 8.2.1 shows $|r_2|$ and $|r_3|$ as functions of $a_{\infty 1}$; it is clear that the solution to (9) is given by $a_{\infty 1} \simeq 0.7$ and that at this point $|r_3| = |r_2|$, with r_3 negative and r_2 positive. Setting $r_3 = -r_2$ we find to four figures

$$a_{\infty 1} = 0.7009.$$

We repeat that, although it is not obvious from this simple example, the L.P. equations (3) usually form the most convenient starting point for a numerical solution.

8.3　Least squares residual minimisation
8.3.1　Setting up the defining equations

The defining equations for the least squares approximation are given by (7.3.14) and have the form (we omit the superfix N and assume real functions)

$$\mathbf{L}_{\mathrm{LS}}\mathbf{a} = \mathbf{y}_{\mathrm{LS}}, \tag{1}$$

$$(\mathbf{L}_{\mathrm{LS}})_{ij} = \int_a^b l_i(s)l_j(s)\,\mathrm{d}s, \quad i,j = 1,\ldots,N,$$

$$(\mathbf{y}_{\mathrm{LS}})_i = \int_a^b l_i(s)y(s)\,\mathrm{d}s. \tag{2}$$

We consider first the cost of setting up these equations. In general the integrals must be evaluated numerically. It is very uneconomic to compute each element $(\mathbf{L}_{\mathrm{LS}})_{ij}$ with a different quadrature rule although at first sight it is tempting to do this since the integrals get 'harder' as i,j increase (consider $h_i(s) = s^{i-1}$, for example). We can estimate the cost of such a procedure as follows. For each i,j:

(a) Introduce a new quadrature rule $Q_q(w_k, \xi_k, k = 1,\ldots,q)$
(b) Evaluate $l_i(\xi_k)$, $l_j(\xi_k)$, $k = 1,\ldots,q$. Cost: $2pq$ operations (see (8.1.4b)).
(c) Evaluate $(\mathbf{L}_{\mathrm{LS}})_{ij}$ from (1). Cost: q operations.

Total cost for the $N \times N$ matrix $\mathbf{L}_{\mathrm{LS}}$: approximately $2N^2pq$ operations where an operation is a multiplication plus an addition. To achieve satisfactory accuracy it is necessary to set $p, q \geq N$; hence the total cost is at least $\mathcal{O}(N^4)$ for this procedure. This can be reduced substantially by using the same quadrature rule for each element $(\mathbf{L}_{LS})_{ij}$. Then step (b) above need only be carried out once and the major cost involved is that of evaluating step (c) N^2 times. This gives a total cost of $(2pq + N^2q) = \mathcal{O}(N^3)$. We may or may not choose the rules Q_p, Q_q to be the same.

The equations (1) are the *normal equations* for the least squares method; provided a single quadrature rule $Q_p(\mathbf{w}_p, \boldsymbol{\xi}_p)$ is used to approximate each l_i in (2) and a single rule $Q_q(\mathbf{w}_q, \boldsymbol{\xi}_q)$ with positive weights $\mathbf{w}_q$ is used to evaluate $(\mathbf{L}_{LS})_{ij}$, for each $i, j = 1, \ldots, N$, the matrix $\mathbf{L}_{LS}$ is symmetric positive definite.

For then

$$(\mathbf{L}_{LS})_{ij} = \sum_{k=1}^{q} l_i(\xi_{qk}) w_{qk} l_j(\xi_{qk}). \tag{3}$$

That is,

$$\mathbf{L}_{LS} = \mathbf{\Lambda}^T \mathbf{W} \mathbf{\Lambda} \tag{4}$$

where $\mathbf{\Lambda}$ is $q \times N$ with elements

$$\Lambda_{ij} = l_j(\xi_{qi}); \quad i = 1, \ldots, q; \quad j = 1, \ldots, N; \tag{5}$$

and $\mathbf{W}$ is diagonal with diagonal elements

$$W_{ii} = w_{qi}, \quad i = 1, \ldots, q. \tag{6}$$

The vector $\mathbf{y}_{LS}$ may also be written in terms of the matrix $\mathbf{\Lambda}$. Introducing the rule Q_q, (2) becomes

$$(\mathbf{y}_{LS})_i = \sum_{k=1}^{q} w_{qk} y(\xi_{qk}) l_i(\xi_{qk}).$$

Hence, writing $y_i = y(\xi_{qi})$, we have

$$\mathbf{y}_{LS} = \mathbf{\Lambda}^T \mathbf{W} \mathbf{y}. \tag{7}$$

In some applications of the least squares method the points and weights $\boldsymbol{\xi}_q$, $\mathbf{w}_q$ used in (3) represent an equally weighted set of pre-chosen points, but are *not* to be interpreted as a quadrature rule. The algorithm which results is then to be interpreted as a direct least squares procedure minimising the residual norm over the given point set. Such an algorithm is in practice less efficient than one based on an approximate representation of the continuous least squares problem. This is demonstrated in the following example.

Table 8.3.1. *Least squares solution of the integral equation (8.2.5) for various point sets*

Choice of Points	q	N 2 ε_{LS}	ε_{max}	4 ε_{LS}	ε_{max}	6 ε_{LS}	ε_{max}
Points equally	5	4.0×10^{-3}	1.4, −2	2.7, −6	1.0, −5		
spaced in $[0, \pi/2]$	10	2.8, −3	8.5, −3	6.2, −7	2.3, −6	7.2, −11	2.5, −10
$w_{qk} = 1, k = 1,\ldots,q$	20	2.5, −3	6.1, −3	2.9, −7	5.7, −7	2.7, −11	5.1, −11
	50	2.5, −3	4.8, −3	2.9, −7	5.2, −7	2.3, −11	4.3, −11
Gauss–Legendre	4	2.5, −3	3.9, −3	9.6, −7	2.3, −6		
points and weights,	8	2.5, −3	3.9, −3	3.3, −7,	5.3, −7	2.3, −11	5.2, −11
q-point rule	12	2.5, −3	3.9, −3	3.3, −7	5.3, −7	2.3, −11	4.7, −11

Example 8.3.1.
We compute yet again the solution of the integral equation of Example 7.3.2 (see equation (8.2.5)) with two choices of point set ξ_{qk} and weights w_{qk}: (a) q equally spaced points on $[0, \pi/2]$, with $w_{qk} = 1$, representing a direct discrete L_2 method; (b) points and weights from a q-point Gauss–Legendre quadrature rule, so that we approximate the results of the continuous L_2 calculation.

The results are shown in Table 8.3.1 where we compare, for the two families of point sets, two standard measures of the achieved error $e_N(s) = x(s) - x_N(s)$:

(i) $\varepsilon_{max} = \max_{s \in P} |e_N(s)|$

(ii) $\varepsilon_{LS} = \left[\frac{1}{10} \sum_{s \in P} |e_N(s)|^2 \right]^{\frac{1}{2}}$

where P is a point set of ten equally spaced points on the given interval $[0, \pi/2]$.

We see that, as N increases for given q, both the least squares error ε_{LS} and the maximum error ε_{max} decrease rapidly. For a given N the errors also decrease with increasing q; this decrease is very slow for the discrete method, but convergence with increasing q is attained extremely rapidly for the approximate L_2 method, so that the point set needed for given N is much smaller than for the discrete approach.

8.3.2 Solving the defining equations

Since L_{LS} is symmetric positive definite, equations (1) may be solved by Gauss elimination without interchanges, in about $\frac{1}{6}N^3$ operations. This is a very satisfactory procedure if the set $\{h_i\}$ is chosen to be orthogonal, since then the condition number $C = \mathrm{Cond}\,(\Lambda)$ of Λ (and hence of $\mathbf{L}$) is usually quite small. However, if the set $\{h_i\}$ and hence $\{l_i\}$ is almost linearly dependent, Λ will be very illconditioned and since $\mathrm{Cond}\,(\mathbf{L}_{\mathrm{LS}}) \sim C^2$, the system (1) may prove quite intractable. Under these circumstances it is usually recommended that $\mathbf{L}_{\mathrm{LS}}$ be not found at all. Instead, we form the weighted overdetermined $q \times N$ system (see (4), (7))

$$\mathbf{W}^{\frac{1}{2}}\Lambda\mathbf{a} = \mathbf{W}^{\frac{1}{2}}\mathbf{y} \tag{8}$$

and directly seek that vector **a** which minimises the vector 2-norm

$$\min_{\mathbf{a}} \| \mathbf{W}^{\frac{1}{2}}\Lambda\mathbf{a} - \mathbf{W}^{\frac{1}{2}}\mathbf{y} \|_2. \tag{9}$$

It is immediately clear that the solution of this problem satisfies the normal equations (1), with $\mathbf{L}_{\mathrm{LS}}, \mathbf{y}_{\mathrm{LS}}$ given by (4), (7); so the problems are equivalent.

Several methods are available for the direct solution of (9). That based on the modified Gram–Schmidt process produces first an orthogonal matrix **R** such that

$$\mathbf{R}^T\mathbf{A} = \begin{pmatrix} \mathbf{U} \\ \mathbf{0} \end{pmatrix} \tag{10}$$

where $\mathbf{A} = \mathbf{W}^{\frac{1}{2}}\Lambda$ and **U** is upper triangular. Such a matrix **R** always exists. Now using the fact that **R** is orthogonal, that is $\mathbf{R}\mathbf{R}^T = \mathbf{R}^T\mathbf{R} = \mathbf{I}$, we may rewrite the normal equations in the form

$$(\mathbf{A}^T\mathbf{R}\mathbf{R}^T\mathbf{A})\mathbf{a} = \mathbf{A}^T\mathbf{R}\mathbf{R}^T\mathbf{y}. \tag{11}$$

That is,

$$(\mathbf{U}^T|\mathbf{0})\begin{pmatrix} \mathbf{U}\mathbf{a} \\ \mathbf{0} \end{pmatrix} = (\mathbf{U}^T|\mathbf{0})\begin{pmatrix} \mathbf{y}' \\ \mathbf{y}'' \end{pmatrix}.$$

Hence

$$\mathbf{U}^T\mathbf{U}\mathbf{a} = \mathbf{U}^T\mathbf{y}'. \tag{12}$$

Now if rank $(\mathbf{A}) = N$ the $N \times N$ matrix $\mathbf{U}^T$ is nonsingular and (12) is equivalent to the triangular system

$$\mathbf{U}\mathbf{a} = \mathbf{y}'. \tag{13}$$

If, on the other hand, rank $(\mathbf{A}) < N$, (12) has no unique solution but any solution of (13) is also a solution of (12) and hence of the least squares problem.

Although orthogonalisation techniques are more stable than the direct solution of the normal equations, they are also slightly more expensive. Given the matrix **A** the costs are:

Normal equations: Form $\mathbf{A}^T\mathbf{A}$: $\frac{1}{2}N^2q$. Solve: $\frac{1}{6}N^3$. Total: $\frac{1}{2}N^2(q + \frac{1}{3}N)$ operations. *Modified Gram–Schmidt*: $\frac{1}{2}N^2(q + N)$.

8.3.3 Rank deficient systems

In both methods it is necessary to make allowances for the possibility that rank $(\mathbf{A}) < N$, either because the set $\{h_i\}$ is in fact linearly dependent, or because round-off errors have lost all significant digits in the solution. It is possible for either method to return a satisfactory solution of the least squares problem in such cases; since such a solution exists but is not unique, any solution of the numerically singular equations can be returned to give the same residual error. Usually the most satisfactory procedure is to return a solution with one or more of the components of **a** set to zero. To see how we can achieve this, let us consider the solution of the normal equations (1) by Gauss elimination. Recall that this process (we describe it for simplicity without row or column interchanges) first subtracts multiples of the first equation from succeeding equations to eliminate the coefficient of variable 1; that is, of a_1; then multiples of the (modified) second equation are subtracted from succeeding equations to eliminate the coefficient of a_2, and so on. After k stages of elimination we obtain a set of equations with the same solution as the original, but with the form

$$\mathbf{L}^{(k)}\mathbf{a} = \left(\begin{array}{c|c} \overbrace{\mathbf{0}}^{\mathbf{U}^{(k)}} & \mathbf{L}_{12}^{(k)} \\ \hline \mathbf{0} & \mathbf{L}_{22}^{(k)} \end{array}\right)\left(\begin{array}{c} \mathbf{a}_1^{(k)} \\ \mathbf{a}_2^{(k)} \end{array}\right) = \left(\begin{array}{c} \mathbf{y}_1^{(k)} \\ \mathbf{y}_2^{(k)} \end{array}\right). \tag{14}$$

After $k = (N-1)$ steps, the matrix $\mathbf{L}^{(N-1)}$ is upper triangular and we can find the solution of this triangular system by back substitution. But now suppose we have stopped after k stages and set arbitrarily $\mathbf{a}_2^{(k)} = \mathbf{0}$; that is,

$$a_{k+1} = a_{k+2} = \cdots = a_N = 0.$$

Then the form of (14) shows that values of $a_1, \ldots, a_k$ satisfying the first k equations exactly could be found by back substitution in the equations

$$\mathbf{U}^{(k)}\mathbf{a}_1^{(k)} = \mathbf{y}_1^{(k)}. \tag{15}$$

Now with

$$\mathbf{a}^* = \left(\begin{array}{c} \mathbf{a}_1^{(k)} \\ \hline \mathbf{0} \end{array}\right)$$

let us see how well we solve the original equations.

Defining the residual $r_i = y_i^{(k)} - \sum_{j=1}^{N} L_{ij}a_j^*$, we find

$$r_i = 0, \quad i = 1, 2, \ldots, k, \tag{16}$$
$$= y_i^{(k)}, \quad i = k+1, \ldots, N.$$

Finally, suppose that we stop at the kth stage because it is apparent at this stage (because we have come across an effectively zero pivot) that the equations are numerically singular. Then, since the least squares equations are always consistent even if they are not of full rank, the vector $y_2^{(k)}$ must be effectively zero apart from noise, and hence so must $\mathbf{a}_2^{(k)}$; the solution $\mathbf{a}^*$ will represent one (of the infinite number) of the solutions to the least squares problem, while reporting to the user that he might as well not have used the additional expansion functions $h_{k+1}, \ldots, h_N$.

Example 8.3.2.
We illustrate these remarks with a simple example. Returning yet again to Example 7.3.2 (see equation (8.2.5)), let us suppose that we seek an approximate least squares solution

$$x(s) \approx x_c(s) = a_{c1}s + a_{c2}s^3 + a_{c3}(s - s^3) \tag{17}$$

where c stands for crazy and it is clear from the start that the problem will be rank deficient, the 3×3 normal equations having rank 2. The normal equations can be constructed directly from the 2×2 matrix in Example 7.3.2. Rounding all entries in this matrix to one decimal place and then scaling by a factor ten for illustrative purposes, we obtain the 3×3 system of equations

$$\begin{pmatrix} 6 & 9 & -3 \\ 9 & 18 & -9 \\ -3 & -9 & 6 \end{pmatrix} \begin{pmatrix} a_{c1} \\ a_{c2} \\ a_{c3} \end{pmatrix} = \begin{pmatrix} 5 \\ 6 \\ -1 \end{pmatrix}. \tag{18}$$

The solution of the 2×2 system obtained by omitting the third row and column is:

$$a_{c1} = \tfrac{4}{3} = 1.3333; \quad a_{c2} = -\tfrac{1}{3} = -0.3333.$$

If we try to solve system (18) by Gauss elimination without interchanges, we find after the first two stages of elimination the systems:
After stage 1

$$\begin{pmatrix} 6 & 9 & -3 \\ 0 & 4.5 & -4.5 \\ 0 & -4.5 & 4.5 \end{pmatrix} \begin{pmatrix} a_{c1} \\ a_{c2} \\ a_{c3} \end{pmatrix} = \begin{pmatrix} 5 \\ -1.5 \\ 1.5 \end{pmatrix}$$

After stage 2

$$\begin{pmatrix} 6 & 9 & -3 \\ 0 & 4.5 & -4.5 \\ 0 & 0 & 0 \end{pmatrix} \begin{pmatrix} a_{c1} \\ a_{c2} \\ a_{c3} \end{pmatrix} = \begin{pmatrix} 5 \\ -1.5 \\ 0 \end{pmatrix}$$

At this stage we recognise a zero pivot: the system clearly has rank 2. Following the procedure given above, we set $a_{c3} = 0$ and back substitute in the second and then the first equation:

$$4.5a_{c2} + 0 = -1.5: \quad a_{c2} = -\frac{1.5}{4.5} = -\tfrac{1}{3};$$

$$6a_{c1} - \tfrac{9}{3} + 0 = 5: \quad a_{c1} = \tfrac{8}{6} = \tfrac{4}{3}.$$

We have recovered an exact solution of the rank deficient equations: that which corresponds to omitting the offending third expansion function.

8.4 The Galerkin method

The defining equations for this method are given by (7.3.18, 19) and we must now produce the coefficient matrix $\mathbf{L_G}$ and vector $\mathbf{y_G}$

$$\mathbf{L_G} : (\mathbf{L_G})_{ij} = (h_i, l_j)$$
$$\mathbf{y_G} : (\mathbf{y_G})_i = (h_i, y). \tag{1}$$

As with the least squares method, we assume that a quadrature rule $Q_p(\mathbf{w}_p, \boldsymbol{\xi}_p)$ is used to approximate the l_i and a rule $Q_q(\mathbf{w}_q, \boldsymbol{\xi}_q)$ is used to approximate the inner products in (1). Again it is not economic to use a different rule for different elements of $\mathbf{L_G}, \mathbf{y_G}$ and we therefore assume that a single pair of rules is used. Then (1) is approximated by

$$(\mathbf{L_G})_{ij} = \sum_{k=1}^{q} w_{qk} h_i(\xi_{qk}) l_j(\xi_{qk}),$$

$$(\mathbf{y_G})_i = \sum_{k=1}^{q} w_{qk} h_i(\xi_{qk}) y(\xi_{qk}). \tag{2a}$$

That is, using the notation of Section 8.3.2,

$$\mathbf{L_G} = \mathbf{H}^T \mathbf{W} \mathbf{\Lambda},$$
$$\mathbf{y_G} = \mathbf{H}^T \mathbf{W} \mathbf{y}, \tag{2b}$$

where $\mathbf{H}$ is a $q \times N$ matrix with elements

$$H_{ij} = h_j(\xi_{qi}). \tag{3}$$

The cost of forming $\mathbf{L_G}$ given $\mathbf{H}, \mathbf{\Lambda}$ is therefore about N^3 operations. Note that, in general and unlike the least squares matrix $\mathbf{L_{LS}}$, $\mathbf{L_G}$ is not symmetric. Thus, solution of the system

$$\mathbf{L_G} \mathbf{a} = \mathbf{y_G} \tag{4}$$

using Gauss elimination will take about $\tfrac{1}{3}N^3$ operations and should use at least row interchanges (partial pivoting). However, if the kernel K is symmetric and we choose the quadrature rule $Q_q \equiv Q_p$, then $\mathbf{L_G}$ is symmetric. For Λ has elements

$$\Lambda_{ij} = h_j(\xi_{qi}) - \int_a^b K(\xi_{qi}, t)h_j(t)dt \quad \text{(exact)},$$

$$= h_j(\xi_{qi}) - \sum_{k=1}^p w_{pk}K(\xi_{qi}, \xi_{pk})h_j(\xi_{pk}) \quad \text{(numerical)}.$$

If we now set $p = q$, $\mathbf{w}_p = \mathbf{w}_q$, $\boldsymbol{\xi}_p = \boldsymbol{\xi}_q$, we see that

$$\Lambda = \mathbf{H} - \mathbf{KWH}, \tag{5}$$

$$\mathbf{L_G} = \mathbf{H}^T(\mathbf{W} - \mathbf{WKW})\mathbf{H}, \tag{6}$$

where

$$K_{ij} = K(\xi_{qi}, \xi_{qj}). \tag{7}$$

If $K(s, t)$ is symmetric so is the matrix $\mathbf{K}$ and hence $\mathbf{L_G}$ and we can achieve a solution time of about $\frac{1}{6}N^3$ operations.

If the basis $\{h_i\}$ is illconditioned then so is $\mathbf{L_G}$ but less so than the least squares system; the solution strategy described for the normal equations in Section 8.3.2, including the treatment of rank deficiency in Section 8.3.3, is then again applicable. If the $\{h_i\}$ are well conditioned and in particular if they are chosen to be orthogonal, then a convergent and rapid iterative scheme can be given for the system (4) (the same scheme applies also to the normal equations (8.3.1)); see Section 9.8 below.

8.5   The Fast Galerkin algorithm

The $\mathcal{O}(N^3)$ operation cost of setting up the normal equations (8.3.1) and the Galerkin equations (8.4.1), compares unfavourably with the $\mathcal{O}(N^2)$ cost of setting up the defining equations for the Nystrom method. If Gauss elimination is used to solve the equations in each case, the overall cost of an $N \times N$ Nystrom solution will be rather less than for an N-term expansion method solution; if the iterative technique described in Section 4.6.2 is used to solve the Nystrom equations, then for large N the Nystrom method will be much faster.

We now discuss a particular version of the Galerkin method, based on an expansion in Chebyshev polynomials, for which the operation counts are reduced to:

Equation set up: $\mathcal{O}(N^2 \ln N)$

Solution: $\mathcal{O}(N^2)$

the first being achieved by use of Fast Fourier Transform techniques, and the second by the introduction of an iterative solution technique.

There have been a number of methods proposed for Fredholm equations based on expansions in Chebyshev polynomials. The method of El-gendi

(1969), which is essentially a modification of the Nystrom scheme, was described in Chapter 4; that described here is given in Delves (1977); see also Delves, Abd-Elal and Hendry (1979).

We suppose without loss of generality that the interval of integration has been mapped onto $[-1, 1]$ so that the equation to be solved has the form

$$x(s) = y(s) + \int_{-1}^{1} K(s,t)x(t)dt \tag{1}$$

and we choose as basis the Chebyshev polynomials $T_i(s)$, $i = 0, 1, \ldots$. To retain the 'natural' suffix, we count from zero and make the following expansion for $x(s)$:

$$x(s) = \sum_{i=0}^{N} a_i T_i(s). \tag{2}$$

The Chebyshev polynomials are orthogonal on $[-1, 1]$ with weight function $(1 - s^2)^{-\frac{1}{2}}$:

$$\bar{D}_{ij} \equiv \int_{-1}^{1} \frac{T_i(s)T_j(s)ds}{(1 - s^2)^{\frac{1}{2}}} = \begin{cases} \pi, & i = j = 0, \\ \dfrac{\pi}{2}, & i = j > 0, \\ 0, & i \neq j. \end{cases} \tag{3}$$

We therefore introduce this weight function into the inner product in the defining equations (8.4.1) to obtain

$$(\bar{\mathbf{D}} - \bar{\mathbf{B}})\mathbf{a} = \bar{\mathbf{y}} \tag{4}$$

where $\bar{\mathbf{D}}$ is the diagonal matrix (3) and

$$\bar{B}_{ij} = \int_{-1}^{1} \frac{T_i(s)}{(1 - s^2)^{\frac{1}{2}}} \int_{-1}^{1} K(s,t)T_j(t)dt\,ds, \quad i,j = 0, 1, \ldots, N, \tag{5}$$

$$\bar{y}_i = \int_{-1}^{1} \frac{T_i(s)y(s)}{(1 - s^2)^{\frac{1}{2}}}ds, \quad i = 0, 1, \ldots, N. \tag{6}$$

We now need to perform these integrals numerically. A suitable quadrature rule for the integration over s is clearly the Gauss–Chebyshev $(p + 1)$-point rule with weights $w_k = \pi/p$, $k \neq 0$, p; $w_0 = w_p = \pi/2p$ and points $\xi_k = \cos(\pi k/p)$, $k = 0, 1, \ldots, p$; see Section 2.4.3. This yields

$$\bar{y}_i \sim y_i = \frac{\pi}{p}\sum_{k=0}^{p}{}'' y\left(\cos\left(\frac{\pi k}{p}\right)\right)T_i\left(\cos\left(\frac{\pi k}{p}\right)\right)$$

$$= \frac{\pi}{p}\sum_{k=0}^{p}{}'' y\left(\cos\left(\frac{\pi k}{p}\right)\right)\cos\left(\frac{\pi i k}{p}\right) \tag{7}$$

where the symbol $\sum''$ implies that the first and last terms are halved and we have used the relation

$$T_k(s) = \cos(k \cos^{-1} s). \tag{8}$$

An attempt to use this rule for the integral over t would lead to large numerical errors because of the 'missing' weight function $(1 - t^2)^{-\frac{1}{2}}$. However, suppose that we evaluate numerically not $\bar{B}_{ij}$, but $\bar{K}_{ij}$, where

$$\bar{K}_{ij} = \int_{-1}^{1} \frac{T_i(s)}{(1 - s^2)^{\frac{1}{2}}} \int_{-1}^{1} K(s,t) \frac{T_j(t)}{(1 - t^2)^{\frac{1}{2}}} \, dt \, ds. \tag{9}$$

Then $\bar{K}$ can be efficiently approximated by $\mathbf{K}$:

$$K_{ij} = \frac{\pi^2}{p^2} \sum_{r,s=0}^{p}{}'' K\left(\cos\left(\frac{r\pi}{p}\right), \cos\left(\frac{s\pi}{p}\right)\right) \cos\left(\frac{ri\pi}{p}\right) \cos\left(\frac{sj\pi}{p}\right),$$

$$i, j = 0, 1, \ldots, N. \tag{10}$$

Now (9), (6) identify $\bar{K}_{ij}$, $\bar{y}_i$ as expansion coefficients in the expansions

$$\frac{\pi^2}{4} K(s,t) = \sum_{i=0}^{\infty}{}' \sum_{j=0}^{\infty}{}' \bar{K}_{ij} T_i(s) T_j(t), \tag{11}$$

$$\frac{\pi}{2}(y(s)) = \sum_{i=0}^{\infty}{}' \bar{y}_i T_i(s), \tag{12}$$

while (5) similarly identifies $\bar{B}_{ij}$ as the (i,j)th coefficient in an expansion of the function $(\pi^2/4) K(s,t)(1 - t^2)^{\frac{1}{2}}$. We can therefore obtain the coefficients $\bar{B}_{ij}$ from the $\bar{K}_{ij}$ and the Chebyshev expansion of $(1 - t^2)^{\frac{1}{2}}$:

$$(1 - t^2)^{\frac{1}{2}} = \frac{2}{\pi} - \frac{4}{\pi} \sum_{r=1}^{\infty} \frac{T_{2r}(t)}{4r^2 - 1} \tag{13}$$

by multiplying the two series together. Using the identity

$$T_p(s) T_q(s) = \frac{1}{2}[T_{p+q}(s) + T_{|p-q|}(s)] \tag{14}$$

we find at once

$$\frac{\pi}{2} \bar{B}_{ij} = \bar{K}_{ij} - \sum_{l=1}^{\infty} \frac{1}{(4l^2 - 1)} (\bar{K}_{i,j+2l} + \bar{K}_{i,|j-2l|}). \tag{15}$$

We can therefore compute an accurate approximation $\mathbf{B}$ to $\bar{\mathbf{B}}$ by replacing in (15) $\bar{K}_{pq}$ by K_{pq}, $0 \le p, q \le N$, and by 0, $p, q > N$.

We now estimate the cost of this two-stage process:

(i) *Calculation of* $\mathbf{K}$. Direct evaluation of $\mathbf{K}$ from (10) would seem at first sight to take $\mathcal{O}(N^2 p^2)$ operations but using the reduction

$$S_{rj} = \frac{\pi}{p} \sum_{s=0}^{p}{}'' K\left(\cos\left(\frac{r\pi}{p}\right), \cos\left(\frac{s\pi}{p}\right)\right) \cos\left(\frac{sj\pi}{p}\right),$$

$$K_{ij} = \frac{\pi}{p} \sum_{r=0}^{p}{}'' S_{rj} \cos\left(\frac{ri\pi}{p}\right),$$

can be achieved in about $Np(p+N)$ operations. However, if we set $p = N$ we can identify (10) as representing a discrete Fourier (cosine) transform of the function $K(\cos\theta, \cos\phi)$. This transform can be carried out using the Fast Fourier Transform (FFT) technique, which produces $\mathbf{K}$ in $\mathcal{O}(N^2 \ln N)$ operations. The FFT procedure can also be used to evaluate $\mathbf{y}$ in $\mathcal{O}(N \ln N)$ operations. This count is almost a factor of N better than the standard Galerkin order count.

It is not in fact necessary to set $p = N$. The choice $p < N$ is not sensible because it guarantees very large errors in the matrix and the solution, but we may wish to set $p > N$ if there is reason to suppose that the kernel and driving term y each have awkward features which make them hard to integrate, but which cancel in the solution (so that the vector $\mathbf{a}$ converges more rapidly than does $\mathbf{y}$ or $\mathbf{K}$). With $p > N$, the FFT procedure yields a $p \times p$ matrix for $\mathbf{K}$ (and then $\mathbf{B}$) which can then be truncated, only the leading $N \times N$ minor being used.

(ii) *Evaluation of* $\mathbf{B}$. The direct use of (15) also takes $\mathcal{O}(N^3)$ operations. However, two one-dimensional Chebyshev series each of $\mathcal{O}(N)$ terms can be multiplied together in $\mathcal{O}(N \ln N)$ operations using the FFT, as we now demonstrate. We introduce the following notation:

Notation	Meaning
$f \to \mathbf{f}(N)$	Given a function $f(s)$, or function values at the $N+1$ Chebyshev points, $\xi_k = \cos(k\pi/N)$, $k = 0, 1, \ldots, N$, evaluate f_i, $i = 0, 1, \ldots, N$, the Chebyshev expansion coefficients of f, from (7). This can be carried out with FFT in $\mathcal{O}(N \ln N)$ operations.
$\mathbf{f}(N) \to \mathbf{f}(M)$	If $N > M$, omit the coefficients f_i, $i > M$. If $N < M$, add coefficients $f_i = 0$, $i = N+1, \ldots, M$.
$\mathbf{f}(N) \to f^{(N)}$	Evaluate $f^{(N)}(\xi_k) = \sum_{i=0}^{\prime N} f_i T_i(\xi_k)$ at the $N+1$ points $\xi_k = \cos(k\pi/N), k = 0, 1, \ldots, N$. This can be carried out using FFT in $\mathcal{O}(N \ln N)$ operations.

Now suppose that we have two sets of Chebyshev coefficients $\mathbf{a}(N)$, $\mathbf{b}(M)$ of length N, M respectively, which represent truncated expansions of the functions $a(s)$, $b(s)$. We suppose that $M \geq N$. We can calculate approximations to the first N expansion coefficients $\mathbf{c}(N)$ of the product function $c(s) = a(s) \times b(s)$ (that is, we can multiply the two Chebyshev series together) in $\mathcal{O}(N \ln N)$ operations as follows:

Algorithm 8.5.1.

Step 1 $\mathbf{a}(N) \to \mathbf{a}(M) \to a^{(M)}$ — Pad out $\mathbf{a}$ with zeros and evaluate $a(\xi_k)$ at the $(M+1)$ Chebyshev points ξ_k, $k = 0, 1, \ldots, M$.

Step 2 $\mathbf{b}(M) \rightarrow b^{(M)}$ Evaluate $b(\xi_k)$ at the same points.

Step 3 Form $c^{(M)}(\xi_k) = a(\xi_k) \times b(\xi_k), k = 0, 1, \ldots, M$.

Step 4 $c^{(M)} \rightarrow \mathbf{c}(M) \rightarrow \mathbf{c}(N)$ Produce the Chebyshev expansion of $c(s)$ and truncate it (if required) to length N.

This algorithm is less complicated than it appears and we make the following comments:

Step 1: Padding out **a** with zeros merely involves inserting zero coefficients $a_{N+1}, a_{N+2}, \ldots, a_M$ to treat $a(s)$ as a polynomial of degree M rather than N. Step 1 then involves computing $\sum_{i=0}^{\prime M} a_i T_i(\xi_k)$ at the Chebyshev points ξ_k, $k = 0, 1, \ldots, M$, and this can be carried out using the FFT.

Step 2: Repeats step 1 for the function $\sum_{i=0}^{\prime M} b_i T_i(\xi_k)$.

Step 3: Multiplies these two functions together at the Chebyshev points, at a cost of $\mathcal{O}(N)$ operations.

Step 4: Recognises that the product function values just calculated are all that is needed to form estimates of the expansion coefficients of the product function, using the FFT with $(M + 1)$ points.

This algorithm is not exact, although it can be made to be so by padding out both **a** and **b** to length $2M$ before starting; however, in the context under discussion, both **a** and **b** represent not exact representations of polynomials, but approximate and truncated representations of (hopefully) smoothly varying functions; then **a** and **b** both contain implicit truncation errors and we require only that the errors caused by performing an inexact product are of the same order as these. It is shown in Delves (1977) that this is the case; see also Chapter 9.

Example 8.5.1.

We illustrate the algorithm by using it to multiply the two Chebyshev series
$$\mathbf{a} = (2.0, 0.1) \quad \text{and} \quad \mathbf{b} = (2.0, -0.1)$$
to form a product expansion
$$\mathbf{c} = (c_1, c_2).$$

These represent the polynomials $1 + 0.1\,s$, $1 - 0.1\,s$; the coefficients of the second terms have been chosen small to reflect the use of the algorithm with a truncated expansion for which the truncation error is about 10%. We hope to find that we have computed an approximation to the function $c(s) = 1 - 0.01\,s^2$, with expansion

$$c(s) = \sum_{i=0}^{2}{}' c_i T_i(s); \quad \mathbf{c} = (1.990, 0.0, -0.005).$$

For this problem, $N = M = 1$, and the Chebyshev points are

$$\xi_0 = 1, \quad \xi_1 = -1,$$

and we go through the following stages:

Step 1: $a(1) = 1 + 0.1 = 1.1$

$$a(-1) = \sum_{i=0}^{1} {}' a_i T_i(-1) = 1 - 0.1 = 0.9$$

Step 2: $b(1) = 1 - 0.1 = 0.9; b(-1) = 1 + 0.1 = 1.1$

Step 3: $c(1) = 0.9 \times 1.1 = 0.99; c(-1) = 0.9 \times 1.1 = 0.99$

Step 4: We use equation (7) with $p = 1$ to find the expansion coefficients of $c(s)$:

$$c_0 = \frac{2}{\pi} \int_{-1}^{1} \frac{T_0(s)c(s)ds}{(1-s^2)^{\frac{1}{2}}}$$

$$\approx \frac{2}{\pi} \cdot \pi [\tfrac{1}{2} c(+1) \cos(0) + \tfrac{1}{2} c(-1) \cos(0)] = 1.98,$$

$$c_1 = \frac{2}{\pi} \int_{-1}^{1} \frac{T_1(s)c(s)ds}{(1-s^2)^{\frac{1}{2}}}$$

$$\approx \frac{2}{\pi} \cdot \pi [\tfrac{1}{2} c(1) \cos(0) + \tfrac{1}{2} c(-1) \cos(\pi)] = 0.0.$$

We have thus computed the approximate truncated product expansion $\mathbf{c} = (1.98, 0.0)$, with an error of only $\tfrac{1}{2}\%$ in c_0.

The algorithm is applicable to equation (15), which represents the explicit result of multiplying the Chebyshev expansion of $(\pi^2/4)K(s, t)$ (see (11)) by that of $(1 - t^2)^{\frac{1}{2}}$. The latter is a function of one variable, so that (see (15)) we can multiply the double series for K a row at a time, taking successively $i = 0, 1, ..., N$ and yielding an overall operation count of $\mathcal{O}(N \ln N \times N) = \mathcal{O}(N^2 \ln N)$. It remains to choose a suitable value for M, the number of terms used in the series for $(1 - t^2)^{\frac{1}{2}}$. The method is equivalent to a direct use of (15) if we set $M = 2N$, for, with only an $(N + 1) \times (N + 1)$ matrix of coefficients K_{ij}, we have to set $\bar{K}_{ij} = 0$ in (15) if either i or j (or both) $> N$. Thus the largest value of l which enters is that for which $|j - 2l| = N$ with $j = N$, leading to $l_{\max} = N$, $2l_{\max} = 2N$; that is, (15) requires $2N$ terms in the series for $(1 - t^2)^{\frac{1}{2}}$. More generally, it is shown (in Delves, Abd-Elal and Hendry, 1979) that, if the series $\mathbf{a}$ converges rapidly (as will be the case here if $K(s, t)$ is reasonably smooth) but the series $\mathbf{b}$ does not (as is certainly the case here), the first N terms of the product vector $\mathbf{c}$ have small error provided that we set $M = 2N$.

We have now shown that the $(N + 1) \times (N + 1)$ Galerkin matrix equations can be set up in $\mathcal{O}(N^2 \ln N)$ operations. An error analysis of the method is given in the next chapter, which shows that the method will converge very rapidly provided that $K(s,t)$, $y(s)$ are smooth and that it is possible to provide cheaply computable error estimates, which take into account both truncation errors (those errors due to truncation of the series after $N+1$ terms) and quadrature errors (the errors due to the use of the quadrature formulae (7), (10)).

In addition, it is shown there that there exists a simple iterative scheme for solving the equations (4), with an overall $\mathcal{O}(N^2)$ operations count. The total cost for the Fast Galerkin method is therefore $\mathcal{O}(N^2 \ln N)$; it is this lower count together with the rapid convergence and rather detailed error estimates which make the scheme attractive. We demonstrate the method with a simple example:

Example 8.5.2.

$$x(s) = y(s) + \lambda \int_0^1 e^{\beta st} x(t) dt,$$

$$y(s) = e^{\alpha s} - \lambda(e^{\alpha+\beta s} - 1)/(\alpha + \beta s).$$

(16)

Solution: $x(s) = e^{\alpha s}$.

For this example, $y(s)$ and $K(s,t)$ are smooth and the error analysis of the next chapter predicts very rapid convergence for the Fast Galerkin method. Table 8.5.1 shows the results obtained with Algorithm 8.5.1 for several values of (α, β, λ), together with the error estimates obtained from the analysis of Sections 9.6 and 9.7. It also shows results obtained using the standard Galerkin algorithm; that is, with the same Chebyshev expansion but with direct use of the quadrature rules (7), (10) and of the expansion (15), rather than the FFT techniques used in the Fast Galerkin scheme, and with a direct rather than iterative solution of the equations. We can make the following comments on these results:

(i) The accuracies achieved by the two algorithms are very similar but not identical; the difference stems from the iterative solution of the equations in the Fast scheme, iterations being terminated when the error remaining is less than that estimated from other sources (truncation, quadrature).

(ii) Similarly, the error estimates differ chiefly because those for the Fast scheme include a component for the iterative solution. For both versions of the algorithm the error estimates are on the whole rather good, reflecting fairly closely the actual error achieved, but occasionally they are very pessimistic.

(iii) The Fast scheme is slower than the standard scheme for small N, but achieves large savings for large values of N, the crossover point being at about $N = 8$.

(iv) For the standard algorithm, the time taken depends only on N; the time taken by the Fast scheme depends also on the difficulty of the problem, iterative solution of the equations taking longer for $\beta = 20$ than for $\beta = 1$.

The rapid convergence demonstrated by the results of Table 8.5.1 cannot be expected for all problems but depends crucially on the smoothness of the kernel and driving term. However, the Fast scheme proves to have unexpected advantages even for singular problems, as the next example shows.

***Example* 8.5.3.**

$$x(s) = y(s) + \int_0^{\pi/2} s^\beta t^{\frac{1}{2}} x(t) \mathrm{d}t, \tag{17}$$

$$y(s) = s^{\frac{1}{2}} - \frac{\pi^3}{24} s^\beta.$$

Table 8.5.1. *Solution of equation* (16) *using the standard and Fast Galerkin schemes.*

The 'achieved' error is the maximum error over the interval $(0,1)$; the 'estimated' error is that reported by the program and includes an estimate of truncation and quadrature errors, and, for the Fast Galerkin method, of the error involved in an iterative solution of the equations. Times are in arbitrary units.

Parameters α, β, λ	Algorithm	N	Accuracy		Time taken
			Achieved	Estimated	
1, 1, 1	Standard	4	2.8×10^{-5}	6.4, -3	48
	Fast	4	1.4, -3	4.5, -3	72
	Standard	8	1.7, -10	2.0, -8	238
	Fast	8	8.3, -8	1.6, -8	253
	Standard	16	1.2, -10	5.2, -9	1 353
	Fast	16	2.3, -10	1.4, -10	722
1, 5, 1	Standard	8	5.0, -6	1.7, -3	238
	Fast	8	2.5, -4	8.7, -4	264
	Standard	16	6.0, -10	4.4, -9	1 352
	Fast	16	3.8, -8	2.7, -9	830
1, 20, 1	Standard	16	1.1, -3	1.0, -3	1 353
	Fast	16	1.2, -3	1.5, -3	1 155
	Standard	64	8.5, -4	3.4, -4	65 089
	Fast	64	5.7, -4	2.1, -3	16 001

Table 8.5.2. *Results for the integral equation* (17) *with* $\beta = 1.5$, *using the standard and Fast Galerkin schemes.*
The notation is as in Table 8.5.1.

N	Algorithm	Accuracy		Time Taken
		Achieved	Estimated	
8	Standard	7.0×10^{-2}	1.5, −2	205
	Fast	2.4, −2	7.4, −2	223
16	Standard	3.7, −2	2.7, −3	1148
	Fast	3.7, −3	2.3, −2	469
32	Standard	1.9, −2	5.0, −4	7613
	Fast	1.3, −3	6.4, −3	1951

Solution: $x(s) = s^{\frac{1}{2}}$
This example has a rank 1 kernel so that it can be solved for an arbitrary driving term (do this!). Both $y(s)$ and $K(s,t)$ are decidedly nonsmooth so that the series (11), (12) will converge only slowly, as will the series (2) for the solution. The results achieved by the standard and Fast Galerkin schemes are shown in Table 8.5.2.

We see that for this example both algorithms converge only slowly, as expected. However, a closer inspection of the results shows that the two algorithms behave rather differently as N increases. The standard algorithm makes only very slow progress, despite reporting error estimates which indicate that it thinks it is progressing quite well. The Fast algorithm however, although also converging slowly, achieves an overall error which is a factor of more than 10 better, as well as a quite realistic error estimate. These results reflect an improved stability on the part of the revised algorithm: the FFT techniques used, because of their low operation count, yield much smaller round-off errors than the direct schemes.

We return to the problem of the efficient solution of singular integral equations in Chapter 11.

8.6   Comparison of methods

There is a very close connection between the Nystrom method described in Chapter 4 and each of the methods (L_∞, L_2, Galerkin) described above. The Nystrom method introduces a rule $Q_q(w_q, \xi_q)$ with N points to discretise the integrals. Let us compare the Nystrom method with an expansion method using an arbitrary basis $\{h_i, i = 1, \ldots, N\}$.

The Galerkin and least squares methods introduce two rules Q_p, Q_q; we choose these to be the same N-point rule used by the Nystrom method.

Finally, we choose the point set over which the L_∞ residual norm is minimised, to be the set of quadrature points $\xi_{qi}, i = 1, \ldots, N$. We recall the form of the defining equations for each method, in the notation of this chapter.

Nystrom: $\qquad$ $(\mathbf{I} - \mathbf{KW})\mathbf{x} = \mathbf{y}$ $\qquad\qquad\qquad\qquad$ (1)

Least squares: $\;\; \mathbf{\Lambda}^T\mathbf{W}\mathbf{\Lambda a} = \mathbf{\Lambda}^T\mathbf{Wy}$ $\qquad\qquad\qquad$ (2)

Galerkin: $\qquad$ $\mathbf{H}^T\mathbf{W}\mathbf{\Lambda a} = \mathbf{H}^T\mathbf{Wy}$ $\qquad\qquad\qquad$ (3)

$L_\infty:$ $\qquad\qquad$ $\min_{\mathbf{a}} \| \mathbf{y} - \mathbf{\Lambda a} \|_\infty$ $\qquad\qquad\qquad$ (4)

Now with the choices we have made $\mathbf{\Lambda}$ and $\mathbf{H}$ are square. We assume they are nonsingular; then (2) and (3) each reduce to the system

$$\mathbf{\Lambda a} = \mathbf{y} \qquad\qquad\qquad\qquad (5)$$

so that the Galerkin and least squares methods yield the same solution. Moreover, the minimal residual in (4) is zero, attained at the solution point of (5), so that the L_∞ method also reduces to the other two. Finally, inserting the form (8.4.5) for $\mathbf{\Lambda}$, (5) becomes

$$(\mathbf{H} - \mathbf{KWH})\mathbf{a} = \mathbf{y}. \qquad\qquad\qquad (5a)$$

But now,

$$\mathbf{x}_i = x_N(\xi_i) = \sum_{j=1}^{N} a_j h_j(\xi_i).$$

That is,

$$\mathbf{x} = \mathbf{Ha} \qquad\qquad\qquad\qquad (6)$$

and (5a) is completely equivalent to (1). Note that this equivalence is independent of the choice of expansion set $\{h_i\}$; the properties of the methods are determined solely by the quadrature rule used.

This identity does not hold in general; however, it suggests that the methods will achieve comparable accuracy if comparable quadrature rules are used in each and this suggestion is largely borne out in practice, although (see Chapter 10) details of the numerical implementations affect the performance of a given method quite significantly. We should therefore choose between them on other grounds, including those of:

$\quad$ (i) Stability

$\quad$ (ii) Ease of obtaining an error estimate

$\quad$ (iii) Cost.

The cost of the Nystrom method is determined by the solution of the equations if direct elimination is used and is then about $\frac{1}{3}N^3$ operations. This is much faster than the costs quoted above for any of the expansion methods and reduces to $\mathcal{O}(N^2)$ if the iterative solution technique of Section 4.6.2 is implemented. Of the expansion methods, the Galerkin method is the cheapest; we saw above that an $\mathcal{O}(N^2 \ln N)$ operation count could be achieved for this method, which is comparable with that for the Nystrom scheme. The least squares and L_∞ residual minimisation schemes do not appear to offer any particular advantages in return for their higher costs and we shall not pursue these further. In the next chapter we see that a rather complete error estimate can be provided for the Galerkin method with orthogonal expansion set, and for the Fast Galerkin scheme in particular, and this adds to the attraction of the Fast Galerkin method. An overall comparison of the various methods is given in Chapter 10.

Exercises

1. Let $\mathbf{A}$ be a real $m \times n$ matrix with $m > n$, and let $\mathbf{x}, \mathbf{y}$ be vectors of length n. Show that $(\mathbf{A}\mathbf{x} - \mathbf{y})^T(\mathbf{A}\mathbf{x} - \mathbf{y})$ achieves its minimum value of $(\mathbf{y}^T\mathbf{y} - \mathbf{x}^T\mathbf{A}^T\mathbf{A}\mathbf{x})$ when $\mathbf{x}$ is a solution of $\mathbf{A}^T\mathbf{A}\mathbf{x} = \mathbf{A}^T\mathbf{y}$.

2. Show that the Chebyshev polynomials satisfy the following relations

$$\int_0^s T_0(s)\mathrm{d}s = T_1(s),$$

$$\int_0^s T_1(s)\mathrm{d}s = \tfrac{1}{4}\{T_2(s) + 1\},$$

$$\int_0^s T_r(s)\mathrm{d}s = \frac{1}{2}\left[\frac{T_{r+1}(s)}{r+1} - \frac{T_{r-1}(s)}{r-1}\right], \quad r \geq 2.$$

3. Show, using the results of Exercise 2, that the Chebyshev expansion of $(1 - t^2)^{\frac{1}{2}}$ is given by

$$(1 - t^2)^{\frac{1}{2}} = \sum_{j=0}^{\infty} {}' a_{2j} T_{2j}(t)$$

where

$$a_{2j} = \frac{4}{\pi(1 - 4j^2)}.$$

4. The integral equation

$$x(s) = y(s) + \int_a^b K(s, t)x(t)\mathrm{d}t$$

is to be solved using the standard Galerkin, collocation (see Chapter 7, Exercise 4) and Nystrom methods, where the expansion-type methods use the N-term approximation

$$x(s) \approx x_N(s) = \sum_{i=1}^{N} a_i^{(N)} h_i(s)$$

and $(P+1)$-point Gauss–Chebyshev formulae are used to approximate any integrals. The overall cost for each method consists of the number of kernel and driving term evaluations, the cost of setting up the equations and their direct solution. Show that when $P = N$ and Gauss elimination is used to solve the resulting set of simultaneous equations, the overall costs for each method are as follows:

Standard Galerkin: $\frac{10}{3} N^3 (\mu + \alpha)$

Collocation $\quad : \frac{7}{3} N^3 (\mu + \alpha)$

Nystrom $\quad\quad : \frac{1}{3} N^3 (\mu + \alpha)$

where μ and α refer to the operations of multiplication and addition respectively.

5. The Fast Fourier Transform method evaluates

$$X(j) = \sum_{k=0}^{N-1} A(k) e^{2\pi i jk/N}, \quad j = 0, 1, \ldots, N - 1,$$

in $N \cdot (\sum_{i=1}^{m} N_i)$ operations where $N = \prod_{i=1}^{m} N_i$, compared to the straightforward method which uses N^2 operations. Show that when $N_i = r$, $i = 1, \ldots, m$, the FFT method evaluates $X(j)$ in $rN \log_r N$ operations and show further that when $m = 2$, $r = 4$ the FFT method reduces the standard operation count by a factor of two. What is the corresponding factor when $m = 7$, $r = 2$? when $m = 3$, $r = 2$? Comment on the importance of the use of FFT techniques within the Fast Galerkin algorithm.

6. Use Algorithm 8.5.1 to square the Chebyshev series $1 + s = T_0(s) + T_1(s)$, first by following the algorithm as given and then by repeating the process for the 'cubic' $T_0(s) + T_1(s) + 0 \cdot T_2(s) + 0 \cdot T_3(s)$. Comment on the difference obtained. Check your result by expanding $(1 + s)^2$ directly as a Chebyshev series.

9

Analysis of the Galerkin method with orthogonal basis

9.1 Structure of the equations

The Galerkin method is the most important of the expansion methods discussed in Chapter 8. We saw there that ignoring quadrature and round-off errors, its accuracy depended not on the basis $\{h_i\}$ but on the space spanned by this basis, the same solution being obtained with the set $\{s^{i-1}, i = 1,\ldots,N\}$ as with the set of Legendre polynomials $\{P_{i-1}(s), i = 1,\ldots,N\}$. This is no longer true when quadrature errors are present; indeed, we saw in Section 8.6 that when we use an N-point quadrature rule with an N-term expansion the resulting approximate solution is completely independent (at the quadrature points) of the basis $\{h_i\}$ and depends only on the quadrature rule used. However, the choice of basis is always important in determining the *conditioning* of the linear equations and hence the stability of the calculation against both quadrature and round-off errors. An orthonormal basis has the advantage that it guarantees the stability of the matrix equations. In addition, the resulting structure of the equations is such that an iterative solution of the equations with $\mathcal{O}(N^2)$ operations count is possible and such that cheap and effective error estimates can be obtained for the solution, provided that error estimates for the quadrature rule used are available in a suitable form. These properties are essentially independent of the normalisation of the basis functions; in this chapter we assume that the basis $\{h_i\}$ is orthogonal with weight function $w(s)$:

$$\int_a^b w(s)h_i(s)h_j(s)\mathrm{d}s = \delta_{ij}w_j \tag{1}$$

where $w_j = 1$ if the basis is in fact orthonormal and we develop the analysis on which these remarks are based. We begin by considering the structure of the Galerkin equations and assume that the weight function $w(s)$ is incorporated into the Galerkin equations (see (6), (7) below). We consider the second kind Fredholm equation

$$x = y + \lambda K x \tag{2}$$

and introduce the exact expansion

$$x(s) = \sum_{i=0}^{\infty} b_i h_i(s) \tag{3}$$

and the approximation x_N:

$$x_N(s) = \sum_{i=0}^{N} \bar{a}_i^{(N)} h_i(s). \tag{4}$$

With an orthonormal basis, the Galerkin equations for $\bar{\mathbf{a}}^{(N)}$ have the form

$$(\mathbf{I} - \lambda \bar{\mathbf{B}}) \bar{\mathbf{a}}^{(N)} = \bar{\mathbf{y}} \tag{5}$$

where $\mathbf{I}$ is the $(N+1) \times (N+1)$ unit matrix and

$$\bar{B}_{ij} = \int_a^b w(s) \int_a^b K(s,t) h_i(s) h_j(t) dt \, ds, \tag{6}$$

$$\bar{y}_i = \int_a^b w(s) y(s) h_i(s) ds. \tag{7}$$

As remarked in Section 8.5, (6) and (7) identify $\bar{B}_{ij}$ and $\bar{y}_i$ as expansion coefficients in the orthonormal expansions of $K(s,t)/w(t)$, $y(s)$:

$$w^{-1}(t) K(s,t) = \sum_{i=0}^{\infty} \sum_{j=0}^{\infty} \bar{B}_{ij} h_i(s) h_j(t), \tag{8}$$

$$y(s) = \sum_{i=0}^{\infty} \bar{y}_i h_i(s). \tag{9}$$

These equations are enough to guarantee (provided that $w^{-1}(t) K(s,t)$ is square integrable) that $\bar{B}_{ij}$ and $\bar{y}_i$ are decreasing functions for large i, j. For we have at once

$$\| y \|^2 = \sum_{i=0}^{\infty} \bar{y}_i^2,$$

$$\| w^{-1} K \|^2 = \sum_{i=0}^{\infty} \sum_{j=0}^{\infty} \bar{B}_{ij}^2. \tag{10}$$

The infinite series in (10) therefore converge; hence there exist positive constants B_0, y_0, p, q, r such that for all i, j:

$$|\bar{y}_i| \le y_0 i^{-r}, \tag{11}$$

$$|\bar{B}_{ij}| \le B_0 i^{-p} j^{-q}. \tag{12}$$

(See Delves and Freeman, 1981, Chapters 9, 11.)

With an orthogonal basis, that is, one satisfying equation (1), the Galerkin equations for $\bar{\mathbf{a}}^{(N)}$ have the form

$$\Lambda^{\frac{1}{2}}(\mathbf{I} - \lambda\bar{\mathbf{B}})\Lambda^{\frac{1}{2}}\bar{\mathbf{a}}^{(N)} = \Lambda^{\frac{1}{2}}\bar{\mathbf{y}} \tag{13}$$

where $\Lambda = \mathbf{diag}(w_i)$ and $\bar{\mathbf{B}}, \bar{\mathbf{y}}$ are defined in equations (6), (7).

Thus in the case of an orthonormal basis, the Galerkin equations for $\bar{\mathbf{a}}^{(N)}$ are given by

$$\mathbf{L}\bar{\mathbf{a}}^{(N)} = \bar{\mathbf{y}} \tag{5a}$$

where $\mathbf{L} = \mathbf{I} - \lambda\bar{\mathbf{B}}$; when the basis is orthogonal but not orthonormal we solve the closely related system

$$(\Lambda^{\frac{1}{2}}\mathbf{L}\Lambda^{\frac{1}{2}})\bar{\mathbf{a}}^{(N)} = \Lambda^{\frac{1}{2}}\bar{\mathbf{y}}. \tag{13a}$$

For the remainder of this chapter we shall make use of either of the forms (5a), (13a) as seems appropriate.

In particular, if we choose as orthogonal basis the Chebyshev polynomials, as in the Fast Galerkin scheme, equation (13a) reduces to

$$(\bar{\mathbf{D}} - \lambda\bar{\mathbf{B}})\bar{\mathbf{a}}^{(N)} = \bar{\mathbf{y}} \tag{14}$$

(see Section 8.5), where

$$\bar{D}_{ij} = \int_{-1}^{1} \frac{T_i(s)T_j(s)\mathrm{d}s}{(1-s^2)^{\frac{1}{2}}} = \begin{cases} \pi, & i = j = 0, \\ \dfrac{\pi}{2}, & i = j > 0, \\ 0, & i \neq j, \end{cases} \tag{14a}$$

$$\bar{B}_{ij} = \int_{-1}^{1} \frac{T_i(s)}{(1-s^2)^{\frac{1}{2}}} \int_{-1}^{1} K(s,t)T_j(t)\mathrm{d}t\,\mathrm{d}s, \quad i,j = 0,1,\ldots,N, \tag{14b}$$

$$\bar{y}_i = \int_{-1}^{1} \frac{T_i(s)y(s)\mathrm{d}s}{(1-s^2)^{\frac{1}{2}}}, \quad i = 0,1,\ldots,N. \tag{14c}$$

The convergence of the series in (10) ensures that $p, q, r > \frac{1}{2}$. (See Delves and Freeman (1981), Chapters 9, 11). In practice, we expect rather larger values for these parameters. The values which obtain depend on the smoothness properties of K, y, and on the basis h_i; we demonstrate this in the next two sections by considering two typical expansion sets, limiting the discussion for simplicity to the one-dimensional expansion (11):

    (i) $h_i(s) = \cos is$: $(a,b) = (0,\pi)$. We show that, with this set, the constraints required for r to be large are very restrictive.

    (ii) $h_i(s) = T_i(s)$: $(a,b) = (-1,1)$. With this set, we show that the value of r in (11) depends directly on the number of continuous derivatives which $y(s)$ has on the interval $(-1,1)$. This result is typical of that obtained for orthogonal polynomial bases.

9.2 Convergence of Fourier half range cosine expansions

Let us consider the expansion (9.1.9) for the case

$$[a, b] = [0, \pi]; h_i(s) = \begin{cases} \left(\dfrac{2}{\pi}\right)^{\frac{1}{2}} \cos is, & i > 0, \\ \left(\dfrac{1}{\pi}\right)^{\frac{1}{2}}, & i = 0. \end{cases} \tag{1}$$

Since we are interested in the behaviour of (9.1.9) for large i, we take $i > 0$. Then

$$\left(\frac{\pi}{2}\right)^{\frac{1}{2}} \bar{y}_n = \int_0^\pi y(s) \cos ns \, ds. \tag{2}$$

We now derive an asymptotic expansion for $\bar{y}_n$ by repeatedly integrating (2) by parts.

Integrating once, we have

$$\left(\frac{\pi}{2}\right)^{\frac{1}{2}} \bar{y}_n = \left[\frac{1}{n} y(s) \sin ns\right]_0^\pi - \frac{1}{n} \int_0^\pi y'(s) \sin ns \, ds$$

$$= -\frac{1}{n} \int_0^\pi y'(s) \sin ns \, ds \tag{3a}$$

which is valid provided that $y'(s)$ is piecewise continuous on $[0, \pi]$. If it is in fact continuous we can integrate by parts again to obtain

$$\left(\frac{\pi}{2}\right)^{\frac{1}{2}} \bar{y}_n = \left[\frac{1}{n^2} y'(s) \cos ns\right]_0^\pi - \frac{1}{n^2} \int_0^\pi y''(s) \cos ns \, ds \tag{3b}$$

and this process can obviously be repeated provided that the relevant derivatives of $y(s)$ exist. If $y(s)$ has p continuous derivatives on $[0, \pi]$ and its $(p + 1)$st derivative exists and is piecewise continuous, we can integrate by parts $(p + 1)$ times to obtain the identity

$$\left(\frac{\pi}{2}\right)^{\frac{1}{2}} \bar{y}_n = \sum_{\substack{k=2 \\ k \, \text{even}}}^{p+1} \left[(-1)^{(k/2)+1} \frac{1}{n^k} y^{(k-1)}(s) \cos ns\right]_0^\pi$$

$$- (-1)^{[p/2]} n^{-(p+1)} \int_0^\pi y^{(p+1)}(s) \begin{cases} \cos ns \, ds, & p \, \text{odd}, \\ \sin ns \, ds, & p \, \text{even}, \end{cases} \tag{4a}$$

that is,

$$\left(\frac{\pi}{2}\right)^{\frac{1}{2}} \bar{y}_n = \sum_{\substack{k=2 \\ k \, \text{even}}}^{p+1} n^{-k} (-1)^{(k/2)+1} [(-1)^n y^{(k-1)}(\pi) - y^{(k-1)}(0)]$$

$$+ (-1)^{[p/2]+1} n^{-(p+1)} R_n \tag{4b}$$

where R_n denotes the remainder integral in (4a).

We now note the following:

(i) since $|\cos ns|$, $|\sin ns| \le 1$, we have for all n

$$|R_n| \le \pi \sup_{0 \le s \le \pi} |y^{(p+1)}(s)|. \text{ Hence the last term in (4b) is } \mathcal{O}(n^{-(p+1)}).$$

(ii) In general, the 'surface terms' in (4b) do not vanish; thus, usually the first nonvanishing surface term yields the dominant behaviour of $\bar{y}_n$ for large n.

Usually, the first term will not vanish, so that we expect in general, for a cosine expansion

$$\bar{y}_n = \mathcal{O}(n^{-2}), \quad \text{provided that } y(s) \in C^{(2)}[0, \pi]. \tag{5}$$

This is a very slow rate of convergence, which suggests that we avoid the use of such an expansion. We can identify classes of functions for which more rapid convergence is attained. For example, if $y(s) \in C^2[0, \pi]$ and $y'(\pi) = y'(s) = 0$, then for every n the first surface term vanishes, and

$$\bar{y}_n = \mathcal{O}(n^{-3}).$$

More generally, define γ to be the largest integer such that

$$y^{(2k-1)}(\pi) = y^{(2k-1)}(0) = 0, \quad k = 1, 2, \ldots, \gamma$$

and let $y(s) \in C^{(p)}[0, \pi]$ and $y^{(p+1)}$ exist almost everywhere in $[0, \pi]$. Then it follows from (4b) that

$$\bar{y}_n = \mathcal{O}(n^{-q}), \quad q = \min[p+1, 2\gamma + 2]. \tag{6}$$

9.3   Expansions in Chebyshev polynomials

The Chebyshev polynomials $T_i(s)$ are orthogonal on the interval $[-1, 1]$ with weight function $w(s) = (1 - s^2)^{-\frac{1}{2}}$:

$$\int_{-1}^{1} \frac{T_i(s)T_j(s)ds}{(1 - s^2)^{\frac{1}{2}}} = \begin{cases} \frac{\pi}{2}\delta_{ij}, & i > 0, \\ \pi\delta_{i0}, & i = 0, \end{cases} \tag{1}$$

and hence for an expansion in Chebyshev polynomials

$$y(s) = \sum_{i=0}^{\infty}{}' \bar{y}_i T_i(s) \tag{2a}$$

we have

$$\left(\frac{\pi}{2}\right)\bar{y}_n = \int_{-1}^{1} \frac{T_n(s)y(s)ds}{(1 - s^2)^{\frac{1}{2}}}. \tag{2b}$$

The convergence properties of $\bar{y}_n$ can be derived quite simply in terms of those for the half range cosine expansion discussed above by noting the relation

$$T_n(s) = \cos(n\cos^{-1}s) \qquad (3)$$

from which we obtain on setting $s = \cos\theta$ in (2b):

$$\left(\frac{\pi}{2}\right)\bar{y}_n = \int_0^\pi \cos n\theta \, y(\cos\theta)\mathrm{d}\theta. \qquad (4)$$

Thus the expansion coefficients $\bar{y}_n$ for the Chebyshev expansion of $y(s)$ are the same as those for the half range cosine series expansion of $Y(\theta) = (2/\pi)^{\frac{1}{2}}y(\cos\theta)$. But this apparently trivial difference in the effective function being expanded makes a very large difference to the properties of the expansion. For

$$\frac{\mathrm{d}Y(s)}{\mathrm{d}s} = \frac{\mathrm{d}y(\cos s)}{\mathrm{d}(\cos s)} \times \frac{\mathrm{d}(\cos s)}{\mathrm{d}s} = -y'(\cos s)\sin s \qquad (5)$$

and hence $\mathrm{d}Y(\pi)/\mathrm{d}s = \mathrm{d}Y(0)/\mathrm{d}s = 0$ provided that $y'(+1)$, $y'(-1)$ are finite; the same is true of the higher derivatives of $Y(s)$.

This means that in the cosine expansion of $Y(s)$ all of the surface terms vanish in (4b) and we have the very simple characterisation:

Let $y(s)\in C^{(p)}[-1,1]$ and $y^{(p+1)}(s)$ exist almost everywhere in $[-1,1]$. Then for some y_0 and all n, the coefficients $\bar{y}_n$ in expansion (2a) are bounded by

$$|\bar{y}_n| \le y_0 \, \hat{n}^{-(p+1)} \qquad (6)$$

where

$$\hat{n} = \begin{cases} n, & n > 0 \\ 1, & n = 0. \end{cases}$$

This result is not the strongest which can be obtained; it is in fact sufficient for the derivatives of $y(s)$ to be continuous on the open interval $(-1,1)$, provided that they do not diverge too violently at the endpoints: see Bain and Delves (1977) or Delves and Freeman (1981, Chapter 9) for details of these weaker restrictions. These references also consider other classical orthogonal polynomial expansions and show that these lead to results similar to (6).

The fact that (6) is not the tightest possible bound is obvious if we consider the function $(1-t^2)^{\frac{1}{2}}$. The exact Chebyshev expansion of this function is given by (8.5.13) and satisfies (9.1.11) with $r = 2$. But $(1-t^2)^{\frac{1}{2}}$ has *no* continuous derivatives on the closed interval $[-1,1]$ since its first derivative is singular at $t = \pm 1$. We must therefore set $p = 0$ in (6), which predicts correctly, but pessimistically, $r = 1$.

9.4   Expansion of functions of two variables

Two-dimensional expansions such as (9.1.8) can be analysed by similar methods; the details are more complicated but the principles are unchanged. We refer the interested reader to Delves and Bain (1977) or to Delves and Freeman (1981), Chapter 10, and give only the corresponding results to (9.3.6) for expansions in Chebyshev polynomials:

Let $K(s,t) \in C^{(p,q)}$; that is, assume that $K(s,t)$ has p continuous partial derivatives with respect to s and q continuous partial derivatives with respect to t on $[-1,1] \times [-1,1]$ and $(p+1)$st, $(q+1)$st partial derivatives defined almost everywhere in $[-1,1] \times [-1,1]$. Then for some K_0 and all n, m the coefficient K_{nm} in the expansion

$$K(s,t) = \sum_{n=0}^{\infty}{}' \sum_{m=0}^{\infty}{}' K_{nm} T_n(s) T_m(t) \tag{1}$$

is bounded by

$$|K_{nm}| \leq K_0 \hat{n}^{-(p+1)} \hat{m}^{-(q+1)}. \tag{2}$$

Similar results can be given for other orthogonal polynomial expansions. The most directly relevant use of (2) is in the analysis of the matrix $\bar{\mathbf{B}}$, the coefficients of which are given by equation (9.1.14b), for the Fast Galerkin method, which corresponds to setting $w(t) = (1-t^2)^{-\frac{1}{2}}$ in (9.1.8). If the kernel $K(s,t)$ is of class $C^{(p,q)}$ then the matrix $\bar{\mathbf{K}}$, given by equation (8.5.9), has elements bounded by an equation of the form (2) above. The corresponding bound on $\bar{\mathbf{B}}_{nm}$ (see equation (9.1.8)) can be obtained from the identity (8.5.15), which yields

$$\frac{\pi}{2}|\bar{B}_{ij}| \leq K_0 \left[i^{-(p+1)} \hat{j}^{-(q+1)} + \sum_{l=1}^{\infty} \frac{1}{4l^2 - 1} (i^{-(p+1)} \widehat{(j+2l)}^{-(q+1)} + i^{-(p+1)} \widehat{|j-2l|}^{-(q+1)}) \right]. \tag{3}$$

The third term in this bound is dominant; it can be estimated in a straightforward manner to yield

$$\frac{\pi}{2}|\bar{B}_{ij}| \leq K_0' \hat{i}^{-(p+1)} \hat{j}^{-2}. \tag{4}$$

A direct use of the rather weak result (2) would predict a $\hat{j}^{-1}$ dependence.

9.5   Estimates for the solution of the Galerkin equations

We now turn to the application of these results to the Galerkin method. For simplicity we assume that the equations have been written in terms of an orthonormal basis.

The structure implied by bounds of the form (9.1.11, 12) allows a rather complete analysis of the solution of equations (9.1.5a) and the prediction of the behaviour of the coefficients $\mathbf{b}$, $\bar{\mathbf{a}}^{(N)}$ of equations (9.1.3), (9.1.4). The implications of this structure are discussed in considerable detail in Delves and Freeman (1981) and the references cited there.

We introduce the following definitions:

Definition 9.5.1. The matrix $\mathbf{L}$ is *asymptotically lower diagonal of type A and degree p* if there exist constants $p \geq 0$, $C > 0$ such that, for all $i > j$,

$$\frac{|L_{ij}|}{(|L_{ii}||L_{jj}|)^{\frac{1}{2}}} \leq Ci^{-p}.$$

Definition 9.5.2. The matrix $\mathbf{L}$ is *asymptotically upper diagonal of type A and degree p* if $\mathbf{L}^T$ is asymptotically lower diagonal of type A and degree p.

Then it follows that the matrix $\mathbf{I} - \lambda\bar{\mathbf{B}}$ is asymptotically lower diagonal of type A and degree p, and asymptotically upper diagonal of type A and degree q. For sets of equations which have coefficient matrices with this structure, it proves possible to derive rather complete results on the behaviour of the solution vector. The relevant analysis is given in Freeman and Delves (1974); Theorems 6 and 7 of this paper yield the following lemmas (for N sufficiently large):

Lemma 9.5.1. For some C_b independent of i, and all i:

$$|b_i| \leq C_b i^{-s}, \quad s = \min(p, r). \tag{1}$$

Lemma 9.5.2. For some C_b' independent of i, N:

$$|b_i - \bar{a}_i^{(N)}| \leq C_b' N^{-(s+q-1)}. \tag{2}$$

Lemma 9.5.3. For some Q independent of N:

$$\|(\mathbf{I} - \lambda\bar{\mathbf{B}})^{-1}\| \leq Q. \tag{3}$$

These results can be proved directly only for sufficiently small λ; they cannot be true for λ a characteristic value since then $(\mathbf{I} - \lambda\bar{\mathbf{B}})^{-1}$ does not exist. However, (1) and (3) effectively follow for all regular values λ from the alternative arguments:

(i) If $K(s, t) \in C^{(p,q)}$ (which implies (9.4.2)) then for all $\mathscr{L}^2$ functions $x(t)$, $\int_a^b K(s, t)x(t)dt \in C^p[-1, 1]$.

Hence, if $y(s) \in C^r[-1, 1]$, the equation

$$x = Kx + y$$

must have a solution $x \in C^s[-1, 1]$, where $s = \min(p, r)$, and (1) follows.

(ii) $\mathbf{I} - \lambda \bar{\mathbf{B}}$ is an $(N + 1) \times (N + 1)$ truncated basis of the operator $I - \lambda K$ for the orthonormal basis $h_i(s)$. Hence if the basis $\{h_i\}$ is complete, $\lim_{N \to \infty} \|(\mathbf{I} - \lambda \bar{\mathbf{B}})^{-1}\| = \|(I - \lambda K)^{-1}\|$ and hence $\|(\mathbf{I} - \lambda \bar{\mathbf{B}})^{-1}\|$ is uniformly bounded at least for N sufficiently large.

Note that the bound for $|b_i|$ is independent of the parameter q (see (9.1.12)) so that the low value $q = 2$ obtained, even for smooth kernels, with the Fast Galerkin method has no effect on the solution, although it does affect the rate at which $(b_i - \bar{a}_i^{(N)})$ tends to zero with N; we see in the next section that this does not materially affect the overall accuracy of the calculation.

9.6   Error estimates

In practice, we do not solve (9.1.5a) or (9.1.14), but the approximate system

$$(\mathbf{I} - \lambda \mathbf{B})\mathbf{a}^{(N)} = \mathbf{y} \tag{1}$$

or

$$(\mathbf{D} - \lambda \mathbf{B})\mathbf{a}^{(N)} = \mathbf{y} \tag{2}$$

where $\mathbf{D}, \mathbf{B}, \mathbf{y}$ contain quadrature errors. In the case where the equations have been written in terms of an orthonormal basis we assume for simplicity that the matrix $\mathbf{I}$ in (1) is exact. This is likely to be the case. Even if it is computed numerically its (i, j) element is

$$I_{ij} = \int_a^b w(s) h_i(s) h_j(s) \, ds, \quad i, j = 0, 1, \dots, N.$$

Now suppose that the basis $\{h_i\}$ is polynomial; then the highest degree polynomial is that given by the integrand of I_{NN}, which is of polynomial degree $2N$. An $(N + 1)$-point Gauss rule with weight function $w(s)$ is of degree $2N + 1$ and hence will integrate exactly every element of the matrix $\mathbf{I}$. Of course, this argument ignores round-off errors but these are normally negligible compared to the quadrature errors.

We write (1) in the form

$$(\mathbf{I} - \lambda \bar{\mathbf{B}} - \lambda \delta \mathbf{B})(\bar{\mathbf{a}}^{(N)} + \delta \mathbf{a}) = \bar{\mathbf{y}} + \delta \mathbf{y} \tag{3}$$

where $\lambda \delta \mathbf{B}, \delta \mathbf{y}$ represent the quadrature errors. The total error in the computed solution is then $e_N(s) = x(s) - x_N(s)$:

$$e_N(s) = \sum_{i=0}^{N} (b_i - a_i^{(N)}) h_i(s) + \sum_{i=N+1}^{\infty} b_i h_i(s)$$

$$= \sum_{i=0}^{N} [(b_i - \bar{a}_i^{(N)})] h_i(s) + \sum_{i=0}^{N} (\bar{a}_i^{(N)} - a_i^{(N)}) h_i(s)$$

$$+ \sum_{i=N+1}^{\infty} b_i h_i(s)$$

$$= e_1(s) + e_2(s) + e_3(s) \tag{4}$$

whence we find at once

$$\| e_N \| \leq \| e_1 \| + \| e_2 \| + \| e_3 \|. \tag{5}$$

Since the set $\{h_i\}$ is assumed orthonormal these individual terms have the form

$$\| e_1 \|^2 = \sum_{i=0}^{N} (b_i - \bar{a}_i^{(N)})^2, \tag{6}$$

$$\| e_2 \|^2 = \sum_{i=0}^{N} (\bar{a}_i^{(N)} - a_i^{(N)})^2 = \| \delta \mathbf{a} \|^2, \tag{7}$$

$$\| e_3 \|^2 = \sum_{i=N+1}^{\infty} b_i^2. \tag{8}$$

We now estimate each of these in turn.

9.6.1 Approximation and truncation errors

From (9.5.2) we have

$$\| e_1 \|^2 \leq C_b'^2 \sum_{i=1}^{N} N^{-2(s+q-1)} = C_b'^2 N^{-(2s+2q-3)} \tag{9}$$

and from (9.5.1) it follows that

$$\| e_3 \|^2 \leq C_b^2 \sum_{i=N+1}^{\infty} i^{-2s} \leq \frac{C_b^2}{2s-1} N^{-2s+1}. \tag{10}$$

Provided that $q > 1$, it follows from (9), (10) that $\| e_1 \|^2$ is negligible for large N compared with $\| e_3 \|^2$. This is a very satisfactory result since e_3 represents the irreducible error made when we truncate the series (9.1.3), while e_1 represents the error due to approximating the b_i, $i = 0, 1, \ldots, N$, with the Galerkin algorithm. We therefore have the overall estimate

$$\| e_1 \| + \| e_3 \| \lesssim \frac{C_b N^{\frac{1}{4}}}{(2s-1)^{\frac{1}{2}}} N^{-s}. \tag{11}$$

We can further relate the estimate (11) to quantities which are available during the calculations.

From (9.5.1), (9.5.2) we have the simple *a posteriori* estimate

$$C_b N^{-s} \sim |b_N| \sim |\bar{a}_N^{(N)}| \sim |a_N^{(N)}| \tag{12}$$

while we obtain an *a priori* estimate by noting that $s = \min(p, r)$ and that hence, within a constant (unknown) factor

$$C_b N^{-s} \sim \max(|\bar{B}_{N0}|, |\bar{y}_N|) \sim \max(|B_{N0}|, |y_N|). \tag{13}$$

The relations (12), (13) are not meant to be taken literally. In any particular calculation, the $a_i^{(N)}$, B_{ij}, y_i may not be smooth functions of their suffixes; for example, a symmetry property of $y(s)$ may imply $y_i = 0$, i even, while $y_i \neq 0$, i odd. Thus, y_N might be identically zero, as might $a_N^{(N)}$, B_{N0}. In practice, we would estimate 'smoothed' values for these quantities by inspecting the last few entries in the vectors $\mathbf{y}$, $\mathbf{a}^{(N)}$ and the first column of $\mathbf{B}$; or (see Chapter 12) we may choose to find a fit to these vectors of the form (9.1.11), (9.5.1), (9.1.12) respectively and hence estimate the parameters C_b, p, q, r.

9.6.2   Quadrature error

If we have estimates for the quadrature errors $\delta\mathbf{B}$, $\delta\mathbf{y}$, $\|e_2\|$ can be estimated in a straightforward manner since we have the standard bound (see (3))

$$\|e_2\| = \|\delta\mathbf{a}\| \leq Q\{\lambda\|\delta\mathbf{B}\|\,\|\bar{\mathbf{a}}^{(N)}\| + \|\delta\mathbf{y}\|\}/(1 - \lambda Q \|\delta\mathbf{B}\|). \tag{14}$$

Here, we have inserted the uniform bound (9.5.3) for $\|(\mathbf{I} - \lambda\bar{\mathbf{B}})^{-1}\|$. Only a crude estimate of Q need be made; in practice, the very simple lower bound on Q

$$Q \sim \frac{\|\bar{\mathbf{a}}^{(N)}\|}{\|\bar{\mathbf{y}}\|} \tag{15}$$

proves quite effective and has the virtue of being cheap. It is not possible to simplify (14) further without specifying the quadrature rule being used; we therefore now consider a specific case, that of the Fast Galerkin method.

9.7   Quadrature errors for the Fast Galerkin method
9.7.1   Quadrature errors in y

For the Fast Galerkin method, the appropriate orthonormal basis is

$$h_i(s) = \alpha_i T_i(s)$$

where

$$\alpha_i = \left(\frac{2}{\pi}\right)^{\frac{1}{2}}, \quad i > 0,$$

$$\left(\frac{1}{\pi}\right)^{\frac{1}{2}}, \quad i = 0.$$

It is more usual however to write this method in terms of the unnormalised Chebyshev polynomials $T_i(s)$; then the defining equations are (9.1.14) rather than (9.1.5a) and we solve the approximate system (9.6.2). The analysis of Section 9.6 should be adjusted to allow for this. However, because the normalisation factor α_i is essentially independent of i, the error estimates of Section 9.6 are effectively unchanged by this adjustment apart from a constant factor which we ignore. Within this framework, the coefficient $\bar{y}_i$ is given by

$$\bar{y}_i = \int_{-1}^{1} \frac{y(s)}{(1-s^2)^{\frac{1}{2}}} T_i(s) ds \tag{1}$$

and is approximated by the $(p+1)$-point Gauss–Chebyshev rule (8.5.7). The error of such a rule was discussed in Chapter 2 and we recall the result given there (equation (2.7.45)):

Let $g(s) = \sum_{k=0}^{\infty}{}' g_k T_k(s)$ and for convenience set

$$I_s(f) = \int_{-1}^{1} \frac{f(s) ds}{(1-s^2)^{\frac{1}{2}}}; \quad Q_s(f) = \frac{\pi}{p} \sum_{k=0}^{p}{}'' f\left(\cos \frac{\pi k}{p}\right).$$

Then

$$\int_{-1}^{1} \frac{g(s) ds}{(1-s^2)^{\frac{1}{2}}} - \frac{\pi}{p} \sum_{k=0}^{p}{}'' g\left(\cos \frac{\pi k}{p}\right) \equiv I_s(g) - Q_s(g) = -\pi \sum_{j=1}^{\infty} g_{2pj}. \tag{2}$$

Comparing (2) with (1), we see that the integrals are identical if we set $g(s) = y(s) T_i(s)$.

But now (see (8.5.12))

$$\frac{\pi}{2} y(s) = \sum_{k=0}^{\infty}{}' \bar{y}_k T_k(s)$$

and hence since

$$T_k(s) T_i(s) = \frac{1}{2}[T_{i+k}(s) + T_{|i-k|}(s)]$$

we find

$$y(s) T_i(s) = \frac{1}{\pi} \sum_{k=0}^{\infty}{}' [T_{i+k}(s) + T_{|i-k|}(s)] \bar{y}_k$$

$$= \frac{1}{\pi} \sum_{k=0}^{\infty}{}' [\bar{y}_{i+k} + \bar{y}_{|i-k|}] T_k(s). \tag{3}$$

Thus (2) yields the identity

$$\delta y_i \equiv I_s(yT_i) - Q_s(yT_i) = -\sum_{k=1}^{\infty} (\bar{y}_{2pk-i} + \bar{y}_{2pk+i}). \tag{4}$$

This sum is usually rapidly convergent and we estimate $|\delta y_i|$ by its leading term which is $\bar{y}_{2p-i}$. Using an N-term expansion, $i \leq N$ and we therefore have

$$|\delta y_i| \leq |\bar{y}_{2p-N}|, \quad i \leq N \tag{5}$$

which should be taken in the same spirit as (9.6.12, 13): we must use smoothed values of **y** to obtain a numerical estimate from (5).

9.7.2   Quadrature errors in B

We estimate the quadrature errors in **B** in two stages. First, we consider the errors in the matrix **K**, equation (8.5.10). The elements in this matrix have been computed with a product Gauss–Chebyshev quadrature rule and we can estimate the error in this product rule from the one-dimensional error expansion (2) as follows. If we denote by δK_{ij} the error in evaluating (8.5.10) we have

$$\delta K_{ij} = [I_{(s)}I_{(t)} - Q_{(s)}Q_{(t)}][K(s, t)T_i(s)T_j(t)].$$

Introducing the intermediate function $P_j(s)$ and its expansion coefficients $\mathbf{p}^{(j)}$:

$$P_j(s) \equiv \int_{-1}^{1} \frac{T_j(t)K(s, t)}{(1 - t^2)^{\frac{1}{2}}} dt = I_{(t)}[K(s, t)T_j(t)]$$

$$= \sum_{r=0}^{\infty}{}' p_r^{(j)} T_r(s) \tag{6}$$

we have

$$K_{ij} = I_{(s)}T_i(s)P_j(s)$$

and it follows from (4) that

$$K_{ij} - Q_{(s)}T_i(s)P_j(s) = -\sum_{k=1}^{\infty} (p_{2Nk-i}{}^{(j)} + p_{2Nk+i}{}^{(j)}). \tag{7}$$

Now inserting the definition of $P_i(s)$ and introducing a second auxiliary function $Q_i(t)$ and its expansion coefficients $\mathbf{q}^{(i)}$:

$$Q_i(t) = \frac{\pi}{N} \sum_{k=0}^{N}{}'' K\left(\cos\left(\frac{k\pi}{N}\right), t\right) \cos\left(\frac{k\pi i}{N}\right)$$

$$= \sum_{l=0}^{\infty}{}' q_l^{(i)} T_l(t) \tag{8}$$

and using again the one-dimensional error expansion (4), we find

$$\delta K_{ij} = - \sum_{k=1}^{\infty} (p_{2Nk-i}^{(j)} + p_{2Nk+i}^{(j)} + q_{2Nk-j}^{(i)} + q_{2Nk+j}^{(i)})$$

$$\equiv -\frac{2}{\pi} \sum_{k=1}^{\infty} (\bar{K}_{2Nk-i,j} + \bar{K}_{2Nk+i,j} + \bar{K}_{i,2Nk-j} + \bar{K}_{i,2Nk+j}) \qquad (9)$$

where we have noted that for example

$$p_i^{(j)} \equiv \frac{2}{\pi} \int_{-1}^{1} \frac{P_j(s)T_i(s)ds}{(1-s^2)^{\frac{1}{2}}}$$

$$= \frac{2}{\pi} \int_{-1}^{1} \int_{-1}^{1} \frac{T_i(s)T_j(t)K(s,t)}{(1-s^2)^{\frac{1}{2}}(1-t^2)^{\frac{1}{2}}} ds\,dt \equiv \frac{2}{\pi} \bar{K}_{ij}.$$

Now introducing a bound of the form (9.4.2) for $\bar{K}_{ij}$:

$$|\bar{K}_{ij}| \le C_k \hat{i}^{-u} \hat{j}^{-v} \qquad (10)$$

we find, on bounding the sums in (9) and using the following result (see (2.7.28a))

$$\sum_{j=N+1}^{\infty} (j-i)^{-r} \le (N+1-i)^{-r+1} \frac{r}{r-1}, \quad \text{provided} \quad r > 1, \qquad (11)$$

that

$$|\delta K_{ij}| \le C_k' \left[\left(1 + \frac{2^{1-u}u}{u-1}\right)N^{-u} + \left(1 + \frac{2^{1-v}v}{v-1}\right)N^{-v} \right] \qquad (12)$$

which shows explicitly that the matrix $\bar{\mathbf{K}}$ is computed accurately provided $K(s,t)$ is smooth. It remains to obtain a bound on δB_{ij}. Since $\mathbf{B}$ is found by multiplying two Chebyshev series together, this requires a detailed analysis of the algorithm used to do this and we quote only the result, in a form which is slightly weaker than necessary for application to Chapter 8, but suitable for later use in Chapter 11 (see Delves, Abd-Elal, Hendry, 1979): Let $\mathbf{B}$ be found by Chebyshev multiplication of the matrix of coefficients $\mathbf{K}$ by the matrix of Chebyshev coefficients H_{ij} of a function $H(s,t)$, using the FFT algorithm (Algorithm 8.5.1) described in Chapter 8; let H_{ij} be known exactly and satisfy

$$|H_{ij}| \le C_H. \qquad (13)$$

Then

$$|\delta B_{ij}| \le CC_k C_H \left(\frac{u-1}{u-2} N^{2-u} + \frac{v-1}{v-2} N^{2-v} \right) \qquad (14)$$

where C is a constant of order unity.

For the Fast Galerkin algorithm, $H(s,t) = (1-t^2)^{\frac{1}{2}}$ and this bound can be improved; however (14) is used in Chapter 11, where $H(s, t)$ will represent a more general ill-behaved function.

From (14) we derive a computable error estimate as follows:

(i) $\dfrac{u-1}{u-2}, \quad \dfrac{v-1}{v-2} \approx 1$

(ii) $C_k N^{-u} \approx |K_{N0}|; \quad C_k N^{-v} \approx |K_{0N}|$

(iii) from (12), $\|\mathbf{H}\|_\infty \approx N C_H$

(iv) Hence, $|\delta B_{ij}| \lesssim N \|\mathbf{H}\|_\infty (|K_{0N}| + |K_{N0}|).$ (15)

The error bounds of this chapter are those used in producing the error estimates of Tables 8.5.1, 8.5.2; see also the error estimates given in Section 11.6.2. These tables show that the estimates reflect rather well the order of the actual error; the ability to provide cheap estimates in this way is one advantage of the Galerkin scheme.

9.8   Iterative solution of the equations $\mathbf{L}\mathbf{a}^{(N)} = \mathbf{y}$

We now turn to solution techniques for the Galerkin equation (9.6.1) or (9.6.2) with orthonormal or orthogonal basis. A direct solution by Gauss elimination will have an operation count of $\mathcal{O}(N^3)$ for an $(N + 1)$-term expansion, and we seek instead to construct an iterative solution algorithm with guaranteed and rapid convergence. Such an algorithm, if one exists, will have an associated cost of $\mathcal{O}(N^2)$. We will show that the structure of the matrix $\mathbf{L}$ (the coefficient matrix of equations (9.6.1) or (9.6.2)), namely that of 'block weakly asymptotic diagonality', along with some extra conditions on $\mathbf{L}$ (see Theorem 9.8.1) guarantees the existence of a rapidly convergent iterative algorithm for systems of this type.

These guarantees can strictly only be given in the absence of quadrature errors in setting up the equations; in practice, however, the stability of the equations implies that the numerical errors do not affect the convergence properties, and for simplicity in this section we assume that the equations are exact.

We shall need the following definitions:

Definition 9.8.1. The matrix $\mathbf{L}$ is (*block*) *weakly asymptotically diagonal* (WAD) if it is partitioned into blocks $\mathbf{L}_{ij}$ in such a manner that, for each i, $\mathbf{L}_{ii}$ is a non-null square matrix, and in some norm $\|\cdot\|$, and for some constant C

$$F_{ij} = \frac{\|\mathbf{L}_{ij}\|}{(\|\mathbf{L}_{ii}\|^{\frac{1}{2}} \|\mathbf{L}_{jj}\|^{\frac{1}{2}})} \leq C A_{ij}$$

where the matrix $\mathbf{A} = \{A_{ij}\}$ satisfies certain inequalities given in Delves and Freeman (1981), Chapter 7. These inequalities form a natural (although not obvious) generalisation of those in Definitions 9.5.1 and 9.5.2. $\mathbf{L}$ is said to be *normalised* if $\|\mathbf{L}_{ii}\| = 1$, $\forall i$. In the case where $\mathbf{L}_{ij}$ is a scalar, $\mathbf{L}$ is said to be *point weakly asymptotically diagonal*.

Note that an asymptotically diagonal matrix can always be normalised by use of the transformation

$$\mathbf{L} \rightarrow \mathbf{\Lambda}^{-1}\mathbf{L}\mathbf{\Lambda}^{-1}$$

where $\mathbf{\Lambda} = \mathbf{diag}\,(\|\mathbf{L}_{ii}\|^{\frac{1}{2}}\mathbf{E}_i)$ is a block diagonal matrix and $\mathbf{E}_i$ is a unit matrix whose size is governed by the (i, i) block of $\mathbf{L}$. Then instead of solving

$$\mathbf{L}\mathbf{a}^{(N)} = \mathbf{y} \tag{1}$$

we may solve the system

$$(\mathbf{\Lambda}^{-1}\mathbf{L}\mathbf{\Lambda}^{-1})\mathbf{\Lambda}\mathbf{a}^{(N)} = \mathbf{\Lambda}^{-1}\mathbf{y}. \tag{1a}$$

***Definition* 9.8.2.** Let $(\mathbf{L})_{ij}$ be positive and semi-monotone increasing in j, bounded above and semi-monotone decreasing in i. Moreover let

$$\|\mathbf{L}\|_{\infty} = \mathscr{C} < \infty$$

and

$$\|\mathbf{L}\|_1 = \Omega < \infty.$$

Then $\mathbf{L}$ *has property* MN (for monotonicity and norm).

The following important theorem provides us with the existence of a convergent block iterative scheme for the solution of (1):

***Theorem* 9.8.1.** Let $\mathbf{L}$ be block WAD, normalised, with property MN and suppose that for some Q which is independent of N, and for i sufficiently large, the diagonal blocks $\mathbf{L}_{ii}$ satisfy

$$\|\mathbf{L}_{ii}\|\,\|\mathbf{L}_{ii}^{-1}\| \leq Q.$$

Then there exists an integer $M(<N)$, independent of N, such that block Jacobi and block Gauss–Seidel iterations on equations (1), with leading super block of order M blocks and subsequent blocks of order 1 block, are convergent. Moreover, for any δ independent of N, M may be chosen such that the spectral radius ρ of the iteration matrix satisfies

$$\rho < \delta.$$

***Proof*.** See Delves (1977).

This theorem leads directly to a rapidly convergent iterative scheme for a Galerkin calculation with orthogonal basis. Consider first the case of an orthonormal basis, with defining equations given by (9.6.1). These equations are already 'almost' normalised in the sense of Definition 9.8.1, and it follows from the bounds of Sections 9.6, 9.7 that the matrix $\mathbf{L} = \mathbf{I} - \lambda\mathbf{B}$ satisfies (in general)[†] the conditions of Theorem 9.8.1. We therefore partition the equations (9.6.1) in the form

$$\mathbf{L} = \mathbf{I} - \lambda\mathbf{B} = \mathbf{L}_0 + \delta\mathbf{L}$$

where $\mathbf{L}_0$ contains the leading $M \times M$ submatrix of $\mathbf{L}$ together with its main diagonal, and iterate using the scheme

$$\mathbf{L}_0\mathbf{a}^{(k+1)} = \mathbf{y} - \delta\mathbf{L}\mathbf{a}^{(k)}$$

or equivalently

$$\mathbf{a}^{(k+1)} = \mathbf{a}^{(k)} + \delta^{(k+1)},$$ (2)
$$\mathbf{L}_0\delta^{(k+1)} = \mathbf{y} - \mathbf{L}\mathbf{a}^{(k)},$$

where we have dropped the superscript N on $\mathbf{a}$.

Theorem 9.8.1 then guarantees that (in general[†]), there exists an M, independent of N, such that the scheme (2) is convergent with spectral radius $\rho \leq \rho_0 < 1$; the number of iterations required to achieve a fixed accuracy is then independent of N, and the total cost of the scheme is $\mathcal{O}(N^2)$.

This result requires of course that we find a suitable M; in practice, the choice of M presents no problems (see Delves (1977) for further details) and M may be increased if convergence proves to be too slow. We require also a direct method to solve the system (2) with matrix $\mathbf{L}_0$; for an initial choice of M Gauss elimination with row interchanges proves appropriate since further rows may be added if necessary without further interchanges.

9.8.1   Iterative scheme for the Fast Galerkin method

The Fast Galerkin scheme requires the solution of equation (9.6.2) where the matrix $\mathbf{B}$ is an approximation to the matrix $\bar{\mathbf{B}}$ of equation (9.1.14). We recall from Section 8.5 that the coefficients $\bar{B}_{ij}$ are given by

$$\frac{\pi}{2}\bar{B}_{ij} = \bar{K}_{ij} - \sum_{l=1}^{\infty}\frac{1}{(4l^2-1)}(\bar{K}_{i,j+2l} + \bar{K}_{i,|j-2l|}), \quad i,j = 0,1,\ldots,$$ (3)

where

$$\bar{K}_{ij} = \int_{-1}^{1}\frac{T_i(s)}{(1-s^2)^{\frac{1}{2}}}\int_{-1}^{1}K(s,t)\frac{T_j(t)}{(1-t^2)^{\frac{1}{2}}}\,dt\,ds, \quad i,j = 0,1,\ldots,N,$$ (4)

can be approximated by K_{ij} using a product $(N+1)$-point Gauss–Chebyshev rule. $\mathbf{K}$ can be evaluated using FFT techniques in $\mathcal{O}(N^2 \ln N)$ operations and an approximation $\mathbf{B}$ to the matrix $\bar{\mathbf{B}}$ may be obtained by

---

[†] If λ is a characteristic value of the integral operator, the theorem breaks down; if λ is 'close to' a characteristic value the theorem may be valid only for large N and is then not very useful.

using equation (3) with the convention $K_{ij} = 0, i,j > N$. The total cost of the solution of equations (9.6.2) includes the cost of kernel and driving term evaluations, the set-up time and the solution time for the equations; production of the matrix $\mathbf{B}$ has an $\mathcal{O}(N^3)$ cost component and we now show that a minor extension of the iterative algorithm avoids the need for the explicit formation of $\mathbf{B}$ and its associated $\mathcal{O}(N^3)$ cost. We note that equation (3) can be written in the form

$$\mathbf{B} = \mathbf{KC}$$

where $\mathbf{C}$ is the matrix with elements

$$C_{2m+1,2n} = C_{2m,2n+1} = 0,$$

$$\frac{\pi}{2}C_{ij} = -\left[\frac{1}{(i-j)^2 - 1} + \frac{1}{(i+j)^2 + 1}\right],$$

$$i,j \text{ both odd or both even, } i > 0, \tag{5}$$

$$C_{00} = \frac{2}{\pi}; \; C_{0j} = \tfrac{1}{2}C_{j0}, \quad j > 0.$$

Thus if we partition the matrices $\mathbf{D}, \mathbf{B} = \mathbf{KC}$ and the vectors $\mathbf{a}^{(N)}$, $\mathbf{y}$ in equation (9.6.2) so that

$$\mathbf{K} = \begin{pmatrix} \mathbf{K}_{00} & \mathbf{K}_{01} \\ \mathbf{K}_{10} & \mathbf{K}_{11} \end{pmatrix}, \text{ etc.} \tag{6}$$

where $\mathbf{K}_{00}$ is the leading $M \times M$ submatrix of $\mathbf{K}$, the following block iterative scheme (where again we drop the superscript N on $\mathbf{a}$)

$$\mathbf{g}^{(k)} = \mathbf{C}\mathbf{a}^{(k)}, \tag{7}$$

$$(\mathbf{D}_{00} - \lambda\mathbf{K}_{00}\mathbf{C}_{00})(\mathbf{a}_0^{(k+1)} - \mathbf{a}_0^{(k)}) = \mathbf{y}_0 - \mathbf{D}_{00}\mathbf{a}_0^{(k)} + \lambda(\mathbf{K}_{00} \; \mathbf{K}_{01})\mathbf{g}^{(k)},$$

$$\mathbf{D}_{11}\mathbf{a}_1^{(k+1)} = \mathbf{y}_1 + \lambda(\mathbf{K}_{10} \; \mathbf{K}_{11})\mathbf{g}^{(k)},$$

is convergent, for sufficiently large M, to the solution of (9.6.2). The order of M required to guarantee convergence is independent of N and the spectral radius of the resulting iteration matrix may be made as small as desired by choosing M sufficiently large. For M sufficiently large, the spectral radius is bounded away from unity so that the number of iterations required to achieve a given accuracy is independent of N.

The main attraction of this scheme is that we obtain the complete solution of (9.6.2) to fixed accuracy in $\mathcal{O}(N^2)$ operations, since we dispense with the explicit formation of $\mathbf{B}$. Only the leading $M \times M$ submatrix of $\mathbf{B}$ need be calculated explicitly, the cost per iteration being approximately $2N^2(M+2) + \mathcal{O}(MN)$ operations.

Exercises

1. A square infinite unsymmetric matrix $\mathbf{L}$ is said to be *asymptotically diagonal of type A and of degree* (p_L, p_U) if the elements L_{ij} of every

finite submatrix $\mathbf{L}_N$ satisfy

$$\frac{|L_{ij}|}{(|L_{ii}||L_{jj}|)^{\frac{1}{2}}} \le \begin{cases} (C_L)_j \, i^{-p_L}, & i \ge j, \\ (C_U)_i \, j^{-p_U}, & i \le j \end{cases}$$

where $(C_L)_j, (C_U)_i, p_L, p_U > 0$.

Show that these properties are invariant under a diagonal transformation of $\mathbf{L}$ and that the product of two asymptotically diagonal matrices of sufficiently high degree (how high?) is asymptotically diagonal.

2. The defining equations for the continuous Galerkin method are

$$\mathbf{L}_G \mathbf{a} = \mathbf{y}.$$

Consider the perturbed system

$$(\mathbf{L}_G + \delta\mathbf{L}_G)(\mathbf{a} + \delta\mathbf{a}) = \mathbf{y} + \delta\mathbf{y}$$

where $\delta\mathbf{L}_G$ is the quadrature error in the matrix $\mathbf{L}_G$ and the computed vectors $\mathbf{a}$, $\mathbf{y}$ have errors $\delta\mathbf{a}$, $\delta\mathbf{y}$ respectively. Show, providing that $Q\|\delta\mathbf{L}_G\| < 1$, where $Q = \|\mathbf{L}_G^{-1}\|$, that

$$\|\delta\mathbf{a}\| \le \frac{Q(\|\delta\mathbf{L}_G\|\,\|\mathbf{a}\| + \|\delta\mathbf{y}\|)}{(1 - Q\|\delta\mathbf{L}_G\|)}.$$

3. The coefficients K_{ij} in the expansion

$$K(s,t) = \sum_{i=0}^{\infty}{}' \sum_{j=0}^{\infty}{}' K_{ij} T_i(s) T_j(t)$$

are bounded by

$$|K_{ij}| \le C_k \hat{i}^{-u} \hat{j}^{-v}$$

where

$$\hat{n} = \begin{cases} n, & n > 0, \\ 1, & n = 0. \end{cases}$$

Show, using the relation (9.7.11), that

$$\left| \sum_{r=1}^{\infty} (K_{2Nr-i,j} + K_{2Nr+i,j}) \right| \le C_k \left(1 + \frac{2^{1-u}u}{u-1} \right) N^{-u}$$

where $i \le N$.

(*HINT.* Write the dominant series as

$$\sum_{r=1}^{\infty} (2Nr - i)^{-u} = (2N - i)^{-u} + \sum_{r=1}^{\infty} \{2N(r+1) - i\}^{-u}$$

and use $i \le N$ to obtain upper bounds for the elements in both series).

10

Numerical performance of algorithms for Fredholm equations of the second kind

10.1   The comparison of algorithms

As we have introduced and discussed the various possible schemes for solving Volterra and Fredholm equations of the second kind, we have made a number of general qualitative and quantitative comparisons of the methods, in terms of their cost (for a given discretisation size N), their likely accuracy (for a given N) and rate of convergence (as N increases), their stability and the ease with which usable error estimates can be provided for the method. It is unfortunately a truism that the performance of a method in practice is also affected by apparently minor details of the implementation (that is, of the program). It is also true that comparisons based on one or two numerical examples can be misleading; an overall picture of the performance of a method depends on its behaviour over a wide class of problems. In this chapter we consider in some detail how objective comparisons between methods can be made and we illustrate the problems by looking at the performance of several specific routines for solving second kind Fredholm equations. These detailed comparisons are very interesting but, when two different algorithms give fairly close results, the reader is warned of the dangers of mentally ranking the algorithms: different implementations of the algorithms might well rank in the reverse order, so that only quite large differences in performance should be taken seriously.

Implementations of a given algorithm can be of two main types, namely *automatic* or *non-automatic* routines. Non-automatic routines require the user to specify the limits of integration, routines for computing the kernel function and driving term, and a parameter N which usually specifies the number of points in the quadrature rule used to replace the integral in (0.3.2) or (0.3.4) according as we solve a Fredholm or Volterra equation of the second kind. The distance between successive quadrature points is referred to as the *step size* or *steplength*; thus an N-point Newton–Cotes rule uses a fixed or constant steplength while the higher order N-point Gauss rule uses a variable steplength. In the case where an expansion method is used to solve problems of the form (0.3.2), N refers to the number of terms used. Non-automatic routines may or may not provide an estimate of the error in the returned N-point solution. However the only way in which results correct to some prescribed accuracy or *tolerance* may be obtained, using such methods, is to re-solve the original problem recursively, each time using a larger value of N, until two consecutive sets of approximate function values agree to the required accuracy. (Usually, in the case of successive applications of the Newton–Cotes rules, the value of N is doubled, that is the steplength is halved, since this enables previously computed kernel values to be used again.)

Ideally the user should be able to specify a tolerance ε which defines the maximum absolute error he is prepared to accept in all computed function values; then, given an estimate of the quadrature error incurred at each step, the routine may use this estimate to choose an optimal value of the steplength which will guarantee that the criterion is satisfied. Clearly, such routines, termed automatic routines, take into account the nature of the integrand; thus in regions where the kernel function has a so-called 'difficult' feature, for example in the form of a strong peak, an automatic routine may choose a relatively small steplength (compared to the length of the interval of integration, or more importantly, to previous steplengths used in regions where such features are not present). Automatic routines thus require the user to specify, in addition to the limits of integration and routines for computing the kernel function and driving term, the parameter ε and an upper bound N on the number of kernel evaluations. The choice of ε must be made with care; if ε is chosen too large and the automatic routine employs an iterative scheme, whereby successive approximations are produced until two agree to within the tolerance ε, spurious agreement may lead the routine to accept a value which is quite unsatisfactory (see Davis and Rabinowitz, 1975); on the other hand if ε is chosen too small, an excessive number of kernel evaluations may lead to automatic termination of the routine when the value of N is exceeded, and in addition round-off errors may accumulate to such an extent that the returned solution is totally unreliable. Of course it is the aim of the routine designer to overcome such problems but in practice his success is usually limited.

The aim of this chapter is to illustrate the comparative performance of currently available routines, both automatic and non-automatic, by using them to solve a series of test problems of the form (0.3.2) and to determine which routine is 'best' in the sense of timing, accuracy and reliability. Of course, to define what we mean by 'best' we need to declare a particular test strategy which can be applied uniformly to each routine and we therefore begin by discussing briefly the difficulties involved in producing a valid comparison between routines.

Test packages for comparing competing routines have proved useful in a number of fields, for example, nonlinear optimisation, numerical quadrature, solution of ordinary differential equations. Most of the test packages which have been developed use a common methodology: the 'Battery' method. This approach considers a set of problems which have known solutions and which vary in difficulty from easy to hard. Each problem is solved by the routine under consideration, perhaps several times with different input parameters (in the case of non-automatic routines we use several different values of N and in the case of automatic routines several values of the input tolerance). Certain data, such as the time taken, the actual error and, usually only in the case of automatic routines, the number of function evaluations, are collected at the end of each run. Finally, these values are averaged out so as to give, hopefully, an indication of how the routine would cope with a 'typical' problem; these average values are then compared with values obtained using other routines which have undergone exactly the same procedure.

Examples of the Battery method in various fields are described by Hull *et al.* (1972) for ordinary differential equations, Enright, Hull and Lindberg (1975) for stiff ordinary differential equations, Casaletto, Pickett and Rice (1969) for automatic quadrature routines and Kahaner (1971) also for automatic quadrature routines.

The Battery method, however, rests on two basic assumptions, namely
  (i) that there is a 'best' routine for solving all problems of the, usually wide, class considered.
  (ii) more importantly, that problems of the same difficulty but with slightly different parameters will deliver similar results (timings, actual error, etc.).

These assumptions are not necessarily valid. For example, in the field of numerical quadrature, Lyness and Kaganove (1977) show that, of currently available automatic routines, those which do well with 'smooth' integrands are based on high-order methods, while those which perform best on 'difficult' integrands are based on low-order methods. Hence it would seem to be better to look for a routine which is the best for solving particular types of problem. The objection to the second assumption is best illustrated by an example.

Consider the Fredholm problem in family 1 defined in Section 10.2. The kernel for this problem has a ridge whose height is governed by the value of c and whose position and orientation are governed by the parameters p_2, p_3 and p_4. For a given value of c we would expect that the difficulty of the problem would be roughly the same whatever the values of p_2, p_3 and p_4. Thus for a particular choice of p_1, p_2, p_4 and c we might expect solution of the problem, for two different values of p_3, to yield actual errors of roughly the same magnitude. In fact, we can plot the actual error against the value of p_3, for a range of such values, to yield the 'performance profile' of the chosen method with this problem (Lyness and Kaganove, 1976). In particular, if we take $p_1 = p_2 = p_4 = 0.5$, $c = 0.1$ and solve the Fredholm problem in family 1 over the range $a \le s \le b$, using the NPL non-automatic routine FRED2B (NAG Library routine DO5ABF) with $N = 15$, we obtain the following results:

$p_3 = 0.32$, actual error $= 1.405 \times 10^{-4}$;

$p_3 = 0.08$, actual error $= 3.976 \times 10^{-6}$.

We note here a factor of about 35 difference in the error. The approximate profile for the integral equation routine FRED2B with the above problem is given in Figure 10.1.1.

Examples of performance profiles for automatic quadrature routines are given in Lyness and Kaganove (1977) where it is shown that the profiles may not even be continuous. The performance profiles of different routines for the same problem are not similar and may cross each other at several places in the range under consideration. Hence with the Battery method one routine would appear better (perhaps much better) than another for certain

Figure 10.1.1. The accuracy achieved by routine FRED2B on problem family 1 with $p_1 = p_2 = p_4 = 0.5$ and p_3 varied over the range $[0,1]$. Note the logarithmic scale for the error.

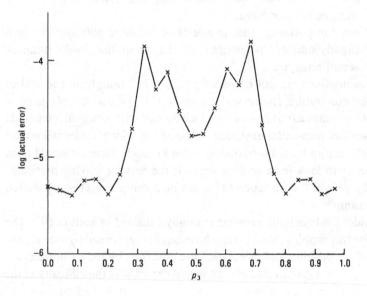

choices of parameter but worse (perhaps much worse) for others, giving misleading comparisons if only one value of parameter is used in a test.

Problems 1–3 of Section 10.2 define a whole class of closely linked problems in which the difficulty (here, the position of the ridge in the kernel) can be moved about the range of interest by varying the values of p_1, p_2, p_3 and p_4. Such problems are called 'problem families'; the basic strategy advocated by Lyness and Kaganove is to carry out statistical tests on single problem families in order to take account of rapid variations in performance profiles, and to seek a 'best' routine only for a single problem family or for a set of related problem families. We give below results based on such comparisons for Fredholm routines.

10.2 Non-automatic routines for second kind Fredholm equations

We apply this strategy to the testing of non-automatic routines for the Fredholm equation

$$x(s) = y(s) + \int_a^b K(s,t)x(t)dt. \tag{1}$$

The routines we shall test are all designed to handle 'smooth' equations; that is, they make no allowance for any analytic singularities in the kernel or driving terms. The following families of test problems reflect this by involving only analytic kernels. Nevertheless, each contains a 'difficult' feature in the form of a strong peak in the kernel and/or driving term. The inherent difficulty of the problem is governed by the ratio

height of peak/half-width

where the 'half-width' is defined as one half of the width of the peak at the point where its height is half the maximum height. This ratio is held constant within a given family but the position and orientation of the peak is allowed to vary. We construct the following families of test problems:

Family 1

$$K(s,t) = c/\{[(1 - p_4)(s - p_2) - p_4(t - p_3)]^2 + c^2\},$$

$$y(s) = p_1 + \frac{p_1}{p_4}\left\{ \arctan\left[\frac{(1 - p_4)(s - p_2) - p_4(b - p_3)}{c}\right] \right.$$

$$\left. - \arctan\left[\frac{(1 - p_4)(s - p_2) - p_4(a - p_3)}{c}\right]\right\}.$$

Solution:

$$x(s) = p_1.$$

Parameter values: $a = 0$, $b = 1$, $c = 0.1$.

The kernel for this problem has a smooth ridge of height $1/c$, whose position and orientation are determined by p_2, p_3, p_4; see Figure 10.2.1(a).

Family 2

$$K(s,t) = c^2/[(s - p_2)^2 + c^2][(t - p_3)^2 + c^2],$$

$$y(s) = p_1 - cp_1\left[\arctan\left(\frac{b - p_3}{c}\right)\right.$$

$$\left. - \arctan\left(\frac{a - p_3}{c}\right)\right]/[(s - p_2)^2 + c^2].$$

Solution:

$$x(s) = p_1.$$

Parameters: $a = 0$, $b = 1$, $c = 0.2$.
The kernel for this problem has a hyperbolic spike of height $1/c^2$; see Figure 10.2.1(b).

Family 3

$$K(s,t) = c/[(1 - p_4)(s - p_2)^2 + p_4(t - p_3)^2 + c^2],$$

$$y(s) = p_1 - p_1 d^{-1}cp_4^{-\frac{1}{2}}\left\{\arctan\left(\frac{b - p_3}{d}\right) - \arctan\left(\frac{a - p_3}{d}\right)\right\},$$

where $d^2 = c^2 + (1 - p_4)(s - p_2)^2$.

Solution:

$$x(s) = p_1.$$

Parameters: $a = 0$, $b = 1$, $c = 0.1$.
This kernel has a parabolic spike; see Figure 10.2.1(c).

Family 4

$$K(s,t) = (s - p_2)(t - p_3),$$

$$y(s) = \frac{c}{(s - p_1)^2 + c^2} - (s - p_2)c\left\{\frac{1}{2}\ln\left[\frac{(b - p_1)^2 + c^2}{(a - p_1)^2 + c^2}\right]\right.$$

$$\left. + \frac{(p_1 - p_3)}{c}\left[\arctan\left(\frac{b - p_1}{c}\right) - \arctan\left(\frac{a - p_1}{c}\right)\right]\right\}.$$

Solution:

$$x(s) = c/[(s - p_1)^2 + c^2].$$

Parameters: $a = 0$, $b = 1$, $c = 0.1$.
This problem has an 'easy' kernel but a strongly peaked driving term leading to a peak in the solution at $s = p_1$ of height $1/c$ and with half-width c.

Figure 10.2.1. Plot of the kernel functions for families 1–3. (a) Family 1: $p_2 = 0.25$, $p_3 = 0.5$, $p_4 = 0.75$, $c = 0.1$. (b) Family 2: $p_2 = 0.25$, $p_3 = 0.5$, $c = 0.2$. (c) Family 3: $p_2 = 0.25$, $p_3 = 0.5$, $p_4 = 0.75$, $c = 0.1$.

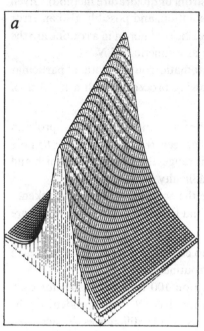

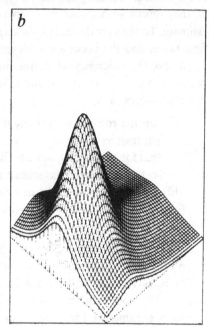

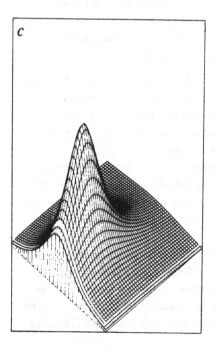

Now non-automatic routines require the user to specify a parameter which refers to either the number of terms used (for an expansion method) or the number of mesh points (for a Nystrom or quadrature method). Given N, they return an N-point or N-term solution, and possibly also an error estimate. To the user the main characteristics of interest in a routine are the time taken and the accuracy achieved as a function of N.

To test the efficiency of a non-automatic routine with a particular problem family we carry out the following procedure with a number of different values of N:

(i) run the routine with many different (random) values of problem parameters $\{p_i, i = 1, \ldots, 4\}$, between the limits $a < p_i < b$; note that for values of p_i outside this range, the peak would vanish and so too would the associated difficulty of the problem.

(ii) after each individual run note the actual error and time taken.

(iii) finally, average out the errors and times. These average values are then plotted against N.

Note that this test procedure gives no credit for extra information (such as an error estimate) returned by the routine.

The results presented here are based on 100 runs per family for each routine. We note that the time taken for fixed N is independent of the problem parameters for most routines, while differences in achieved accuracy of 20% say, are unlikely to be significant (or even noticed) in practice, so that high statistical accuracy is not required. We give results for the four problem families using five routines which are representative of the main techniques described in Chapters 2, 4 and 8:

(i) *FE2SR* (K.S. Thomas, 1976). This implements the Nystrom method with a repeated Simpson's rule, but produces also a 'corrected' solution and an error estimate based on a comparison of the uncorrected and corrected solutions. The parameter N is the total number of quadrature points used.

(ii) *D05CAB* (NAG Algol 68 Library). This also uses the Nystrom method, with a quadrature rule specified by the user from a library of rules. The results shown here were obtained with the N-point Gauss–Legendre rule. This routine produces no estimate of the error.

(iii) *D05AAA/F* (NAG Algol 60/Fortran Library). This routine implements the El-gendi method (see Section 4.5). This is a method which exploits the equivalence between the Nystrom and expansion methods (demonstrated in Chapter 8) with an N-point quadrature rule. It solves the Nystrom equations for a Clenshaw–

Curtis N-point quadrature rule and then converts the solution to a Chebyshev expansion form. No error estimate is returned.

(iv) *FFTNA* (Delves and Abd-Elal, 1977). This implements the Fast Galerkin scheme described in Chapter 8, and the error estimate of Chapter 9, returning an N-term Chebyshev expansion and an error estimate.

(v) *PMODNA* (Riddell, 1981). This is a revised version of FFTNA using an improved Fast Fourier Transform module and with improved termination strategy during the iterative solution of the linear equations.

To be fair to the routines it is important to use the same criterion for each in evaluating the error. We use the following:

$$\text{Error} = \frac{1}{20}\left(\sum_{i=1}^{20} |x_t(\xi_i) - x_N(\xi_i)|^2 \right)^{\frac{1}{2}}$$

where $x_t(\xi_i)$ is the true, and $x_N(\xi_i)$ the approximate solution at ξ_i; the ξ_i, $i = 1,\ldots,20$, are equally spaced points over the range $[a,b]$. In FFTNA, PMODNA and DO5AAA/F the solution is given as a set of coefficients of a Chebyshev series and we can compute

$$x_N(\xi_i) = \sum_{j=0}^{N-1}{}' a_j T_j(\xi_i).$$

For DO5CAB and FE2SR the solution is given only at the quadrature points and for these routines we use the 'natural' interpolation formula

$$x_N(\xi_i) = y(\xi_i) + \sum_{j=1}^{N} \bar{w}_j K(\xi_i, \bar{\xi}_j)\bar{x}_N(\bar{\xi}_j)$$

where $\bar{\xi}_j$ are the quadrature points, $\bar{w}_j$ the quadrature weights and $\bar{x}_N(\bar{\xi}_j)$ the solutions delivered by the routine.

The results we obtain for these routines are shown in Figures 10.2.2–10.2.6.

Figure 10.2.2 gives a plot of time (in arbitrary units) against N for the routines tested. The times taken are essentially independent of family, except for routines FFTNA and PMODNA which use an iterative solution of the equations. Figures 10.2.3–10.2.6 show the achieved average accuracy against N, on a logarithmic scale, for the problem families tested. We now comment on these results.

10.2.1 Timings

The results for routines FFTNA and PMODNA are consistent with the theoretical estimate $T = \mathcal{O}(N^2 \ln N)$. The routines are initially slower than the other routines tested but there is a crossover point at about $N = 27$ where FE2SR becomes slower and by extrapolation the curves of both DO5AAA/F and DO5CAB cross the FFTNA curve at about $N = 80$. In practice FFTNA and PMODNA spend a large proportion of their time in the FFT module and the difference in time between them stems from this.

Figure 10.2.2. Ridge kernel, problem family 1, $c = 0.1$.

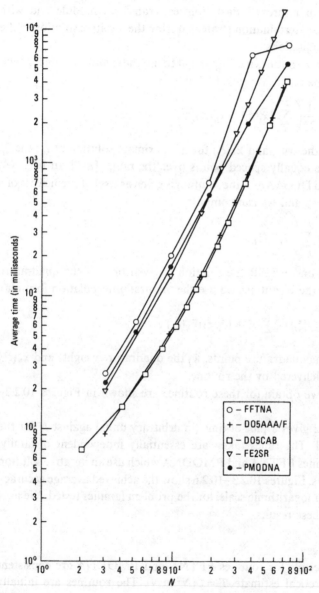

DO5CAB, DO5AAA/F and FE2SR are all based on the Nystrom method but use different quadrature rules. As N increases the time taken in solving the Nystrom equations becomes increasingly dominant; thus, since the same routine is used to solve the equations in each case (NAG routine FO4CAB), there is little difference in the timings.

Figure 10.2.3. Ridge kernel, problem family 1, $c = 0.1$.

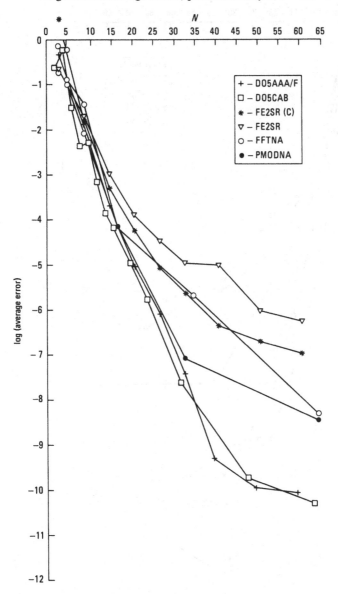

It is apparent from these results that, over the range of N considered, the 'Fast Galerkin' scheme is slower by about a factor two than the Nystrom routines for a given value of N. However, this ignores the cost of producing an error estimate with the Nystrom method. This cost is borne by routine

Figure 10.2.4. Hyperbolic spiked kernel, problem family 2, $c = 0.2$.

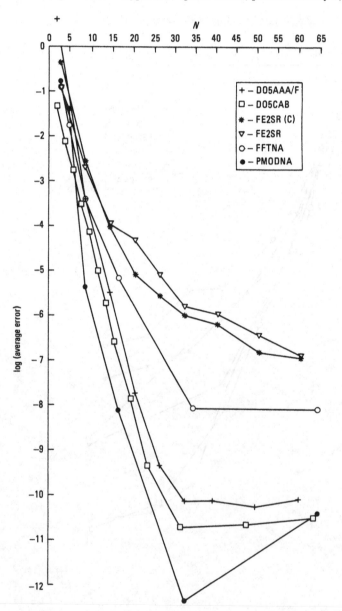

FE2SR, which takes uniformly twice as long as the 'bare' Nystrom routines DO5AAA/F and DO5CAB. We conclude that the Nystrom method and the Fast Galerkin scheme are about equally fast if error estimates are demanded.

Figure 10.2.5. Parabolic spiked kernel, probem family 3, $c = 0.1$.

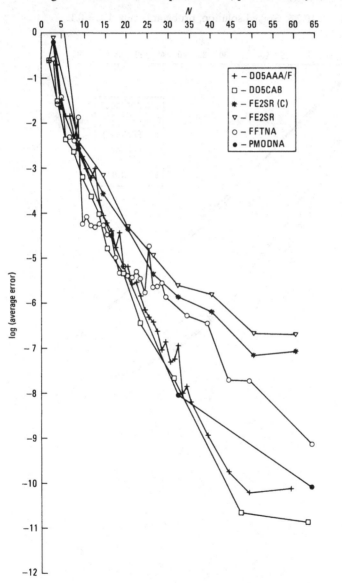

10.2.2 Accuracy

Routines DO5CAB and DO5AAA/F, based respectively on Gauss–Legendre and Gauss–Chebyshev quadrature rules, have similar error curves for each of the problems containing difficult kernels (that is, those of families 1, 2 and 3) whilst FE2SR is the least successful and even the

Figure 10.2.6. Spiked solution, problem family 4, $c = 0.1$.

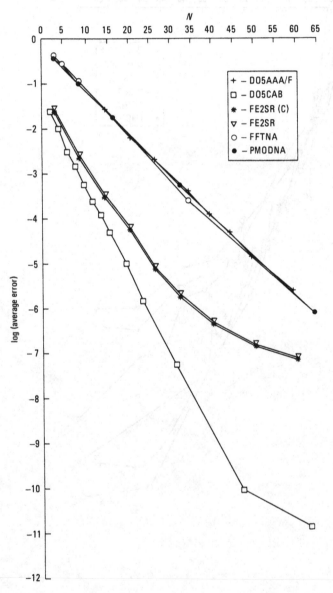

Table 10.2.1. *Routine rankings*

Criterion	Family 1	Family 2	Family 3	Family 4
	DO5CAB	PMODNA	DO5CAB	DO5CAB
Achieved	DO5AAA/F	DO5CAB	DO5AAA/F	FE2SR(C)
accuracy/N	PMODNA	DO5AAA/F	PMODNA	FE2SR
	FFTNA	FFTNA	FFTNA	FFTNA
	FE2SR(C)	FE2SR(C)	FE2SR(C)	PMODNA
	FE2SR	FE2SR	FE2SR	DO5AAA/F

correction mechanism (denoted FE2SR(C) in the graphs) fails to make it competitive here. For these problem families FFTNA compares favourably with DO5CAB and DO5AAA/F only while N is less than about 25. For larger values of N its performance appears to be affected by premature termination of the iterative solution of the equations. The termination criterion used by FFTNA and PMODNA is based on an *a priori* error estimate derived from inspection of the Galerkin matrix and right hand side vectors. The problem families considered here all have 'difficult' kernels and 'easy' solutions, and hence show considerable cancellation effects between the kernel and driving terms which are not detected by FFTNA. However, the revised estimator used in PMODNA overcomes this difficulty. These results demonstrate well how the performance of a routine can be affected by apparently minor details of the implementation.

For family 4, which contains a spiked solution, an entirely different situation exists. Again DO5CAB does well, and in fact, is considerably better than any other routine tested. Routines DO5AAA/F, FFTNA and PMODNA perform similarly, but surprisingly badly, in this instance. The behaviour of FE2SR is similar to that described for problem families 1, 2 and 3 although the effect of the correction is less marked.

We can now try to make an overall assessment of these results. Inspection of Figures 10.2.3–10.2.6 shows that, judged on the basis of achieved accuracy for a given discretisation parameter N, the routines may be ordered as in Table 10.2.1.

As expected, no routine is uniformly best for every problem class but we see that, despite the theoretical and practical care lavished on the more complicated El-gendi and Fast Galerkin algorithms, the clear winner of this contest has been the Nystrom routine DO5CAB with the N-point Gauss–Legendre rule. This routine is extremely simple; it includes a call to a routine which provides the points and weights for the quadrature rule, about twelve lines of code to set up the Nystrom equations and a call to the routine which solves these equations. Such results are enough to make a numerical analyst weep. However, at least the even simpler Simpson's rule routine was out-performed. Further, this verdict is based on the assumption that achieved accuracy (in a given time) is the appropriate figure of merit. To demonstrate the difficulties of reaching uniformly agreed ordering of routines let us change the criterion to one which is in a sense more plausible, and then re-run the contest. In a practical calculation, we shall not know the

Figure 10.2.7. Spiked solution, problem family 4, $c = 0.1$.

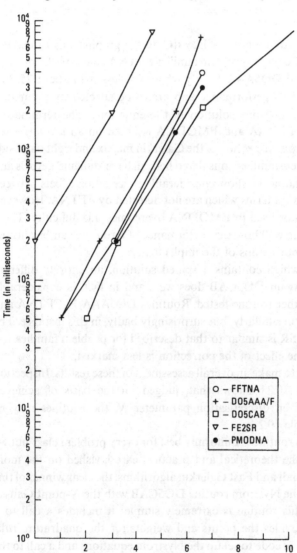

Table 10.2.2. *Routine rankings*

Criterion	Family 1	Family 2	Family 3	Family 4
	DO5CAB	PMODNA	PMODNA	DO5CAB
Accuracy + error	DO5AAA/F	DO5CAB	DO5CAB	PMODNA
estimate/time	PMODNA	FFTNA	DO5AAA/F	FFTNA
	FFTNA	DO5AAA/F	FFTNA	DO5AAA/F
	FE2SR	FE2SR	FE2SR	FE2SR

achieved accuracy. We therefore demand an error estimate and use as a figure of merit the achieved estimated error in a given time. Naturally, we require that the error estimate be a 'reasonable' one; the winner of the contest may now depend however on how the error estimation is carried out. Routines FFTNA, PMODNA and FE2SR already return error estimates which we shall use, while DO5CAB and DO5AAA/F do not. For these routines, we assume that the error estimate is obtained by re-running with a value of N which is 50% larger, and comparing the two results. Different strategies might give different results for the tests but such a strategy would be not unreasonable.

Figure 10.2.7 gives the results for problem family 4, run under these rules, while Table 10.2.2 gives the overall rankings for the routines on each problem family. Note that there is only one entry for FE2SR, since the 'corrected' solution has been used to provide an error estimate; this routine is now uniformly last in the rankings, suggesting strongly that a low-order rule forms a poor basis for such a routine. Routine DO5CAB clearly suffers in ranking by being forced to spend time producing an error estimate; however an overall assessment would probably make it, on these rules, 'joint winner' with the Fast Galerkin routine PMODNA.

A comparison of Tables 10.2.1 and 10.2.2 clearly illustrates how the routine rankings may alter as a result of changing the criterion on which the success of a particular routine is measured.

10.3 Comparing automatic routines

An automatic routine is one which, given an input tolerance ε_{inp}, attempts to adjust the discretisation order N until some measure of the error is less than ε_{inp}; it then returns an approximate solution x_N and possibly other information such as its own estimate of the achieved accuracy, and whether the result appears reliable. Again we give no credit for this extra information (or penalty for its absence) but analyse only the way in which the routine satisfies its basic task. The user has two main questions to ask of such a routine:

    (i) Speed: For a given accuracy, how fast is it?

    (ii) Reliability: How often will it achieve the requested input accuracy?

It is traditional to measure speed in terms of the number of function evaluations used; here, the number of times the kernel $K(s,t)$ is evaluated provides such a measure. However, unlike the situation in numerical quadrature, optimisation and (usually) ordinary differential equations, 'red tape' operations can completely dominate the time taken; on the examples given here, routines typically spend less than 10% of their time in evaluating the kernel $K(s,t)$. It would take a very complicated kernel to change this situation significantly and we therefore quote also the time taken in arbitrary units.

Requirements (i) and (ii) are inter-dependent: a routine writer can trade speed for reliability at the stroke of a keypunch (by setting $\varepsilon_{\text{test}} := 0.1 \times \varepsilon_{\text{inp}}$, for example). The Lyness and Kaganove (1977) analysis for automatic quadrature routines, which we follow here, attempts to disentangle the intrinsic speed of a routine from its intrinsic reliability by factoring out the effect of such possible internal scalings; they achieve this in the following manner.

Suppose we wish to evaluate a routine with a particular problem family; then for each value of input tolerance (ε_{inp}) we run the routine a number of times with random values of problem parameters. At the end of each run we note the following:

$\varepsilon_{\text{act}}(\varepsilon_{\text{inp}}, \mathbf{p}_i)$ — the actual error. This is related to the input tolerance and the values of the problem parameters $\mathbf{p} = (p_1, p_2, p_3, p_4)$

$v(\varepsilon_{\text{inp}}, \mathbf{p}_i)$ — the number of kernel evaluations. Again this is related to ε_{inp} and $\mathbf{p}$

$t(\varepsilon_{\text{inp}}, \mathbf{p}_i)$ — the time taken.

Here i is the number of the run. This information is used to construct the following statistics:

$$v(\varepsilon_{\text{inp}}) = \frac{1}{M} \sum_{i=1}^{M} v(\varepsilon_{\text{inp}}, \mathbf{p}_i),$$

$$t(\varepsilon_{\text{inp}}) = \frac{1}{M} \sum_{i=1}^{M} t(\varepsilon_{\text{inp}}, \mathbf{p}_i),$$

$$\phi(e; \varepsilon_{\text{inp}}) = \frac{1}{M} \{\text{number of values of } i \text{ for which}$$

$$|\varepsilon_{\text{act}}(\varepsilon_{\text{inp}}, \mathbf{p}_i)| \le e\},$$

where M is the total number of runs for each value of ε_{inp}; $\phi(e;\varepsilon_{inp})$ is a statistical distribution function and gives the proportion of the time an actual error e is achieved for a given input tolerance.

We now put ourselves mentally in the place of a knowledgeable (or cynical) user who will reason as follows:

(i) Let ε_{req} be the accuracy I require. No automatic routine can achieve the accuracy I want every time. I will nominate a reliability, s, and ask that it achieve the desired accuracy exactly $(100s)$ per-cent of the time. To achieve more will slow the routine down; to achieve less will be to fail.

(ii) I will make sure the routine does this by taking as input tolerance the value $E_{inp}(s,\varepsilon_{req})$, which is the value of ε_{inp} for which the tolerance of ε_{req} is obtained exactly $(100s)$ percent of the time. We find $E_{inp}(s,\varepsilon_{req})$ as the smallest value of ε_{inp} for which $s = \phi(\varepsilon_{req};\varepsilon_{inp})$.

Plots of $v(E_{inp}(s,\varepsilon_{req}))$ and $t(E_{inp}(s,\varepsilon_{req}))$ against ε_{req} then indicate the intrinsic efficiency of the routine, while a plot of $E_{inp}(s,\varepsilon_{req})$ against ε_{req} indicates the quality of the error control in the routine. The plots depend upon s, the chosen reliability parameter, but in practice the relative orderings of different routines are not sensitive to s (if they were, it would be difficult to interpret the results) and here we take the standard value $s = 0.85$ (that is, 85% reliability).

10.4   Numerical comparison of automatic routines

Given this methodology we use the same problem families as in Section 10.2 to generate comparisons of the following routines:

(i) *FFTA* (Delves, unpublished). A simple extension of the non-automatic routine FFTNA considered previously. A value of the error is estimated for a starting value of N and if this is not sufficiently small then the value of N is stepped up and the procedure is repeated.

(ii) *SIMP* (K.E. Atkinson, 1976). An Algol 68 translation of the routine IESIMP based on the Nystrom method using Simpson's rule and iterative solution of the Nystrom equations (see Section 4.6.2).

(iii) *BOOLE* (K.E. Atkinson, 1976). As SIMP with higher order quadrature rule.

(iv) *GALERKIN*. A version of FFTA based on the standard Galerkin rather than the Fast Galerkin (Delves and Abd-Elal, 1977) technique.

(v) *GAUSS*. A routine also by K.E. Atkinson similar in approach to SIMP and BOOLE, but using a sequence of Gauss–Legendre rules to provide basic approximations.

Performance results for the above routines, with problem families 1–4 are shown in Figures 10.4.1–10.4.9.

(*a*) *Problems with smooth solutions: families* 1–3. Several features of these

Figure 10.4.1. Problem family 1, $c = 0.1$, $s = 0.85$.

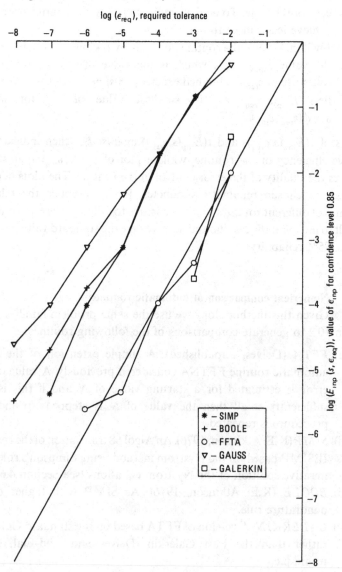

results stand out clearly:

(i) The Fast Galerkin procedure used in FFTA is much more economic in practice than the standard Galerkin procedure; indeed tests with GALERKIN were discontinued because of the costs involved.

Figure 10.4.2. Problem family 1, $c = 0.1$, $s = 0.85$.

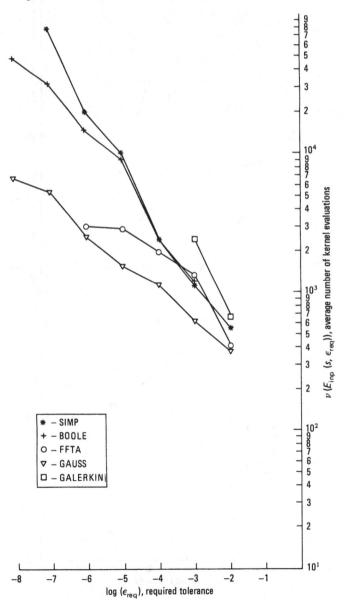

(ii) Routines based on high-order rules, namely FFTA, GALERKIN and GAUSS, are much more economical of kernel evaluations for these families than those based on low-order rules (SIMP, BOOLE).

Figure 10.4.3. Problem family 1, $c = 0.1$, $s = 0.85$.

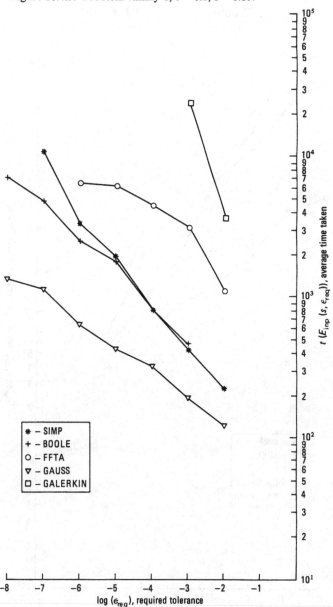

(iii) Nonetheless, the low overheads of SIMP and BOOLE make them competitive with FFTA (but not with GAUSS) for accuracies up to $\varepsilon_{req} = 10^{-6}$.

(iv) Routine GAUSS emerges as a clear overall winner for these families of problems. Note that this is basically the same con-

Figure 10.4.4. Problem family 2, $c = 0.2$, $s = 0.85$.

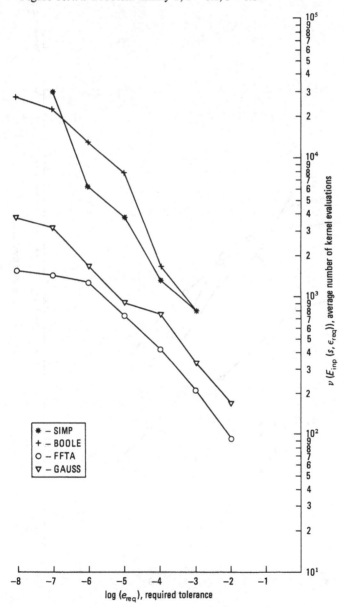

clusion as we reached for the non-automatic routine DO5CAB. For some reason not clear to us, BOOLE fails to improve on SIMP.

(b) *Problems with 'difficult' solutions: family 4.* Only results for BOOLE

Figure 10.4.5. Problem family 2, $c = 0.2$, $s = 0.85$.

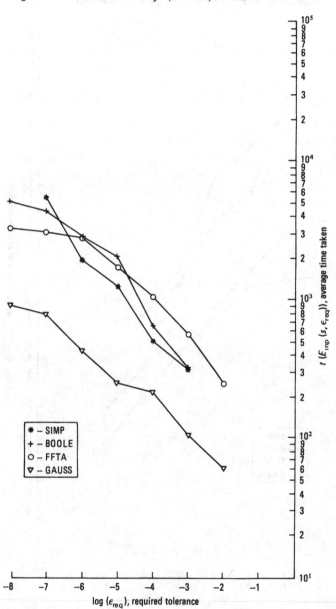

and SIMP are given here. Routine FFTA returned results which it predicted correctly were of too low accuracy to appear on the graphs, while routine GAUSS failed to compute any solution at all. Routine SIMP is therefore declared a winner in this class; the poor results exhibited by the

Figure 10.4.6. Problem family 3, $c = 0.1$, $s = 0.85$.

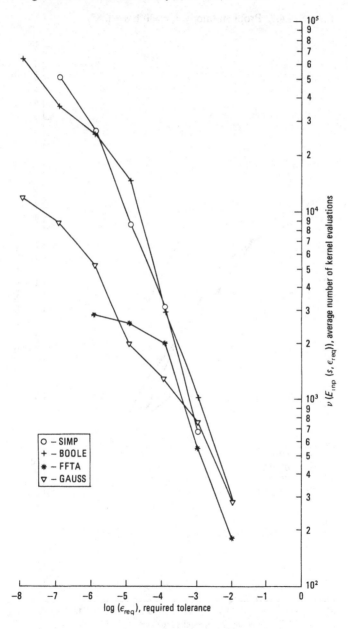

other routines, viewed against the really rather modest difficulties of this problem class, indicate the rather fragile nature of current automatic routines. This in turn is a symptom of their rather primitive nature, compared with currently available routines for automatic quadrature or for initial value ordinary differential equations.

Figure 10.4.7. Problem family 3, $c = 0.1$, $s = 0.85$.

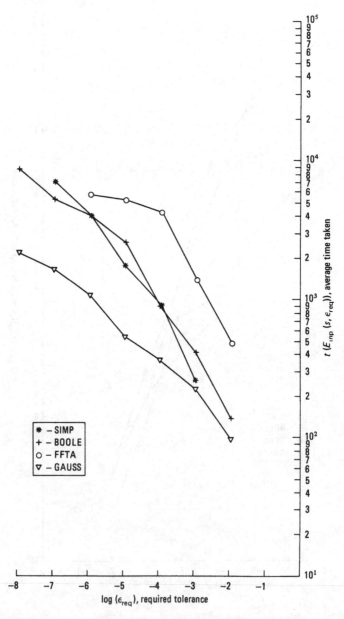

(c) Error Control. Finally, we comment on the error control results. Here, we have shown only one set of results, those for problem family 1, in Figure 10.4.1. This graph typifies very well the results for other problem families. We see that the routines each control the error rather well, but in general achieve a greater accuracy than that requested by the user. The results of

Figure 10.4.8. Problem family 4, $c = 0.1$, $s = 0.85$.

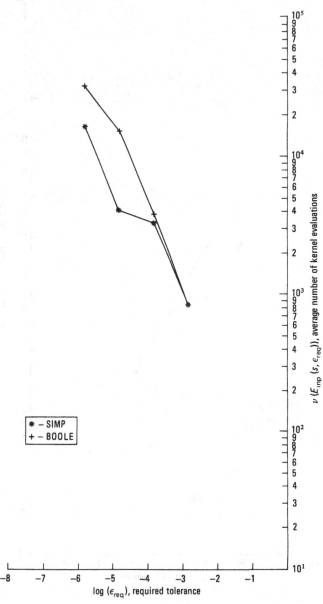

Figure 10.4.1 can in fact be satisfactorily summarised by quoting the number of additional decimal digits of accuracy achieved; the error control results for this and the other problem families are summarised in this way in Table 10.4.1.

Figure 10.4.9. Problem family 4, $c = 0.1$, $s = 0.85$.

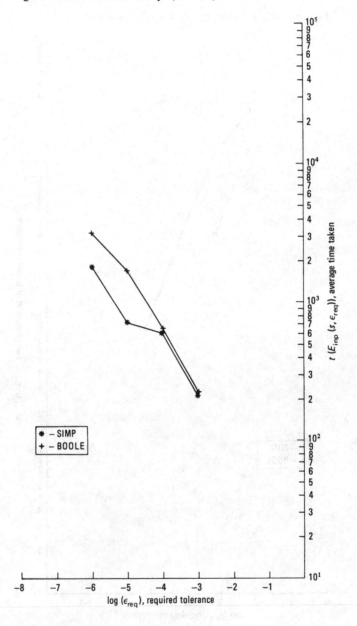

Table 10.4.1. *Error control of automatic routines for problem families 1–4. The entries in the table list the value n = (average number of correct significant figures achieved) – (number requested by the user); this number is observed in practice to be approximately independent of the requested accuracy*

Routine/problem family	1	2	3	4
FFTA	0.0	1.5–4.0	0.0–1.0	—
SIMP	1.5–2.0	1.5–2.0	1.5–2.2	0.0–0.5
BOOLE	2.0	1.2–2.5	1.5–2.2	0.2–1.0
GALERKIN	0.0	—	—	—
GAUSS	2.0–3.0	2.0–4.0	1.0–3.5	—

This table show little overall qualitative difference between the routines. On most problem families, all of the routines err on the side of caution by a considerable margin, with routine SIMP being the least cautious and routine GAUSS the most cautious. This reflects the difficulty of stopping a rapidly convergent process exactly at a particular accuracy: the process is very likely to 'overshoot'. Note however that 'caution' is a statistical concept here: the routines are cautious at least 85% of the time, but certainly not all of the time, and it appears that the routine writers all feel the user will dislike the occasional over-optimistic estimate much more than he will dislike an average pessimistic estimate of the accuracy. Note also that these graphs take no cognisance of the error estimate returned to the user (if any). It is possible for an automatic routine to obtain an accuracy greater than requested (hence appearing 'cautious' on these graphs) but to recognise this, and to return a smaller error estimate to the user. Both the FFTA and GALERKIN routines in fact do this.

The particular routines for which we report results will doubtless (we hope) be soon superseded; the purpose of the chapter is to show the difficulties of reaching objective conclusions. We can summarise our findings as follows:

(i) Comparisons based on one or two examples can be misleading and statistical tests are essential to yield a valid comparison.
(ii) No routine is uniformly good for all classes of problems, and users, faced with a choice of several routines for the solution of a particular problem, should use any knowledge which they have of the problem to influence their choice.
(iii) Important features of the chosen routine are its accuracy for a

given value of the discretisation size N and its rate of convergence as N is increased.

(iv) In addition the cost of the routine for a given value of N is an important consideration.

(v) Equally important is the practical provision of error estimates. Methods which do not yield such estimates naturally should be penalised by recognising the costs which they incur when estimates are required.

(vi) Finally we conclude that high-order methods behave better overall than low-order methods.

11
Singular integral equations

11.1   Introduction

So far we have been mainly concerned with methods for the solution of Fredholm and Volterra equations under the assumption that the kernel and driving terms are smooth functions of their arguments. When this assumption breaks down, the methods we have discussed will either fail completely (if they try to evaluate the kernel at a point where it is singular, for example) or at best converge only slowly. Singular integral equations are very common in practice, and we discuss in this chapter methods for dealing with them; the aim is, or should be, to produce not merely a method which converges, but one which takes sufficient account of the singularity that it converges as fast as methods for smooth problems.

This aim cannot always be met; when it can, it is clear that the resulting method must be tailored in some way to the form of the singularity. Fortunately a few standard types of singularity appear very common in integral equations. We shall use the term 'singularity' rather widely to refer to any lack of analyticity in the problem. Thus, for example, the following features make a problem singular:

    (i) An infinite or semi-infinite range $[a, b]$ for the integral operator.
   (ii) A discontinuous derivative in the kernel or driving term. Hence, a Green's function kernel (see for example (0.3.16)) is singular in this sense.
  (iii) An infinite or non-existing derivative of some finite order. Thus, $y(s) = (1 - s^2)^{\frac{1}{2}}$ is singular on the range $[-1, 1]$ because $y'(s)$ is unbounded at $s = \pm 1$. It is nonsingular on the range $[-\frac{1}{2}, \frac{1}{2}]$.

Methods which ignore this singularity in $y(s)$ will probably work, but converge slowly. The kernel $K(s, t) = \ln |s - t|$ is also singular on the line $s = t$ since it is unbounded on this line. Methods which evaluate $K(s, t)$ on this line will fail disastrously.

11.2   Infinite intervals

The presence of an infinite interval of integration as in the example

$$x(s) = y(s) + \int_a^\infty K(s,t)x(t)\mathrm{d}t, \quad a \le s \le \infty, \tag{1}$$

makes (1) formally singular, but rarely need cause serious problems in practice. Since any square integrable solution $x(s)$ of (1), and also $y(s)$ and $K(s,t)$, must vanish at infinity, one crude approach is to replace the upper limit in (1) by some finite value R. Although this approach is often used in practice it is *not advocated* here, since the results often converge only slowly as R is increased, while (see the cautionary examples in Section 2.6.1) the number of quadrature points needed to cover a large finite interval may be many more than would be needed were proper account taken of the infinite range. Two suitable techniques are available for this:

11.2.1   Direct treatment of the infinite range

It is not difficult to deal directly with the infinite range. For a Nystrom method, we can use a quadrature rule constructed for this range and examples of such rules are given in Section 2.6.2. As noted there, it is important to choose a rule whose annihilation class reflects as closely as possible the structure of the equation; in particular, a rule suitable for functions which decay rapidly at infinity (such as the Gauss–Laguerre rule) should not be used for problems having kernels or driving terms with long tails.

Similarly a direct treatment for an expansion method requires only that we choose a set of expansion functions defined on the infinite interval. Possible and obvious choices for the interval $[0, \infty)$ include

$$h_n(s) = \mathrm{e}^{-\alpha s} L_n(\alpha s), \tag{2}$$

$$h_n(s) = (1+z)^q T_n(z), \tag{3}$$

where

$$z = \left(\frac{2\alpha}{s+\alpha}\right) - 1 \tag{3a}$$

and α, q are parameters.

The first of these might be appropriate if the solution $x(s)$ is known to decay rapidly for large s. The latter expansion set behaves like $(s+\alpha)^{-q}$ for large s and hence might be appropriate if the solution is known to decay only slowly; note that, since we seek a square integrable solution $x(s)$, it is guaranteed to decay for large s faster then $s^{-\frac{1}{2}}$.

11.2.2   Mapping onto a finite interval

An alternative to a direct calculation is to map the integral equation onto a finite interval and then solve the finite interval equation.

For example, setting the lower limit $a = 0$ in (1) and introducing the change of variables

$$\left.\begin{aligned} z &= \frac{2\alpha}{s+\alpha} - 1, \\ \tau &= \frac{2\alpha}{t+\alpha} - 1, \end{aligned}\right\} \tag{4}$$

we find that (1) takes the form

$$X(z) = Y(z) + \int_{-1}^{1} \mathscr{K}(z,\tau) X(\tau) d\tau, \quad -1 \le z \le 1, \tag{4a}$$

where

$$X(z) = x(s),$$

$$Y(z) = y(s),$$

$$\mathscr{K}(z,\tau) = \frac{2\alpha K(s,t)}{(\tau+1)^2}. \tag{4b}$$

We can now use a method designed for the finite interval $[-1,1]$; but note:

(i) The kernel $\mathscr{K}(z,\tau)$ is almost certainly singular at $\tau = -1$, due to the Jacobian factor $(\tau+1)^{-2}$. Hence closed rules cause obvious problems, while, whatever rule is used, the singularity may affect the accuracy achieved.

(ii) The use of a finite interval rule on (4a) is always equivalent to the use of the infinite interval rule formed from that rule by carrying out the inverse mapping, directly on the original equation (1). See Section 2.6.3 for a discussion of such mapped rules; the mapping introduced above is (the inverse of) the Gauss–rational mapping introduced there. We would always advocate the use of a mapped rule, rather than first mapping the problem onto a finite interval: not because there is any essential difference in these approaches, but because physical insight (or rather knowledge) concerning the solution is always useful, and that seems to be most easily expressed in the original coordinate space.

11.2.3 An example

We illustrate the comment made above with the integral equation

$$x(s) = y(s) + \int_{0}^{\infty} K(s,t) x(t) dt, \tag{5}$$

$$y(s) = (1+s^2)^{-\frac{1}{2}} \left(\frac{\pi}{2} e^{-s} + 1 \right),$$

$$K(s,t) = \frac{\cos(st)}{(1+s^2)^{\frac{1}{4}}(1+t^2)^{\frac{1}{4}}},$$

Table 11.2.1. *Solution of the integral equation (5) using the Gauss–Laguerre and Gauss–rational rules with parameter values as shown. The entries are the values of* $\|e_N\|$ *obtained (see equation (6))*

	Gauss–Laguerre			Gauss–rational		
$N \diagdown \alpha$	1	$\frac{1}{4}$	$\frac{1}{16}$	1.0	4.0	16.0
2	0.07	0.13	0.02	0.14	0.27	0.08
4	0.34	0.076	0.016	0.14	0.51	0.43
8	0.10	0.17	0.05	0.15	0.14	0.56
16	0.063	0.41	1.4	0.10	0.035	0.07
32	0.055	0.08	0.33	0.026	0.017	0.035
64	0.033	0.07	0.18	0.014	0.011	0.009

Table 11.2.2. *Solution of the integral equation (5) using an N-point Gauss–Legendre rule on the truncated interval* [0, R]. *The entries are the values of* $\|e_N\|$ *obtained (see equation (6))*

	N					
R	2	4	8	16	32	64
1.0	0.35	0.55	0.63	0.65	0.65	0.67
2.0	0.08	0.27	0.41	0.46	0.47	0.47
4.0	0.49	0.06	0.14	0.21	0.24	0.25
8.0	103(!)	0.24	0.14	0.06	0.11	0.12
16.0	0.05	0.24	0.24	0.06	0.03	0.05
32.0	0.08	0.034	0.25	0.16	0.04	0.02
64.0	0.08	0.056	0.44	0.14	0.07	0.03
128.0	0.004	0.038	0.14	0.15	0.17	0.05
256.0	0.0003	0.002	0.055	1.3	0.31	0.07

which has been constructed to have the exact solution

$$x(s) = (1 + s^2)^{-\frac{1}{2}}. \tag{5a}$$

We solve this using the Nystrom method with three different rules:

Rule 1: the N-point shifted Gauss–Laguerre rule (equation (2.6.8b)) with scale parameter α

Rule 2: the N-point 'Gauss–rational' rule (equation (2.6.15)) with scale parameter α

Rule 3: We truncate the interval $[0, \infty)$ to $[0, R]$ and solve the truncated equation using the N-point Gauss–Legendre rule.

The use of Rule 2 is completely equivalent to mapping the interval $[0, \infty)$ to $[-1, 1]$ using the map (4) and using the Gauss–Legendre rule on the mapped equation. If we follow the arguments of Section 2.6.3, we would comment that the kernel K and driving term y decay only slowly for large s, and hence that the Gauss–rational rule should be the most effective of these three, with the 'truncated region' approach the least effective.

We give the results obtained in Table 11.2.1 and Table 11.2.2, using as an estimate of the error the measure

$$\|e_N\| = \max_{i=1}^{N} |x(\xi_i) - x_N(\xi_i)| \tag{6}$$

where $x_N(s)$ is the computed solution and the ξ_i are the quadrature points of the rule used. For rules 1 and 2 the results depend on the choice of the parameter α, and we quote results for two α values. For rule 3 the results depend both on N and on the choice of R, and we give results for a range of R; the tables represent in brief the process which a user has to go through in practice, experimenting with the available parameters and looking at the results. It is clear from these tables that none of the rules finds the equation particularly easy and the results for small N are rather ragged. For $N > 16$, the Laguerre and rational rules show steady convergence for the quite wide range of parameter values tried, with the Gauss–rational rule, as expected, a fairly clear winner.

The results for the 'truncated region' calculation are particularly interesting. The apparently low values for $\|e_N\|$ with small N stem purely from the definition of $\|e_N\|$ used in (6); for N small and R large, the quadrature points all lie in regions where the solution itself is small. For fixed R and increasing N, the results converge slowly to yield a constant residual stemming from the truncation of the region. As noted in Section 2.6.1, the value of N required to reach this 'stable' region increases with R; the best results in this table are those with $R = 32$, $N = 64$, achieving an accuracy about as high as the 64-point Gauss–Laguerre rule with $\alpha = 1.0$.

11.2.4 Another example

Finally, lest it be thought that all infinite range equations are straightforward to solve numerically, we give a second example, constructed as a slight modification to (5).

The equation

$$x(s) = y(s) + \int_0^\infty K(s,t)x(t)dt, \tag{7}$$

$$K(s,t) = -\cos(st),$$

$$y(s) = \frac{\pi}{2}e^{-s} + \frac{1}{(1+s^2)},$$

Table 11.2.3. *Computed solutions for equation (7) using the Gauss–Laguerre and Gauss–rational rules with parameter* α. *The entries are the values of* $\|e_N\|$ *obtained (see equation (6))*

	Gauss–Laguerre			Gauss–rational		
$N \setminus \alpha$	1	$\frac{1}{4}$	$\frac{1}{16}$	1.0	4.0	16.0
2	0.09	0.14	0.05	1.0	0.3	0.05
4	9.4	0.46	0.04	7.4	0.3	0.05
8	3.4	0.6	0.09	1.0	2.0	0.5
16	0.44	1.95	0.4	0.7	2.1	1.8
32	15.7	1.0	4.5	0.8	0.9	5.4
64	8.3	1.1	0.9	0.3	2.0	4.7

has the square integrable solution

$$x(s) = \frac{1}{(1 + s^2)}. \tag{7a}$$

The driving term $y(s)$ is also an $\mathscr{L}^2$ function on $[0, \infty)$ but the kernel is clearly not. An attempt to solve the equation numerically using the Nystrom method with Gauss–Laguerre and Gauss–rational rules yields the results shown in Table 11.2.3.

These results show no sign of convergence.

11.3   Product integration for singular integrals

We now consider the treatment of singularities (or in general any awkward feature) in the kernel $K(s, t)$. Within the framework of the Nystrom method, the most direct method of attack is to tailor the quadrature rule used to the particular difficult feature at hand. The approach used (product integration) was discussed in Section 4.4 in the context of Fredholm equations and in Section 5.5 in the context of Volterra integral equations.

We observe that, when $s = t_i$, we need an approximation to the integral

$$\int_a^b K(t_i, t)x(t)\,\mathrm{d}t. \tag{1}$$

If K is a rapidly changing function of t, then a low-degree rule will not be particularly successful even if $x(t)$ is smooth. Since K is known, we should be able to produce a rule of the form

$$\int_a^b K(t_i, t)x(t)\,\mathrm{d}t \approx \sum_{j=1}^N w_{ij}x(t_j) \tag{2}$$

which is *exact* for $x(t)$ a polynomial of low degree (rather than for $K(t_i, t)x(t)$ a polynomial of low degree). For fixed t_i, we can always in principle construct such a rule, given a knowledge of the basic integrals on the left side of the equation:

$$\int_a^b K(t_i, t)t^n dt = \sum_{j=1}^{N} w_{ij}t_j^n. \tag{3}$$

Indeed, we can solve these equations for the weights $w_{ij}, j = 1, \ldots, N$, for $n = 0, 1, \ldots, N - 1$. We derive in this way a weighted rule of Newton–Cotes type, together with the disadvantage of such rules if N is large: the weights w_{ij} may not be all positive and the numerical stability may therefore be poor. This will probably matter little in practice. More important is the difficulty of constructing the rule. We have noted explicitly that the weights w_{ij} will depend on the points $t_i, i = 1, \ldots, N$. We therefore have N rules to construct (which must all use this same set of points $t_1, \ldots, t_N$) and these require a knowledge of the N^2 integrals which appear in (3). If K is not of simple form it will be impossible to construct these analytically; we shall then have to approximate them numerically. This is not such a circular process as may appear, especially if we want to solve the integral equation for a wide range of inhomogeneous terms $y(s)$, but it is an undoubted addition to the burden of solution. It is sometimes possible to avoid numerical approximation of the moments by factorising out the singularity in K, in the form

$$K(s, t) = p(s, t)\bar{K}(s, t) \tag{4}$$

where p is singular but of 'standard' form and $\bar{K}$ is (hopefully) smooth; we may then absorb p into the weight function leaving $\bar{K}(s, t)x(t)$ to be integrated explicitly.

Efficient methods are available for computing the weights for the rules involved, provided that the basic integrals ('moments')

$$I_n(s) = \int_a^b p(s, t)t^n dt \tag{5}$$

are known; see Young (1954).

Note that we need not use t^n as the underlying expansion functions for x. If for some set of functions $\{\phi_1(t), \ldots, \phi_N(t)\}$ we know the 'modified moments'

$$I_n'(s) = \int_a^b p(s, t)\phi_n(t)dt, \quad n = 1, \ldots, N, \tag{6}$$

then we can construct a rule which is exact if $x(s)$ has the finite expansion

$$x(s) = \sum_{i=1}^{N} b_i\phi_i(s) \tag{7}$$

Table 11.3.1.

±s	$x(s)$ (i)	(ii)	(iii)	(iv)
0	0.6581	0.6574	0.6576	0.657 41
0.2	—	—	—	0.661 51
0.25	0.6645	0.6638	—	—
0.4	—	—	—	0.673 89
0.5	0.6838	0.6832	0.6827	—
0.6	—	—	—	0.694 48
0.75	0.7152	0.7149	—	—
0.8	—	—	—	0.722 49
1.0	0.7554	0.7557	0.7552	0.755 72

and such an expansion may converge faster than the Taylor series expansion.

As an example, consider the equation (Young, 1954a)

$$x(s) = 1 - \frac{1}{\pi} \int_{-1}^{1} \frac{1}{1 + (s - t)^2} dt.$$

For this we find the following results (see Table 11.3.1) using the rules given below:

   (i) 9-point Trapezoid rule
  (ii) 9-point Trapezoid rule with difference correction
 (iii) 5-point product integration formula
 (iv) 5 panels of a 3-point product integration formula

Clearly, the product integration method is quite efficient for this well behaved integral equation.

11.4   Subtraction of the singularity

Dealing with singular kernels $K(s, t)$ is in general complicated by the fact that the position of the singularity may be a function both of s and of t. There are two cases however in which the singularity can be weakened, by a suitable subtraction technique.

Case (a) *Singularity at a fixed value of t for all s.* This is the simplest case; it arises in quantum scattering problems, for example. We suppose that the singularity can be factored out of $K(s, t)$; that is, taking for definiteness the case of a pole of order α, we write

$$K(s, t) = \frac{K_0(s, t)}{(t - q)^\alpha}, \tag{1}$$

where K_0 is assumed regular and then rewrite the integral operator as

$$\int_a^b K(s,t)x(t)dt = \int_a^b \frac{[K_0(s,t)x(t) - K_0(s,q)x(q)]dt}{(t-q)^\alpha}$$

$$+ K_0(s,q)x(q)\int_a^b \frac{dt}{(t-q)^\alpha}. \tag{2}$$

If $\alpha \leq 1$ the first term is now *regular* at $t = q$, and the second can be evaluated exactly; if $\alpha > 1$ then neither integral exists in the Riemann sense and nor does the original integral operator. In that case, the integral operator will normally have been defined in some extended sense (such as a principal value integral); then (2) remains valid and weakens the form of the singularity. Note that the revised equation (2) includes the function value $x(q)$. If we consider a Nystrom solution of (2) with a rule containing points $\xi_1, \ldots, \xi_N$, we would normally write down the approximating equations at these N points. It is now necessary to write the equation also at the point $s = q, x(q)$ appearing as the $(N+1)$st unknown in the equations (unless q happens to be a quadrature point).

Case (b) *Singularity along the line* $s = t$. A second common type of singular integral has a kernel which is unbounded when $s = t$; a standard Nystrom method will certainly fail for such a kernel, since if ξ_j is a quadrature point, the kernel value $K(\xi_j, \xi_j)$ is always required by such a method. Again we can weaken the singularity by subtraction, by writing

$$\int_a^b K(s,t)x(t)dt = \int_a^b K(s,t)[x(t) - x(s)]dt + \int_a^b K(s,t)x(s)dt$$

$$= \int_a^b K(s,t)[x(t) - x(s)]dt + q(s)x(s) \tag{3}$$

where $q(s) = \int_a^b K(s,t)dt$ is known or can be computed without too much difficulty. The first term on the right side of (3) is now regular at $s = t$ provided that the original integral existed in the Riemann sense; it is in any case less singular than the original form, since $x(t) - x(s) = 0$ at the assumed singular point $t = s$. If we assume the subtracted form is regular, we can now introduce a standard quadrature rule $Q(w_j, \xi_j, j = 1, \ldots, N)$ and derive the modified Nystrom equations for a second kind equation:

$$x(\xi_i) = y(\xi_i) + \sum_{\substack{j=1 \\ j \neq i}}^N w_j K(\xi_i, \xi_j)[x(\xi_j) - x(\xi_i)] + q(\xi_i)x(\xi_i) \tag{4}$$

where the term $j = i$ is omitted in recognition of the identity

$$K(s,s)[x(s) - x(s)] = 0.$$

Note that these subtraction techniques weaken but do not entirely remove the singularity. The resulting equations after subtraction may be regular, but still, in general, some low-order derivative (usually the first derivative) of the modified kernel remains singular. It is possible to continue the subtraction process further in a systematic way, to smooth out the equation to any degree desired.

The basic problem is that of dealing with a function of the general form

$$\frac{f(s)}{\prod_{i=1}^{n}(s - q_i)^{m_i}} = \frac{f(s)}{g(s)}$$

which has poles of order m_i at the points q_i, $i = 1, \ldots, n$. Our aim is to construct a function P_n so that $f(s) - P_n(s)$ has a factor $g(s)$. It is possible to produce a formalism which achieves this, if not neatly then at least systematically. The formalism depends on three rather general identities and we give these first.

11.4.1   Basic subtraction identities

Let $f(s)$ be regular on the possibly complex contour C and let $\{q_i, i = 1, \ldots, n\}$ be a set of points on C. To each q_i we associate an integer m_i, and define a sequence of functions $f_k^{(j)}(s)$ as follows:

$$f_1^{(0)}(s) = \begin{cases} f(s) & \text{if } n = 1, \\ f(s) \Big/ \prod_{i=2}^{n}(s - q_i)^{m_i} & \text{if } n \geq 2, \end{cases} \tag{5}$$

$$f_k^{(j)}(s) = [f_k^{(j-1)}(s) - f_k^{(j-1)}(q_k)]/(s - q_k),$$
$$j = 1, 2, \ldots, m_k; \quad k = 1, 2, \ldots, n;$$

$$f_{k+1}^{(0)}(s) = f_k^{(m_k)}(s)(s - q_{k+1})^{m_{k+1}}, \quad k = 1, 2, \ldots, n-1.$$

We then find the following identities:

I1 $$f(s) = L(s)f_n^{(m_n)}(s) + P(s) \tag{6a}$$

where

$$L(s) = \prod_{i=1}^{n}(s - q_i)^{m_i},$$

$$P(s) = L(s) \sum_{k=1}^{n} \sum_{j=0}^{m_k - 1} f_k^{(j)}(q_k)(s - q_k)^{j - m_k}; \tag{6b}$$

I2 $$j! f_k^{(j)}(q_k) = \left[\frac{d^j f_k^{(0)}(s)}{ds^j}\right]_{s=q_k} = \left[\frac{d^j}{ds^j}\{f(s)S_k(s)\}\right]_{s=q_k} \tag{7a}$$

where

$$S_k(s) = \frac{1}{G_k(s)} \quad \text{and} \quad G_k(s) = \prod_{\substack{i=1 \\ i \neq k}}^{n}(s - q_i)^{m_i}. \tag{7b}$$

Proof. See Abd-Elal and Delves (1976).

I3

$$\left[\frac{d^{\mu}}{ds^{\mu}}f(s)\right]_{s=q_l}$$

$$= \sum_{k=1}^{p} \sum_{j=0}^{m_k-1} f_k^{(j)}(q_k)\left[\frac{d^{\mu}}{ds^{\mu}}\{G_k(s)(s-q_k)^j\}\right]_{s=q_l},$$

$$l = 1, 2, \ldots, n; \quad \mu = 0, 1, \ldots, m_l - 1. \tag{8}$$

Proof. The proof follows on differentiating I1.

Identities I1–I3 form the basis for a generalised subtraction process. In subsequent sections we shall sometimes (but not always) treat the term $f_n^{(mn)}(s)L(s)$ as an error term, a procedure suggested by the following observations.

Lemma 11.4.1. For $\mu = 0, 1, \ldots, m_k - 1$ we have from (6b) the identity

$$f^{(\mu)}(q_k) = P^{(\mu)}(q_k), \quad k = 0, 1, \ldots, n. \tag{9}$$

This relation identifies $P(s)$ as a Hermite interpolating polynomial for $f(s)$, of order m_k at the point q_k. The recurrence relations (5) then represent a convenient numerical scheme for generating both this polynomial and the remainder term $f_n^{(mn)}(s)L(s)$.

Lemma 11.4.2. Let $f(s) \in P_Q(s)$, the class of polynomials of degree $\leq Q$. Then $f_n^{(mn)}(s) \in P_N(s)$ with $N = Q - M$ where for convenience here and later we define

$$M = \sum_{i=1}^{n} m_i. \tag{10}$$

Proof. This follows immediately from (5).

Corollary. If $Q \leq M - 1$, $f_n^{(mn)} = 0$.

Lemma 11.4.3. Let $f(s)$ have at least R continuous derivatives on C. Then $f_n^{(mn)}(s)$ has at least $R - m$ continuous derivatives, where $m = \max_{i=1}^{n} m_i$.

Proof. From (6a) we write $f_n^{(mn)}$ in the form

$$f_n^{(mn)}(s) = \frac{f(s) - P(s)}{L(s)}. \tag{11}$$

Now if $s \neq q_l, l = 1, \ldots, n$, it is clear that $f_n^{(mn)}(s)$ has as many derivatives as

$f(s)$. In the neighbourhood of the point q_l we write a finite Taylor expansion of $f(s)$ and of $P(s)$; then the result follows from Lemma 11.4.1.

Lemma 11.4.4. Let $f(s)$ have at least M continuous derivatives in an interval I containing $s, q_1, \ldots, q_n$. Then for some point $\xi \in I$

$$|f(s) - P(s)| = |L(s)f_n^{(m_n)}(s)| \leq |L(s)|\frac{|f^{(M)}(\xi)|}{M!}. \tag{12}$$

Proof. This follows as a standard result from the observation in Lemma 11.4.1; see Davis (1963, p. 67).

In practical applications, one might apply I1 over a subdivision R of width h of a given interval $[a, b]$; then some of the points q_k (corresponding to genuine singularities) may be 'global', that is, outside R but in $[a, b]$, while others will be 'local', within R. Let S_1, S_2 be the number of local and global points, each counted according to its multiplicity m_i, so that

$$S_1 + S_2 = M. \tag{13}$$

Then from (12), for any s in R:

$$|f(s) - P(s)| \leq \frac{D^{(M)}}{M!}h^{S_1}(b - a)^{S_2} \tag{14}$$

where $D^{(M)} = \sup_{s \in [a, b]} |f^{(M)}(s)|$.

11.4.2   Application to the evaluation of singular integrals
Integrals containing poles
As a first example, we consider the application of I1 to the evaluation of singular integrals of the form

$$If = \oint \left\{ f(s) \Big/ \prod_{i=1}^{n} (s - q_i)^{m_i} \right\} ds \tag{15}$$

where $\oint$ refers to any definition of this integral such that (15) exists, such as the Hadamard finite part, the principal value, or a suitable contour integral around the singularities. We assume that $f(s)$ is regular on the region of integration. In the simplest case when the integrand has a single, simple pole in the region of integration, a number of quadrature schemes have been proposed in the literature; see, for example, Delves (1968) or Davis and Rabinowitz (1975, p. 140). More generally, product integration schemes (see Section 11.3), or if $f(s)$ is analytic the contour deformation scheme of Hetherington and Schick (1965) (see also Section 11.5), are available. We apply I1 to yield

$$If = \oint f_n^{(mn)}(s)ds - \sum_{k=1}^{n} \sum_{j=0}^{m_k-1} f_k^{(j)}(q_k)E(q_k, m_k - j) \tag{16}$$

where

$$E(q_k, l) = -\oint \frac{ds}{(s - q_k)^l}. \tag{17}$$

Comments

(a) In any particular case, (17) can be evaluated trivially analytically. For example, for the principal value integral

$$\oint = P \int_{-R}^{R}$$

we have

$$E(q_k, l) = \begin{cases} \ln \dfrac{|R + q_k|}{|R - q_k|}, & l = 1, \\[2mm] \dfrac{1}{l-1}[(-1)^l(R + q_k)^{1-l} + (R - q_k)^{1-l}], & l > 1, \end{cases} \tag{18}$$

and

$$\lim_{R \to \infty} E(q_k, l) = 0, \quad \forall l, \tag{19}$$

so that infinite intervals cause no difficulty. For the semi-infinite interval $[0, \infty)$ some of the individual terms of (16) diverge although the sum may not; this case may be mapped onto $(-\infty, \infty)$ using the map $t = (s - 1/s)$ before integration.

(b) The integrand $f_n^{(mn)}(s)$ is regular; hence the first term in (16) can be evaluated numerically using a straightforward quadrature rule. If the rule used is of degree Q then (16) is exact whenever $f(s) \in P_R(s)$, $R = Q + M - 1$. This suggests that if M is large enough, we may set $Q = 0$ and neglect the first term in (16) completely. We may make M as large as we please by using in identity I1 more points than are required to cancel the singularities in (15). If we do this, the error bound (12) of Section 11.4.1 can be trivially integrated to bound the total quadrature error, and thus show explicitly that the resulting scheme is of degree M.

(c) Neglecting the first term in (16) completely, our result reduces in the case $m_i = 1$, $i = 1, \ldots, n$ to the product integration rule (Young, 1954), and more generally is, like that rule, of Newton–Cotes type. We therefore expect such a procedure to share the disadvantages of Newton–Cotes rules if

M (see (10)) becomes large. We do *not* recommend seeking increased accuracy in this way, but recommend instead, either

(i) evaluating the first term in (16) numerically, as suggested above, or
(ii) neglecting this term, but subdividing the interval of integration.

If we subdivide in this way, some of the points q_i will in general vary from interval to interval (if none varies, nothing is gained by the subdivision), while others, those corresponding to genuine singularities of the integrand, will remain constant. We may therefore refer to the former as 'local' points and the latter as 'global'.

Which approach is preferable in practice depends upon the rule used for approach (i). A direct comparison may be made if we use approach (i) with an n-panel Newton–Cotes rule of degree p. Then if h is the panel width in each case, the errors are $\mathcal{O}(h^p)$ (method (i)) and $\mathcal{O}(h^{S_1})$ (method (ii)) where S_1 is the number of points 'local' to each panel (see (14)).

(*d*) Nothing in (16) restricts the points q_k to the region of integration. The effect of singularities close to, but outside, the path of integration on the numerical evaluation of regular integrals is well known (see, for example, Lyness, 1969) and it is an advantage of this method that it yields a simple formalism for dealing directly with multiple singularities.

(*e*) The need to evaluate the derivatives (8) may be a nuisance numerically in some circumstances; it seems unavoidable with the subtraction technique used here. We see in the next section that the presence of these derivative terms does not preclude the use of (6) in the solution of integral equations.

An example. As an example we evaluate the contour integral

$$If = \int_{-1+i\varepsilon}^{1+i\varepsilon} \frac{e^s(s-2.5)}{(s-0.5)^3}\,ds = 10.709\,625 \tag{20}$$

using two techniques as follows.

Method R: equation (16) with the first term evaluated with the Trapezoid rule.

Method C (Hetherington and Schick, 1965): we note that the integral is unchanged if the contour is altered to a semicircle, centre the origin, in the upper half plane, and use the Trapezoid rule around this revised contour, but directly on (15).

The results of these two methods with roughly equal steplengths, are given in Table 11.4.1.

We note that method R has the great advantage of requiring functional evaluations only on the region of integration, and of making only weak analyticity assumptions on $f(s)$. Further, at least in this example, although

Table 11.4.1. *Evaluation of the integral* (20)

	Steplength h			
	0.4	0.2	0.1	0.05
|Max error|, method R	0.001 70	0.000 43	0.000 11	0.000 03
|Max error|, method C	2.284 97	0.502 02	0.122 61	0.030 49

both methods are exhibiting, as expected, $\mathcal{O}(h^2)$ convergence, the error for the subtraction method is much the smaller; this is because only the 'error term' $\oint f_1^{(3)}(s)ds$, is integrated numerically, and this term is itself small.

Application to integrands with other singularities

A wide class of singularities can be treated via (16) by 'smothering' them with a suitably high-degree zero. We consider an integrand of the form

$$I = \int_a^b f(s)g(s)ds \qquad (21)$$

where $g(s)$ has some type of singularity at the points q_i, $i = 1, \ldots, n$. Using I1 we have the identity

$$I = \int_a^b f_n^{(m_n)}(s)\bar{g}(s)ds + \sum_{k=1}^{n} \sum_{j=0}^{m_k-1} f_k^{(j)}(q_k)E_{k,j} \qquad (21a)$$

where

$$E_{k,j} = \int_a^b g(s)L(s)(s - q_k)^{j-m_k}ds \qquad (22)$$

and

$$\bar{g}(s) = L(s)g(s). \qquad (23)$$

We assume that the $E_{k,j}$ can be evaluated exactly; then only the first term in (21a) remains for numerical evaluation, and $\bar{g}(s)$ is in general a much smoother function of s than is $g(s)$, the smoothness depending now in general upon the choice of the m_i.

This procedure has an obvious connection with the product rule approach described in Section 11.3 above. It differs in being easier to apply (only a few integrals $E_{k,j}$ need be evaluated analytically), while allowing the singularities to be smoothed as much as proves necessary. We illustrate the technique with a numerical example considered also by Young (1954). We take $g(s) = (1 - s^2)^{-\frac{1}{2}}$, and evaluate

$$I = \int_{-1}^{1} \frac{f(s)}{(1 - s^2)^{\frac{1}{2}}}ds = 1.266\,065\,878 \qquad (24)$$

Table 11.4.2. *Evaluation of the integral* (24) *using the Trapezoid rule*

	h				
m	0.5	0.25	0.125	0.062 5	Order of error
1	1.303 945	1.279 638	1.270 896	1.267 779	$\mathcal{O}(h)$
2	1.236 456	1.259 312	1.264 452	1.265 672	$\mathcal{O}(h^2)$
3	1.276 206	1.268 413	1.266 632	1.266 205	$\mathcal{O}(h^2)$

with $f(s) = (1/\pi) \cosh s$ and singularities at the two points $q_1 = -1$, $q_2 = +1$. We apply I1 with $m_1 = m_2 = m$; then

$$\bar{g}(s) = (1 - s^2)^{m - \frac{1}{4}}$$

and the degree of smoothing clearly depends upon the choice of m. Using the Trapezoid rule to evaluate the first term in (21), we find the results given in Table 11.4.2.

Clearly, the results increase in accuracy with increasing m. Further, as expected, the results for $m = 1$ have an error $\mathcal{O}(h)$, while those for $m > 1$ have an error $\mathcal{O}(h^2)$.

11.4.3 Application to the solution of Fredholm integral equations

We now show how the basic identity I1 can be used in the numerical solution of various kinds of Fredholm integral equations.

Basic method

We consider first the linear equation

$$x(s) = y(s) + \int_a^b K(s,t)x(t)\mathrm{d}t. \tag{25}$$

Let q_i, $i = 1, \ldots, n$, be points on the interval $[a, b]$, and m_i, $i = 1, \ldots, n$ a sequence of associated integers. Then the application of I1, with x replacing f, yields the identity

$$\int_a^b K(s,t)x_n^{(m_n)}(t) \prod_{i=1}^n (t - q_i)^{m_i}\mathrm{d}t + \sum_{k=1}^n \sum_{j=0}^{m_k - 1} x_k^{(j)}(q_k)C_{k,j}(s)$$

$$= x_n^{(m_n)}(s) \prod_{i=1}^n (s - q_i)^{m_i} - y(s) \tag{26}$$

where

$$C_{k,j}(s) = \int_a^b K(s,t) \prod_{\substack{i=1 \\ i \neq k}}^n (t - q_i)^{m_i}(t - q_k)^j\mathrm{d}t$$

$$- \prod_{\substack{i=1 \\ i \neq k}}^n (s - q_i)^{m_i}(s - q_k)^j. \tag{27}$$

As in Section 11.4.2 we now choose one of two alternatives.

(i) We choose some 'appropriate' quadrature rule to evaluate the first term in (26). If $K(s,t) \prod_{i=1}^{n} (t-q_i)^{m_i}$ is a regular function of s, t, any standard quadrature rule will do. This procedure is considered further below.

(ii) We drop the first term in (26), noting that it is zero if $x(s) \in P_N(s)$, where $N \leq M - 1$. As in Section 11.4.2, additional accuracy can be obtained if required by increasing M, or better, by subdividing the interval and applying an identity of the form I1 over each subdivision. We refer to this approach as a 'block' method.

We spell out here the steps involved in choice (i), when we approximate the first term (26) with a quadrature rule with abscissae ξ_i, $i = 1, \ldots, p$. Collocating at these abscissae, we obtain p equations in the $(p + M)$ unknowns $x_n^{m_n}(\xi_i)$ and $x_k^{(j)}(q_k)$. We obtain the additional equations by differentiating (26) μ times before collocating at $s = q_l$, $l = 1, \ldots, n$, where

$$\mu = \begin{cases} 0, 1, 2, \ldots, m_j - 1 & \text{if no point} \quad \xi_i = q_l, \\ 1, 2, \ldots, m_j & \text{if some abscissa} \quad \xi_i = q_l. \end{cases}$$

The rth set of differentiated equations then has the form

$$\int_a^b \left[\frac{\partial^\mu K(s,t)}{\partial s^\mu} \right]_{s=q_r} \prod_{i=1}^{n} (t - q_i)^{m_i} x_n^{(m_n)}(t)\mathrm{d}t + \sum_{k=1}^{n} \sum_{j=0}^{m_k - 1} x_k^{(j)}(q_k)$$

$$\times \left[\frac{\mathrm{d}^\mu}{\mathrm{d}s^\mu} C_{k,j}(s) \right]_{s=q_r} = a_r x_n^{(m_n)}(q_r) - \left[\frac{\mathrm{d}^\mu}{\mathrm{d}s^\mu} g(s) \right]_{s=q_r} \quad (28)$$

where

$$a_r = L^\mu(q_r)$$

$$= \begin{cases} 0, & \mu = 0, 1, \ldots, m_r - 1, \\ (-1)^{m_r - 1} \, m_r! \displaystyle\prod_{\substack{i=1 \\ i \neq r}}^{n} (q_r - q_i)^{m_i}, & \mu = m_r. \end{cases} \quad (29)$$

The collocated equations (26) and (28) can now be solved for the unknowns. Finally, $x(s)$ can be evaluated from I1 at any of the collocation points used, or at other points from (26), which may be written, using I1, in the form

$$x(s) = y(s) + \int_a^b K(s,t) x_n^{(m_n)}(t) \prod_{i=1}^{n} (t - q_i)^{m_i} \mathrm{d}t$$

$$+ \sum_{k=1}^{n} \sum_{j=0}^{m_k - 1} x_k^{(j)}(q_k) \int_a^b K(s,t) \prod_{\substack{i=1 \\ i \neq k}}^{n} (t - q_i)^{m_i} (t - q_k)^j \mathrm{d}t. \quad (30)$$

Integral equations with fixed singularities

We now consider the somewhat special case of an integral equation whose kernel has fixed singularities, typified by the equation

$$x(s) = y(s) + \oint_a^b K_0(s,t)x(t) \prod_{i=1}^n (t - q_i)^{-m_i} dt \tag{31}$$

where $K_0(s,t)$ is a regular kernel.

Equations of this type arise for example in scattering problems of quantum mechanics, and have been considered by Wiener (1971). For equations of this form we may remove the singularities completely by applying I1 with the obvious choice of points q_i and integers m_i, $i = 1, \ldots, n$. The resulting equation (26) contains only regular terms and method (i) outlined above is obviously applicable with any standard integration rule.

Integral equations with variable singularities

If the kernel $K(s,t)$ has variable singularities, method (i) is not in general applicable since then for no choice of q_i, m_i is the first term in (26) regular for all s. Then we recommend method (ii) above: the terms in $x_n^{(m_n)}(t)$ in (26), (28) are dropped and the resulting approximate equations collocated at the points $s = q_i$, $i = 1, \ldots, n$. We note that in this case the interval may be subdivided, with the choice of the points q_i depending wholly or partly on the subdivision. The form taken by (26) after subdivision is clear; for brevity we omit the details.

Comments on the method

(a) It follows from the Corollary to Lemma 11.4.2 that the method is exact if $x(s) \in P_{Q+M}(s)$, where Q is the degree of the quadrature rule used (method (i)) or zero (method (ii)).

(b) In addition to approximating $x(s)$, the method yields simultaneous approximations to the derivatives

$$x^{(\mu)}(q_r), \quad \mu = 1, \ldots, m_r - 1, \text{ at the points } q_1, \ldots, q_n.$$

This follows trivially from identity I3.

(c) If $m_i = 1$, $i = 1, \ldots, n$, then method (ii) reduces to the solution of (28) with $\mu = 0$ only, and is identical in content with the method of product integration (Section 11.3). For then $M = n$, and this method makes the replacement

$$\int_a^b K(q_r,t)h(t)dt = \sum_{i=1}^n w_i(q_r)h(q_i), \quad r = 1, 2, \ldots, n, \tag{32}$$

with the $w_i(q_r)$ chosen so that the result is exact if $h(t) \in P_{n-1}(t)$. The two schemes may be written in the form

$$[\mathbf{W} - \mathbf{I}]\mathbf{x}_n = -\mathbf{g} \quad \text{(product rule)} \tag{33}$$

where

$$W_{ij} = w_j(q_i); \quad x_i = x(q_i)$$

and

$$[\mathbf{L} - \mathbf{D}]\mathbf{X} = -\mathbf{g} \quad \text{(equation (28), } \mu = 0) \tag{34}$$

where

$$X_i = x_i^{(0)}(q_i); \tag{35}$$

$\mathbf{D}$ is diagonal with

$$D_{ij} = \prod_{\substack{k=1 \\ k \neq j}}^{n} (q_i - q_k) \tag{36}$$

and

$$L_{ij} = \int_a^b K(q_i, t) \prod_{\substack{k=1 \\ k \neq j}}^{n} (t - q_k) \mathrm{d}t. \tag{37}$$

Now from I2 with $m_i = 1$, $\forall i$

$$\mathbf{x} = \mathbf{D}\mathbf{X}. \tag{38}$$

Hence $\mathbf{x} = \mathbf{x}_n$ provided that $\mathbf{L} = \mathbf{W}\mathbf{D}$. But this follows straightforwardly from the form of (37) and the definition (32) of the weight $w_i(q_r)$.

More generally, the method shares with the product integration scheme the advantage that it is applicable when $K(s, t)$ is not well behaved or when $\int_a^b$ refers to a contour or other generalised integral. It differs in producing both a higher order scheme for a given number of points (by using more information per point) and a simultaneous approximation to $x(t)$ and its low-order derivatives.

(*d*) *Evaluation of the moments $C_{k,j}(s)$*. The moments $C_{k,j}(s_r)$ of equation (27) and their derivatives $[(\mathrm{d}^\mu/\mathrm{d}s^\mu)C_{k,j}(s)]_{s=s_r}$ required by (28), may in simple cases be evaluated analytically. More generally, numerical quadrature will be required and we comment that, for fixed s_r, the integrands involved are apropriate for the use of the technique of Section 11.4.2 on the kernel $K(s, t)$. Using this technique, we have a uniform and effective method of treating a wide class of singular kernels.

(*e*) *Application to Volterra integral equations*. It is clear that the methods of this section apply also to Volterra equations of the form

$$x(s) = y(s) + \int_a^s K(s, t)x(t)\mathrm{d}t; \tag{39}$$

we have only to replace b by s everywhere in (26), (27), (28), the additional terms appearing in (28) for $\mu > 0$ due to the variable limit s in fact vanishing identically because of the form of the integrand. The stability and convergence properties of a block method, based on approach (ii) above remain to be determined; for approach (i) they clearly depend upon the numerical quadrature rule used. We note however that even if the kernel $K(s,t)$ is positive or an increasing function of t (see Fox, 1962, Section 13.14), the zeros induced into (26), (28) ensure that the kernel of these equations is not.

A numerical example. As a numerical example, we consider the strongly singular equation

$$\int_{-1+i\varepsilon}^{1+i\varepsilon} \frac{K(s,t)}{(t-q)^3} x(t)dt = x(s) + y(s) \tag{40}$$

with separable kernel $K(s,t) = \sum_{i=0}^{3} C_i(s)t^i$,

$$C_0(s) = s(-2.5 + 0.75s - 0.125s^2), \tag{41}$$
$$C_1(s) = s(1 - 2s + 0.75s^2),$$
$$C_2(s) = s^2(-1.5s),$$
$$C_3(s) = s^3.$$

We take $y(s) = \sum_{i=1}^{3} a_i s^i - e^s$, with

$$a_1 = 4\left(e - \frac{1}{9e}\right), \quad a_2 = 2\left(e + \frac{1}{3e}\right), \quad a_3 = e - \frac{1}{e}; \quad q = 0.5; \tag{42}$$

and then the exact solution is $x(s) = e^s$.

The results obtained using method (i) with the repeated Trapezoid rule are given in Table 11.4.3, which shows that the singularities have no effect on the convergence obtained.

Table 11.4.3. Numerical solution of equation (40)

	$n = 5$ $h = 0.5$	$n = 9$ $h = 0.25$	$n = 17$ $h = 0.125$	$n = 33$ $h = 0.0625$	Exact
$x(-1.0)$	0.390 002	0.373 747	0.369 369	0.368 253	0.367 879
$x(-0.5)$	0.610 743	0.607 651	0.606 815	0.606 602	0.606 531
$x(0.0)$	1.000 000	1.000 000	1.000 000	1.000 000	1.000 000
$x(0.5)$	1.648 190	1.648 578	1.648 685	1.648 712	1.648 721
$x(1.0)$	2.710 884	2.716 324	2.717 785	2.718 157	2.718 282
$x'(0.5)$	1.644 663	1.647 647	1.648 449	1.648 653	1.648 721
$x''(0.5)$	1.623 380	1.642 038	1.647 027	1.648 296	1.648 721

11.5 Cauchy integral equations

One wide class of equation for which an elegant solution to the singularity problem exists is the following. Consider a kernel $K(s, t)$ which has a pole at $t = q$ in the interval $[a, b]$: $K(s, t) = K_0(s, t)/(t - q)$. Then the equation

$$x(s) = y(s) + \int_a^b K(s, t)x(t)dt$$

is singular as it stands: the integral is not well defined. Such integral equations arise quite generally from problems involving the scattering of radiation, and the integral is then defined as

$$\int_a^b = \lim_{\varepsilon \to 0} \int_{C(\varepsilon)}$$

where $C(\varepsilon)$ is the contour shown in Figure 11.5.1.

The resulting integral may be evaluated wholly on the real axis. Suppose the pole is at q, $a < q < b$; then the integral has the form

$$\int_a^b \frac{K_0(s, t)}{t - q} x(t)dt$$

and one may in a straightforward manner derive rules for treating the pole exactly. The simplest procedure is to subtract the singularity as described in Section 11.4 above; that is, to write the integral as

$$\int_a^b \frac{[K_0(s, t)x(t) - K_0(s, q)x(q)]dt}{t - q} + K_0(s, q)x(q) \int_a^b \frac{1}{t - q}dt$$

where the first term is now regular at $t = q$ (for a first order pole) and the second can be computed exactly. The exact value depends on the meaning given to this singular integral and with the definition above we find

$$\int_a^b \frac{1}{t - q}dt = \lim_{\varepsilon \to 0} \int_{C(\varepsilon)} \frac{1}{t - q}dt = -i\pi.$$

We can now use a rule for regular functions to represent the first integral; but note in the integral equation context that we have introduced the function value $x(q)$. If q is a constant little difficulty arises because of this; we

Figure 11.5.1. The contour $C(\varepsilon)$ is the path 1–2–3–4.

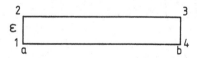

The contour $C(\varepsilon)$ is the path 1-2-3-4

merely introduce the value q into the list of points for which the quadrature rule must be applied. But, if K is symmetric then $K(q,s)$ is singular whenever $K(s,q)$ is singular; more generally, if the singular point depends upon s we appear to be stuck.

An alternative in these cases is to note the following. Let $K(s,t)$ and $y(s)$ be meromorphic in a region $s, t \in R$ which includes the interval $[a,b]$ and let us assume that the only poles of K or y in R are on the real axis. Then by Cauchy's theorem (see Figure 11.5.2)

$$\int_C K(s,t)x(t)\mathrm{d}t = \int_{C'} K(s,t)x(t)\mathrm{d}t$$

and we can solve the integral equation on the contour C' instead of C, by introducing a quadrature rule for the contour integral. A convenient contour for C' is a circle of radius r, since then we may write (see Figure 11.5.2)

$$\int_{C'} f(z)\mathrm{d}z = \mathrm{i}r \int_0^\pi \mathrm{e}^{\mathrm{i}\theta} f(r\mathrm{e}^{\mathrm{i}\theta})\mathrm{d}\theta.$$

In some cases however, a semicircle of diameter $(b-a)$ may not lie wholly in R; we must then adopt a contour which does. We have in this way avoided passing near the singularities on C, but at two costs:

(i) We must use complex arithmetic but this is a minor price to pay.
(ii) We have solved for $x(z)$ for z on C' whereas we require the solution on C.

There are two standard ways by which we can extend the solution to C:

(a) Via the integral equation. We use the notation $s = C$ to denote a point

Figure 11.5.2. We can solve the integral equation along C' instead of C. The contour C'' is the reflection of C' in C.

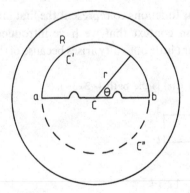

on C, and note that

$$x(C) = y(C) + \int_{t \in C'} K(C, t) x(t) dt.$$

This is the simplest extension; for each point on C it involves a contour integration over C', for which the necessary information is available. This method breaks down near the singular points on C (if these are constant with respect to q) or more generally if the resultant quadrature rule requires the evaluation of $K(C, t_i)$ at or near a singularity.

(*b*) Via Cauchy's theorem. Suppose that the region R also includes the contour C'', the reflection of C' in the real axis. Then the values of $x(s)$ are given on C'' by

$$x(s) = x^*(s^*)$$

and we have the identity

$$x(z) = \frac{1}{2\pi i} \oint_{C + C'} \frac{x(s)}{s - z} ds.$$

This method remains stable at the singularities of K, but fails instead near the edges $z = a$ and $z = b$. However, over these points we can either use (*a*) or else interpolate in terms of the neighbouring points to a and b.

11.5.1 Examples

We consider here two examples.

Example **11.5.1.**

$$x(s) = s \left[1 - \frac{(0.5 - 0.125 \ln 3)}{(4s^2 - 1)} \right] + \int_{-1}^{1} \frac{st x(t) dt}{[(4s^2 - 1)(4t^2 - 1)]}.$$

Solution: $x(s) = s$.

For this problem, the kernel has a pole at $s, t = \pm \frac{1}{2}$ although the solution is well behaved. We take C' to be the semicircle centre 0, radius 1 in the upper half plane. The integral equation has been solved on C' with three rules:

(i) a 32-point Gauss–Legendre rule
(ii) a 33-point Trapezoid rule
(iii) a 33-point Simpson's rule.

These solutions have then been extended to the real axis using either the integral equation or the Cauchy theorem, and one of these three rules.

The results are shown in Figure 11.5.3, from which we derive the following insights:

(i) The Cauchy extension appears much more accurate overall than the integral equation extension. This is of course particularly true near $s = \pm\frac{1}{2}$ but remains true over nearly all the interval.

(ii) As expected, the Gauss rule is better than the Trapezoid rule; but, in this simple example, Simpson's rule does about as well as

Figure 11.5.3.

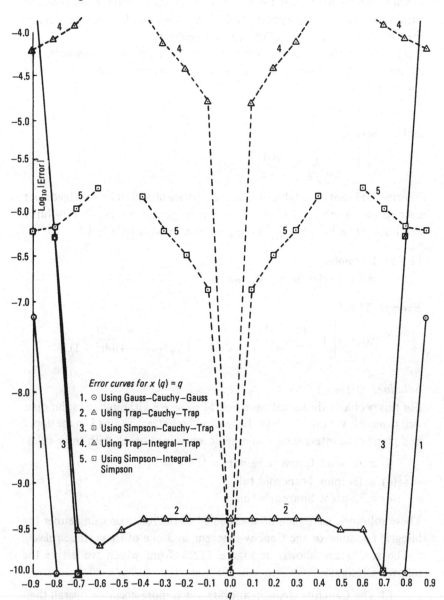

Error curves for $x(q) = q$
1. ⊙ Using Gauss–Cauchy–Gauss
2. △ Using Trap–Cauchy–Trap
3. ⊡ Using Simpson–Cauchy–Trap
4. △ Using Trap–Integral–Trap
5. ⊡ Using Simpson–Integral–Simpson

Gauss (both plunge into the machine noise level) as far as solving
the equation on C' is concerned.

Example 11.5.2.

$$x(s) = b - ia(\pi + 1) - c\left[2\ln(\sqrt{5} - 1) - \ln\sqrt{5} - \frac{i\pi}{2} \right]$$

$$- d(2\ln\sqrt{5} + 1) + \ln\left(\sqrt{5} + \frac{i\pi}{2}\right) + \frac{1}{s^2 + 5}$$

$$+ \int_0^\infty \frac{(t - s)x(t)dt}{(t - 1)^2(s - 1)^2},$$

$$a = -\frac{(2 + s)}{18}; \ b = i(1 - 4s)/18;$$

$$c = (\sqrt{5} + is)/(4\sqrt{5}(2 + i\sqrt{5}));$$

$$d = (\sqrt{5} - is)/(4\sqrt{5}(2 - i\sqrt{5})).$$

Solution: $x(s) = 1/(s^2 + 5)$.

This equation is singular both because of the double poles and because of
the infinite range involved. We cannot now choose C' to be a semicircle; we
take it to be a rectangle of unit height (see Figure 11.5.4). The solution on C'
has then been found by integrating along AB with a 32-point Gauss–
Legendre rule and along BD using a 32-point Gauss–Laguerre or Gauss–
rational formula. The former integrates exactly functions of the form
$\int_0^\infty e^{-s}s^m ds$, the latter, functions of the form $\int_0^\infty (s^2 + 1)^{-m} ds$; we might
expect the latter to yield a more accurate rule. The extension to the real axis
has been carried out using the Cauchy method, and the results are shown in
Figure 11.5.5. They show in fact no conclusive difference between the
Gauss–Laguerre and Gauss–rational rules, but they do show that the
method can be applied successfully to integral equations of this type.

Figure 11.5.4.

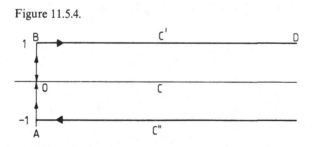

Figure 11.5.5.

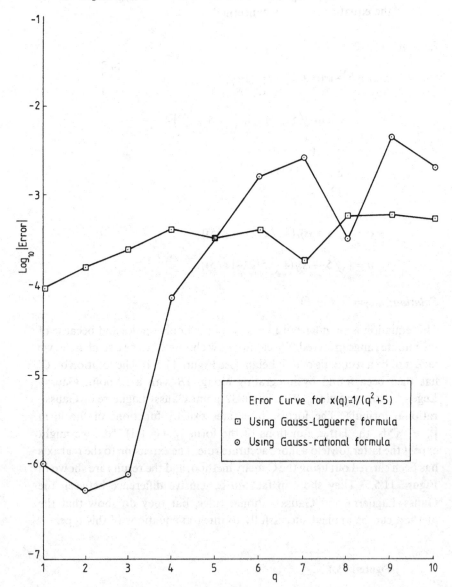

Error Curve for $x(q)=1/(q^2+5)$
□ Using Gauss-Laguerre formula
⊙ Using Gauss-rational formula

11.6   The Singular Galerkin method
11.6.1   Introduction

The product integration scheme described in Section 11.3 gives a method capable of rapid convergence even for singular problems, provided that

   (i) the singularity can be factored out (see (11.3.4))

and

   (ii) some analytic information is available for the singularity (the moments (11.3.5) or modified moments (11.3.6) must be known).

The Fast Galerkin scheme described in Section 8.5 can also be extended rather easily to singular equations and we describe this extension here. It too requires that the singularities can be factored out and that suitable analytic information (their Chebyshev expansions – see below) is available. If this information is available, it yields an expansion method which can handle one or a number of singularities in either the kernel or driving terms, and yield very rapid convergence.

Viewed from the context of the Fast Galerkin (or other expansion) method, singularities in the kernel or driving term may be *benign* – the solution $x(s)$ is smooth, with a rapidly convergent Chebyshev expansion, despite the singularities; or *malignant* – the solution itself contains some analytic singularity and hence has a slowly convergent expansion. Some examples of benign singularities are given by Green's function kernels, and by linear Volterra equations treated as Fredholm equations with discontinuous kernels.

Other examples which occur in practice have singular kernel and driving term, the singularities cancelling to yield a smooth solution. We treat first in Section 11.6.2 the analysis of singular kernels and driving terms without regard to the nature of the solution. We assume here for simplicity that the interval is $-1 \leq s \leq 1$ and we recall that, with a Chebyshev polynomial basis, the Galerkin method approximates the solution $x(s)$ by the truncated expansion

$$x(s) = \sum_{i=0}^{N} a_i T_i(s).$$

The aim in this section is to produce an accurate approximation to the Galerkin equations as cheaply as possible. Solution of the Galerkin equations will yield an accurate approximation to the $(N+1)$-term expansion of $x(s)$ if (and essentially only if) this expansion is rapidly convergent; that is, the singularities are benign. We show by example that in such cases very rapid convergence is obtained. A simple extension of the method to treat also malignant singularities is discussed in Section 11.6.3

and an example given showing that the extension indeed leads to rapidly convergent results.

11.6.2   The basic algorithm

The Fast Galerkin method described in Section 8.5 computes the coefficients a_i as the solution of the equations

$$[\bar{\mathbf{D}} - \bar{\mathbf{B}}]\mathbf{a} = \bar{\mathbf{y}} \tag{1}$$

where

$$\bar{\mathbf{D}} = \mathrm{diag}(\bar{D}_{ii}), \quad \bar{D}_{ii} = \int_{-1}^{1} \frac{T_i^2(s)}{(1-s^2)^{\frac{1}{2}}} ds = \begin{cases} \pi, & i = 0, \\ \dfrac{\pi}{2}, & i > 0, \end{cases} \tag{2a}$$

$$\bar{B}_{ij} = \int_{-1}^{1} \frac{T_i(s)}{(1-s^2)^{\frac{1}{2}}} \int_{-1}^{1} K(s,t)T_j(t)\,dt\,ds, \tag{2b}$$

$$\bar{y}_i = \int_{-1}^{1} \frac{T_i(s)y(s)}{(1-s^2)^{\frac{1}{2}}} ds, \quad i,j = 0,1,\ldots,N. \tag{2c}$$

The integrals appearing in equations (2a, b, c) must be approximated numerically; the technique of Section 8.5 does this by relating $\bar{B}_{ij}$, $\bar{y}_i$ to the Chebyshev coefficients in expansions of (apart from a constant factor) $K(s,t)$, $y(s)$, and then evaluating these coefficients numerically using Fast Fourier Transform techniques. This procedure leads to large quadrature errors if either expansion converges slowly and the techniques described below are designed to overcome this difficulty while retaining an $\mathcal{O}(N^2 \ln N)$ operation count for setting up the matrix $\bar{\mathbf{D}} - \bar{\mathbf{B}}$. In practice singular kernels seem more common than singular driving terms, but for simplicity we consider the latter first.

Singular driving terms

As in Section 8.5 we can identify $\bar{y}_i$ as expansion coefficients in the Chebyshev expansion:

$$\frac{\pi}{2} y(s) = \sum_{i=0}^{\infty}{}' \bar{y}_i T_i(s). \tag{3}$$

We consider the case when $y(s)$ has the form

$$y(s) = w(s)h(s) \tag{4}$$

where $w(s)$ is well behaved but $h(s)$ may not be. We introduce the expansions

$$w(s) = \sum_{i=0}^{\infty}{}' \bar{w}_i T_i(s), \tag{5}$$

$$h(s) = \sum_{i=0}^{\infty}{}' h_i T_i(s), \tag{6}$$

where we assume that the coefficients h_i are known exactly; a number of commonly occurring expansions are given in the Appendix. Since $w(s)$ is assumed smooth, the coefficients $\bar{w}_i$ can be efficiently approximated using a p-point Gauss–Chebyshev rule, $p \geqslant N$ and the Fast Fourier Transform; see Section 8.5. Accurate approximations y_i to the coefficients $\bar{y}_i$, $i = 0, 1, \ldots, N$, can then be computed by multiplying the two Chebyshev series together using the fast multiply algorithm (Algorithm 8.5.1) and $2N$ terms in the expansion of $h(s)$. Both stages of this process take $\mathcal{O}(N \ln N)$ operations; we write formally

$$\mathbf{y}_N = \mathbf{w}_N \otimes \mathbf{h}_{2N}. \tag{7}$$

Singular kernels

We next consider the evaluation of $\bar{B}_{ij}$ in equation (2b) and suppose that the kernel $K(s,t)$ may be factorised in the form

$$K(s,t) = W(s,t)Q(s,t) \tag{8}$$

where W is smooth and Q may be singular. We introduce the function H:

$$H(s,t) = Q(s,t)(1-t^2)^{\frac{1}{2}}. \tag{9}$$

Then equation (2b) takes the form

$$\bar{B}_{ij} = \int_{-1}^{1} \frac{T_i(s)}{(1-s^2)^{\frac{1}{2}}} \int_{-1}^{1} \frac{T_j(t)}{(1-t^2)^{\frac{1}{2}}} H(s,t) W(s,t) \mathrm{d}t \, \mathrm{d}s \tag{10}$$

which, apart from a constant factor, we recognise as the (i, j)th coefficient in the double Chebyshev expansion of the function

$$B = HW:$$

$$\frac{\pi^2}{4} B(s,t) = \sum_{i=0}^{\infty}{}' \sum_{j=0}^{\infty}{}' \bar{B}_{ij} T_i(s) T_j(t). \tag{11}$$

We now proceed in an obvious manner.

We introduce the Chebyshev expansions of W, H:

$$W(s,t) = \sum_{i=0}^{\infty}{}' \sum_{j=0}^{\infty}{}' \bar{W}_{ij} T_i(s) T_j(t), \tag{12}$$

$$H(s,t) = \sum_{i=0}^{\infty}{}' \sum_{j=0}^{\infty}{}' H_{ij} T_i(s) T_j(t) \tag{13}$$

and assume that the coefficients of H_{ij} are known exactly, while approximate coefficients W_{ij} are generated for $\bar{W}_{ij}$ using the FFT as in (8.5.10). Finally, we multiply together the two Chebyshev series to produce accurate approximations B_{ij} to the coefficients $\bar{B}_{ij}$, $i, j = 0, 1, \ldots, N$. The procedure described in Section 8.5 for this multiplication corresponds to the case $W(s, t) = (1 - t^2)^{\frac{1}{2}}$ and applies directly only to functions $W(s, t) \equiv W(t)$ of one variable; however, with an obvious extension of the notation introduced in Section 8.5 we can extend the multiplication algorithm to produce the $N \times N$ matrix $\mathbf{B}(N, N) = \mathbf{W}(N, N) \otimes \mathbf{H}(2N, 2N)$ as follows:

$$
\left.
\begin{aligned}
\textit{Step } 1 \quad &\mathbf{W}(N, N) \to \mathbf{W}(2N, 2N) \to W^{(2N, 2N)}, \\
&\mathbf{H}(2N, 2N) \to H^{(2N, 2N)}
\end{aligned}
\right\}
\begin{aligned}
&\text{at the } (2N + 1)^2 \\
&\text{Chebyshev points} \\
&(\xi_k, \xi_l), \quad k, l = 0, 1, \ldots, 2N. \\
&\xi_k = \cos(k\pi/2N)
\end{aligned}
$$

Step 2 Form

$$
B^{(2N, 2N)}(\xi_k, \xi_l) = W^{(2N, 2N)}(\xi_k, \xi_l) H^{(2N, 2N)}(\xi_k, \xi_l), \quad k, l = 0, 1, \ldots, 2N
$$

Step 3 $\quad B^{(2N, 2N)} \to \mathbf{B}(2N, 2N) \to \mathbf{B}(N, N)$

The cost of this algorithm, and also that for producing $\mathbf{W}(N, N)$, is $\mathcal{O}(N^2 \ln N)$.

Solution of the equations. To complete the description of the algorithm, we note that the matrix $\bar{\mathbf{D}} - \bar{\mathbf{B}}$ still has the structure assumed in Section 9.8, so that the iterative solution described there still applies, although the value of the constant M is likely to be larger for a singular problem because of the slower convergence of the expansions involved.

Error estimate. Finally, we consider the production of error estimates for the method. The discussion given in Chapter 9 is applicable here, including that given for the quadrature errors $|\delta B_{ij}|$, (9.7.14). The only extension required is that for the quadrature errors δy_i in the vector $\mathbf{y}$, since these now include a Chebyshev multiplication. The errors δW_i in the intermediate vector $\mathbf{W}$ are given by (9.7.4, 5); including the effect of the Chebyshev multiplication $\mathbf{y} = \mathbf{W}_N \otimes \mathbf{H}_{2N}$ is straightforward and leads to the final result

$$
\| \delta \mathbf{y} \| \sim N \| \mathbf{H} \|_\infty | W_N |. \tag{14}
$$

Examples

We give two examples of the solution of singular equations with 'benign' singularities, and of the error estimates resulting from the analysis of Chapter 9. These include:

Example **11.6.1.** A Fredholm equation with Green's function kernel.

Example **11.6.2.** A Volterra equation treated as a Fredholm equation with discontinuous kernel, but having in addition two distinct singularities in the kernel K and two in the driving term y.

Example **11.6.1.** Green's function kernel

$$x(s) = y(s) + \int_0^1 K(s, t)x(t)dt,$$

$$K(s, t) = -t(1 - s), \qquad t < s,$$
$$ = -s(1 - t), \qquad t \geqslant s,$$

$$y(s) = s^2.$$

Solution: $x(s) = Ae^s + Be^{-s} - 2,$

$$A = (3e - 2)/(e^2 - 1), \quad B = e(2e - 3)/(e^2 - 1).$$

The computed results are shown in Table 11.6.1 below, where σ_N denotes the computed error for an $(N + 1)$-term expansion.

Example **11.6.2.**

$$x(s) = y(s) + \int_{-1}^s K(s, t)x(t)dt,$$

$$K(s, t) = \ln(s - t)\cdot(s - t - 1) + \frac{(s - t - 0.5)}{(s - t)^{\frac{1}{2}}},$$

$$y(s) = \frac{1}{e}[(s + 1)^{\frac{1}{2}} + 1 + (s + 1)\ln(s + 1)].$$

Solution: $x(s) = e^s.$

Table 11.6.1. *Computed results for Example 11.6.1 using the Fast Galerkin algorithm described above*

N	σ_N	estimated error
2	1.2×10^{-1}	6.0, -1
3	5.8, -4	3.6, -4
4	1.2, -5	2.3, -4
5	1.2, -6	8.4, -7
6	1.8, -8	3.9, -7
7	1.3, -9	1.7, -9
8	1.1, -11	1.1, -9

The computed errors σ_N in these tables are defined to be

$$\sigma_N = \left[\sum_{j=0}^{N} e_N^2(\xi_j)/N \right]^{\frac{1}{2}} \left[\int_{-1}^{1} e_N^2(s)\mathrm{d}s \right]^{\frac{1}{2}} \tag{15}$$

where $\xi_j = \cos(j\pi)/N$, $j = 0, 1, \ldots, N$, and $e_N = x_N - x$.

We can make the following comments concerning these results:

(i) The very rapid convergence obtained shows that the method indeed successfully treats both singular kernels and singular driving terms.

(ii) The error estimates obtained reflect the actual error extremely well, especially when their zero cost is considered.

(iii) Example 11.6.2 has a kernel with two separate singular terms, while the driving term also contains both singular and regular terms. The matrices or vectors for each such term can be separately calculated with the algorithm given here; thus, equations such as that of Example 11.6.2 cause no particular difficulty.

The product integration method can also be extended to cover sums of singular factors by constructing a separate rule for each singular term in K. The method then requires however that the same quadrature points be used for each term, thereby precluding the use of rules of Gauss type (for which the points would differ). Alternatively, rules with different points could be used, but then with M distinct terms and an N-point rule for each, it would be necessary to collocate at each of the MN quadrature points so that even with two terms the advantage of a Gauss rule would be lost. In this respect, the Singular Galerkin method seems rather more flexible; in addition, no

Table 11.6.2. *Computed results for Example 11.6.2 using the Fast Galerkin algorithm described above*

N	σ_N	estimated error
2	4.3×10^{-2}	$1.1, -1$
3	$3.8, -3$	$1.5, -2$
4	$4.2, -4$	$1.5, -3$
5	$3.4, -5$	$1.3, -4$
6	$2.4, -6$	$1.0, -5$
7	$1.5, -7$	$6.5, -7$
8	$8.3, -9$	$4.1, -8$
9	$3.7, -10$	$5.1, -9$
10	$8.0, -11$	$3.7, -9$
11	$1.6, -10$	$3.9, -9$

comparable error estimates to those given in Tables 11.6.1, 11.6.2 are available for the product integration (or the Nystrom) scheme.

11.6.3 Extensions of the algorithm
Singular solutions

Finally, we consider two extensions of the algorithm. The first is to equations with malignant singularities; that is, equations whose solutions are in some sense singular. Then the Chebyshev expansion of the solution is slowly convergent; further, although the algorithm described above produces an accurate approximation to the $(N+1) \times (N+1)$ Galerkin equations, the solution of these equations does not yield an accurate approximation to the first $(N+1)$ expansion coefficients of the solution.

We assume that $x(s)$ may be written in the form

$$x(s) = x_R(s)x_S(s) \tag{16}$$

where x_S, the singular part, is known in advance and x_R is unknown but smooth. Then the integral equation may be rewritten as an equation for x_R:

$$x_S(s)x_R(s) = y(s) + \int_{-1}^{1} K'(s,t)x_R(t)\mathrm{d}t \tag{17}$$

where

$$K'(s,t) = K(s,t)x_S(t). \tag{18}$$

Let $x_S(s)$ have the known expansion

$$x_S(s) = \sum_{i=0}^{\infty}{}' \phi_i T_i(s) \tag{19}$$

and $x_R(s)$ be expanded as before:

$$x_R(s) = \sum_{i=0}^{N} a_i T_i(s). \tag{20}$$

Then the Galerkin equations for the coefficients **a** of x_R are given by

$$(\bar{\mathbf{D}} - \bar{\mathbf{B}})\mathbf{a} = \bar{\mathbf{y}} \tag{21}$$

where

$$\bar{D}_{ij} = \frac{\pi}{4}[\phi_{i+j} + \phi_{|i-j|}],$$

$$\bar{B}_{ij} = \frac{1}{2}\sum_{l=0}^{\infty}{}' \phi_l(\bar{B}_{i,j+l} + \bar{B}_{i,|j-l|}),$$

and $\bar{y}_i$, $i = 0, 1, \ldots, N$, retain the form (11.6.2c). We may now proceed as in Section 8.5 to approximate the coefficients $\bar{B}_{ij}$.

Provided that the singularities in x_S and K' are of a standard type, we may therefore solve for the rapidly convergent expansion of x_R. The error analysis given in previous sections remains valid, but the form of (18) prompts one refinement to the analysis which we now consider.

Fixed singularities in the kernel

If the singular function $H(s, t)$, equation (9), is in fact a function of t only:

$$H(s, t) \equiv H(t) = \sum_{i=0}^{\infty}{}' \alpha_i T_i(t) \tag{22}$$

then the matrix coefficients H_{ij} may be computed from α_i via the relation (21). In this case the error analysis leading up to (14) may be repeated to yield a somewhat tighter error estimate. We find the final result

$$\| \delta \mathbf{B} \| \sim N \| \boldsymbol{\alpha} \|_{\infty} (|W_{0,N}| + |W_{N,0}|) \tag{23}$$

which has the same from as (9.7.15) but with the matrix norm $\| \mathbf{H} \|_{\infty}$ replaced by the vector norm $\| \boldsymbol{\alpha} \|_{\infty}$.

Products of singular functions

A limitation on the use of the algorithm is that, given a singular factor in the kernel, driving term or solution, its Chebyshev expansion must be available. When the factor is the sum of two standard singularities, no difficulties of course arise; it would be nice to be able to handle also the product of standard singularities. What is required is an algorithm for computing accurately the first $2N$ terms of the product of two known but slowly convergent Chebyshev series. No such algorithm appears to be known. In its absence, it is possible to use the heuristic strategy of using the fast multiply algorithm described in Section 8.5 to produce $2N$ terms of the product expansion from $2N$ terms of each factor. An error analysis indicates that this procedure is only slowly convergent; the numerical example below suggests that its use is considerably better than the possible alternative of ignoring the singularity altogether.

An example with malignant singularities

We illustrate these extensions with

Example **11.6.3.**

$$x(s) = y(s) + \int_{-1}^{1} K(s, t)x(t)dt,$$

$$K(s, t) = e^{st}[2 + \alpha + t(s + 1) + s],$$

$$y(s) = (1 + s)^s e^s - e^{s+1}2^{1+\alpha},$$

with solution $x(s) = e^s(1 + s)^{\alpha}$.

We set $\alpha = \frac{1}{2}$ and produce results with two different algorithms:

Algorithm A: We treat the singularities exactly; that is, we set

$$y(s) = y_S(s)y_{R1}(s) + y_{R2}(s),$$
$$x(s) = x_S(s)x_R(s)$$

with

$$x_S(s) = y_S(s) = (1 + s)^{\alpha}; \quad y_{R1}(s) = e^s; \quad y_{R2}(s) = -e^{s+1}2^{1+\alpha}$$

and (see equation (9))

$$H(s, t) = (1 - t^2)^{\frac{1}{2}}(1 + t)^{\alpha}.$$

Algorithm B: As algorithm *A*, but we write $H(s, t)$ as the product of two singular functions

$$H = H_1 H_2; \quad H_1(s, t) = (1 - t^2)^{\frac{1}{2}}; \quad H_2(s, t) = (1 + t)^{\alpha}$$

and evaluate the expansion of $H_1 H_2$ from those of H_1 and H_2 using the 'heuristic procedure' suggested above.

The results obtained are given in Table 11.6.3 and we make the following comments:

(i) Method *A* converges very rapidly as expected: the method successfully handles the singular solution.

(ii) Method *B* behaves surprisingly well initially, but for large *N* converges only slowly as expected; the heuristic strategy used is useful if not ideal. The error estimates for method *B* make no explicit allowance for the error involved in multiplying the singular expansions, and must therefore be considered only heuristic. However, they reflect quite well the initial rapid decrease and final slow convergence of the results.

Table 11.6.3. *Results for Example* 11.6.3

N	σ_N method *A*	estimated error method *A*	σ_N method *B*	estimated error method *B*
2	1.0×10^0	—	1.0, 0	—
3	2.6, -1	5.1, -1	2.6, -1	2.2, 0
4	4.4, -2	7.7, -2	4.4, -2	2.9, -1
5	2.8, -3	6.2, -3	2.5, -3	1.9, -2
6	1.8, -4	7.2, -4	7.1, -5	2.6, -3
7	9.0, -6	4.2, -5	1.4, -4	1.4, -4
8	4.9, -7	3.6, -6	1.0, -4	3.7, -5
9	2.3, -8	1.7, -7	8.1, -5	2.0, -5
10	1.7, -9	1.1, -8	6.2, -5	9.2, -6
11	3.0, -9	5.7, -10	5.0, -5	7.0, -6
12	1.7, -9	1.1, -9	4.0, -5	5.4, -6

Exercises

1. The following are examples of singular integral equations. Identify
the source(s) of difficulty in each case and verify that each has the
stated solution.

(a) $x(s) = e^s + \dfrac{(s+1)^{\frac{3}{2}}}{e} + \displaystyle\int_{-1}^{s} (s-t-1.5)(s-t)^{\frac{1}{2}}x(t)\mathrm{d}t.$

Solution: $x(s) = e^s$.

(b) $x(s) = (s+1)^{\frac{1}{2}} + \dfrac{1}{e} + \dfrac{(s+1)\ln(1+s)}{e}$

$\qquad + \displaystyle\int_{-1}^{s} \{(s-t-1)\ln(s-t) + (s-t-0.5)(s-t)^{-\frac{1}{2}}\}x(t)\mathrm{d}t.$

Solution: $x(s) = e^s$.

(c) $x(s) = s - 0.5[s^2\ln s + (1-s^2)\ln(1-s) - (s+0.5)]$

$\qquad + \displaystyle\int_{0}^{1} \ln|s-t|x(t)\mathrm{d}t.$

Solution: $x(s) = s$.

(d) $x(s) = (1-s^2)^{\frac{3}{4}} - \dfrac{\pi\sqrt{2}}{4}(2-s^2) + \dfrac{2}{3}\displaystyle\int_{-1}^{1} |s-t|^{-\frac{1}{2}}x(t)\mathrm{d}t.$

Solution: $x(s) = (1-s^2)^{\frac{3}{4}}$

2. The partial wave Schrödinger differential equation

$$x_l''(s) = \left[\frac{l(l+1)}{s^2} - k^2 + V(s)\right]x_l(s)$$

satisfies the boundary conditions

$x_l(s) \sim \alpha_l(k)s^{l+1}$ as $s \to 0,$

$\quad \sim ks[j_l(ks) - \tan\delta_l(k)n_l(ks)]$ as $s \to \infty,$

where j_l, n_l are the spherical Bessel, Neumann functions respectively and $\delta_l(k)$ is the so-called phase shift introduced by the
potential $V(s)$, with $s^2V(s) \to 0$ as $s \to 0$ and $V(s) \to 0$ rapidly as
$s \to \infty$.

The above can be reformulated as a Fredholm integral equation
of the second kind:

$$\bar{x}_l(s) = y(s) + \int_{0}^{\infty} K_l(s,t)\bar{x}_l(t)\mathrm{d}t$$

where

$\bar{x}_l(s) = x_l(s)V(s),$

$y(s) = ksj_l(ks)V(s),$

$$K_l(s,t) = \frac{V(s)}{V(t)} G_l(s,t),$$

$$G_l(s,t) = \begin{cases} kst\, j_l(ks)n_l(kt), & s \le t, \\ kst\, j_l(kt)n_l(ks), & s \ge t. \end{cases}$$

In what sense is this integral equation singular?

Use the substitution $t = (1+u)(1-u)^{-1}$ to map the infinite interval of integration to one which is finite and discuss how the problem might be solved numerically for a specified value of l.

3. Show that the singular factors
(i) $h(s) = (1-s^2)^{\frac{1}{2}}$ of the driving term $y(s) = w(s)h(s)$
(ii) $Q(s,t) = 1$ of the Fredholm kernel $K(s,t) = W(s,t)\, Q(s,t)$
   have the following Chebyshev expansion coefficients:

(i) $h_0 = \dfrac{4}{\pi}$; $h_1 = 0$; $h_i = -\dfrac{2(1+(-1)^i)}{\pi(i^2-1)}$, $i \ge 2$,

where

$$h(s) = \sum_{i=0}^{\infty}{}' h_i T_i(s).$$

(ii) $H_{ij} = 0$, $i > 0$;

$$H_{00} = \frac{8}{\pi}; \quad H_{01} = 0; \quad H_{0j} = \frac{-4(1+(-1)^j)}{\pi(j^2-1)}, \quad j > 1,$$

where

$$H(s,t) = \sum_{i=0}^{\infty}{}' \sum_{j=0}^{\infty}{}' H_{ij} T_i(s) T_j(t).$$

4. The Fast Galerkin method for the numerical solution of

$$x(s) = y(s) + \int_{-1}^{1} K(s,t)x(t)\,\mathrm{d}t$$

approximates the solution by the expansion

$$x(s) \approx x_N(s) = \sum_{i=0}^{N} a_i T_i(s)$$

and computes the coefficients a_i as the solution of the equations

$$(\bar{\mathbf{D}} - \bar{\mathbf{B}})\mathbf{a} = \bar{\mathbf{y}}$$

where the coefficients of $\bar{\mathbf{D}}, \bar{\mathbf{B}}, \bar{\mathbf{y}}$ are given in equations (11.6.2).

Show that if the solution contains a known singular factor $x_S(s)$ with the Chebyshev expansion

$$x_S(s) = \sum_{i=0}^{\infty}{}' \phi_i T_i(s)$$

then the Galerkin equations for the coefficients a_i of

$$x_R(s) = \sum_{i=0}^{N} a_i T_i(s),$$

the smooth unknown factor in $x(s)$, are given by

$$(\bar{\bar{D}} - \bar{\bar{B}})a = \bar{y}$$

where

$$\bar{\bar{D}}_{ij} = \frac{\pi}{4}(\phi_{i+j} + \phi_{|i-j|}),$$

$$\bar{\bar{B}}_{ij} = \frac{1}{2} \sum_{l=0}^{\infty}{}' \phi_l(\bar{B}_{i,j+l} + \bar{B}_{i,|j-l|}).$$

12

Integral equations of the first kind

12.1   General discussion

We consider in this chapter integral equations of the form

$$\int_a^b K(s,t)x(t)\,dt = y(s) \qquad \text{(Fredholm)}, \tag{1}$$

$$\int_a^s K(s,t)x(t)\,dt = y(s) \qquad \text{(Volterra)}, \tag{2}$$

where y is known but x is not. It is not at first sight clear why such equations need special discussion. All of the methods discussed previously for Fredholm equations of the second kind apply formally also to first kind equations, with the single exception of the Neumann iteration. For example, introducing an N-point quadrature rule $Q(\mathbf{w}, \boldsymbol{\xi})$ we derive immediately the Nystrom equations

$$\sum_{j=1}^N w_j K(\xi_i, \xi_j)x(\xi_j) = y(\xi_i), \quad i = 1, \ldots, N \tag{3}$$

or

$$\mathbf{Kx} = \mathbf{y}$$

where

$$x_i = x(\xi_i), \quad y_i = y(\xi_i),$$
$$K_{ij} = w_j K(\xi_i, \xi_j).$$

To see the kind of difficulties that arise, we consider a simple example.

Example 12.1.1.

$$\int_0^1 e^{s^\beta t} x(t)\,dt = \frac{(e^{s^\beta + 1} - 1)}{(s^\beta + 1)}. \tag{4}$$

Solution: $x(s) = e^s$.

Table 12.1.1. *Solution of* (4) *by the Nystrom method*

N	2	3	4	5	6	7	8	9	10	20
$\|\text{error}\|_\infty$, $\beta = 1$	8.7×10^{-1}	4.7, −1	3.9, −1	1.5, −1	1.0, −1	5.4, 0	1.3, 2	8.6, 2	3.3, 3	1.1, 5
$\|\text{error}\|_\infty$, $\beta = 2$	9.0, −1	4.5, −1	3.7, −1	1.3, −1	3.8, 0	3.8, 1	5.8, 2	2.3, 2	1.8, 2	7.3, 2

Table 12.1.2. *Solution of Example* 12.1.2

		p	
N	3	6	12
2	4.8×10^{-3}	3.8, −4	3.8, −4
3.	4.7, −3	4.9, −6	6.7, −7
4	4.6, −3	4.7, −6	6.0, −10
5	4.6, −3	4.7, −6	4.6, −12
6	4.6, −3	4.7, −6	4.5, −12

Solving this equation using the Nystrom method (3) with an N-point Gauss–Chebyshev rule, we find the results in Table 12.1.1 for $\beta = 1$, $\beta = 2$.

Clearly, after a reasonable start, something has gone drastically wrong. The reason is that such problems as (1), (2) are in general *ill-posed*. That is, small changes to the problem can make very large changes to the answers obtained. To illustrate this, we consider a second example taken from Barrodale (1974).

Example **12.1.2.**

$$\int_0^1 e^{st} x(t)\,\mathrm{d}t = y_p(s), \quad 0 \le s \le 1,$$

where $y_p(s) = (e^{s+1} - 1)/(s + 1)$ rounded to p significant figures.
The exact solution (for $y_\infty(s)$) is $x(s) = e^s$.
Using the approximating function $x_N(s) = \sum_{i=1}^N a_i s^{i-1}$ and minimising the L_∞ norm of the residual r_N, we find the values of $\|r_N\|_\infty$ given in Table 12.1.2.

These look very satisfactory: the residual reaches a level consistent with the noise in $y(s)$ and then remains steady. However, the solution corresponding to these results, although reasonably satisfactory for 'small' $N(N = 2)$, is certainly not so for larger values, as Table 12.1.3 shows.

Table 12.1.3. *Computed expansion coefficients a_i for Example* 12.1.2

N	i	$p = 3$	$p = 6$	$p = 12$	$p = \infty$
2	1	8.6022×10^{-1}	8.5779, − 1	8.5780, − 1	1.0000, 0
	2	1.7155, 0	1.7202, 0	1.7202, 0	1.0000, 0
5	1	2.5639, 2	3.0487, − 1	1.0001, 0	1.0000, 0
	2	− 4.8397, 3	1.4166, 1	9.9823, − 1	1.0000, 0
	3	− 2.1622, 4	− 5.6650, 1	5.1155, − 1	5.0000, − 1
	4	− 3.1878, 4	8.6802, 1	1.3818, − 1	1.6667, − 1
	5	− 1.5637, 4	− 4.2434, 1	7.0217, − 2	4.1667, − 2

Although $\|r_N\|_\infty$ remains small, $\|e_N\|_\infty = \|x - x_N\|_\infty$ unfortunately does not. Small changes to $y(s)$ have led to very large changes in $x_N(s)$. For $N = 5$, only the solution for y_{12} bears any resemblance to the true solution.

The next two examples show that first kind equations may have no solution at all.

Example 12.1.3.

$$\int_{-1}^{1} ((s + 1)^2 + (t + 1)^2)^{\frac{1}{4}} x(t) dt = \ln(1 + s), \quad -1 \leqslant s \leqslant 1. \qquad (5)$$

For this example, the driving term is $\mathscr{L}^2$ but unbounded at $s = -1$. But the kernel is bounded in $[-1, 1] \times [-1, 1]$ and hence if $x(t)$ is integrable the left side of (5) cannot be singular anywhere.

Example 12.1.4.

$$\int_{-1}^{1} e^{(s+1)(t+1)/4} x(t) dt = (1 - s^2)^{\frac{1}{4}}, \quad -1 \leqslant s \leqslant 1. \qquad (6)$$

This problem can have no solution because the kernel is analytic in s over the whole plane and hence cannot have an infinite derivative at $s = \pm 1$.

Even if a solution exists to (1), a small change δy to y may make an arbitrarily large change δx to the solution x. But round-off errors in the solution process correspond to just such small changes in y, and hence we find in practice large errors creeping (or leaping – see Table 12.1.1) into the numerical solution. In terms of the Nystrom method (3), we find that the condition number of the matrix **K** becomes very large. This is a reflection of the result proved in Chapter 6, that the eigenvalues of a Fredholm kernel have an accumulation point at zero. But the matrix **K** is an approximation to the operator K, and hence for large N **K** must be very illconditioned.

We illustrate the delicate nature of (1) further by considering the case that the kernel K is of finite rank.

Theorem 12.1.1. Let K be of finite rank n and have the representation:

$$K = \sum_{v=1}^{n} a_v \otimes b_v. \tag{7}$$

Then (1) has no solution unless $y \in \text{span}\,\{a_1, \ldots, a_n\}$; that is, unless for some constants $\alpha_1, \ldots, \alpha_n$

$$y = \sum_{v=1}^{n} \alpha_v a_v. \tag{8}$$

Further, if y satisfies (8), then (1) has the solution

$$x = \sum_{v=1}^{n} \beta_v a_v \tag{9a}$$

if and only if the equations

$$\sum_{v=1}^{n} (b_\mu, a_v)\beta_v = \alpha_\mu, \quad \mu = 1, \ldots, n, \tag{9b}$$

are nonsingular.

Proof. The first part follows immediately since

$$Kx = \sum_{v=1}^{n} (b_v, x)a_v \tag{10}$$

has the form (8) for any x. The second part follows on substituting (9a) into (10), which yields (9b) as a set of necessary conditions for the solution x. If (9b) has an unique solution β, it is then simple to show that (9a) satisfies (1).

Note that nothing is said in this theorem about the uniqueness of the solution; this is because the solution, even if it exists, cannot be unique. For we saw in Chapter 6 that, for a kernel K of finite rank, there exist infinitely many zero eigenvalues and corresponding eigenfunctions ϕ_i such that

$$K\phi_i = 0.$$

But now if x is a solution of (1), so is $x + \alpha\phi_i$ for any α. Hence, there is an infinity of linearly independent solutions of (1); by solving (9b) we pick out not *the* solution but *a* solution (which will usually have physical meaning, the $\{\phi_i\}$ corresponding in applications usually to unwanted noise).

Any numerical method for the solution of (1), (2) must take cognisance of its ill-posed nature. It should certainly take steps to damp down the instabilities which occur, or at least to give a clear warning to quit when they become dangerous; it would be nice if the method of solution could also recognise problems which had no solution at all. In the next sections we consider the major methods which have been suggested to deal with first kind equations.

12.2 Eigenfunction expansions

The first method we consider (Baker *et al.*, 1964; Turchin, Kozlov and Malkevich, 1971; Hanson, 1971) is a direct extension of the numerical method for finite rank kernels implied by Theorem 12.1.1 (that is :expand x in the form (9a); set up and solve (9b)).

12.2.1 Hermitian kernels

Suppose first that the kernel K is Hermitian; then it has real eigenvalues λ_i, with corresponding eigenfunctions v_i, satisfying the equation $\int_a^b K(s,t)v_i(t)dt = \lambda_i v_i(s)$, or in operator form, $Kv_i = \lambda_i v_i$. If the $\{v_i\}$ form a complete set, any function, and in particular the solution x and driving term y, has an expansion of the form

$$x = \sum_{i=1}^{\infty} \beta_i v_i, \tag{1a}$$

$$y = \sum_{i=1}^{\infty} \alpha_i v_i. \tag{1b}$$

Substituting (1a) into (12.1.1) we find

$$Kx = \sum_{i=1}^{\infty} \beta_i Kv_i = \sum_{i=1}^{\infty} \beta_i \lambda_i v_i. \tag{2}$$

Comparing (2) with (1b), we have the explicit solution $\beta_i = \alpha_i/\lambda_i$; that is,

$$x = \sum_{i=1}^{\infty} (v_i, y)\lambda_i^{-1} v_i \tag{3}$$

where we have recalled the orthonormality of the eigenfunctions v_i and hence the identity $\alpha_i = (v_i, y)$.

To make (3) the basis of a numerical method we must in general find the eigenfunctions v_i numerically. But this can be done in many ways, for example from the Nystrom equations

$$(\mathbf{K} - \lambda_i \mathbf{I})\mathbf{v}_i = 0, \tag{4}$$

we can obtain approximations to N eigenvalue–eigenvector pairs at the quadrature points ξ_i, $i = 1,\ldots,N$, in $\mathcal{O}(N^3)$ operations, and we can substitute these into the discretised form of (3):

$$\mathbf{x} = \sum_{i=1}^{N} \mathbf{v}_i^T \mathbf{y} \lambda_i^{-1} \mathbf{v}_i. \tag{3a}$$

Moreover, the solution of (4) is a well posed problem, so that we have (apparently) succeeded in replacing a difficult by an easy problem.

Alternatively, expansion methods can be used to approximate the eigenvectors and eigenvalues; see Chapters 7 and 8.

Of course, we have not avoided the difficulties associated with (12.1.1). The eigenvalues λ_i tend to zero as i increases, giving the sum (3) a dangerous look. Now if $\lim_{i \to \infty}(v_i, y)\lambda_i^{-1} \neq 0$, the sum (3) does not represent an $\mathscr{L}^2$ function; that is, the equation has no solution (and we could use this relation to test numerically for consistency; see Section 12.4.4 for a detailed discussion of such tests). However, even if $(v_i, y)\lambda_i^{-1} \to 0$, the series (3) is very unstable. Suppose we replace y by $(y + \varepsilon v_k)$. Then the solution x changes to $(x + \varepsilon\lambda_k^{-1}v_k)$, and the 'response ratio', defined as $\| \delta x \|/\| \delta y \|$, is $|\lambda_k^{-1}|$. Since $\lim_{k \to \infty} |\lambda_k| = 0$, this response ratio can be made arbitrarily large, illustrating the comment made in Section 12.1 that the computed solution x is very sensitive to 'noise' in the function y. This noise is sure to show up if we take arbitrarily many terms in (3). In practice therefore, we must add terms cautiously, stopping when it seems that the decreasing truncation error is outweighed by the increasing noise.

Examples
We illustrate the behaviour typically exhibited with two examples

1)   Example 12.1.1 with $\beta = 1$.

2)   *Example 12.2.1.*
$$\int_0^1 (s^2 + t^2)^{\frac{1}{2}}x(t)\mathrm{d}t = \tfrac{1}{3}((1 + s^2)^{\frac{3}{2}} - s^3).$$

Exact solution: $x(s) = s$.

These two problems were solved by Lewis (1975) using an eigenfunction expansion method retaining N terms; the eigenfunctions themselves were calculated using an expansion in cubic splines. Lewis used a number of variations of the method; we omit details of these and give the results for his 'method a' and 'method d' in Table 12.2.1, together with the results for Example 12.2.1 using a straightforward Nystrom method with an N-point Gauss–Chebyshev rule.

Looking at these results, and those given in Table 12.1.1, we see that:

  (i) The eigenfunction expansion methods work very much better than does the Nystrom method.

  (ii) Although 'method d' seems rather better than 'method a' (a

Table 12.2.1. *Solution of Examples 12.1.1 and 12.2.1, using the eigenfunction expansion methods (a) and (d) of Lewis (1975) with N terms, and the Nystrom method with N-point Gauss–Chebyshev quadrature rule. The entries are the error norm* $\|e_N\|_\infty$

	Example 12.1.1 ($\beta = 1$)		Example 12.2.1		
N	method (a)	method (d)	method (a)	method (d)	Nystrom
2	2.0×10^{-2}	2.0, -2	7.3, -2	7.2, -2	4.2, -1
3	5.5, -4	5.8, -4	2.6, -2	2.6, -2	1.1, -1
4	6.4, -4	1.1, -5	1.3, -2	1.0, -2	2.1, -1
5	3.8, -2	2.4, -6	1.2, -2	3.4, -3	1.3, -1
6	1.1, $+1$	5.2, -5	8.6, -3	7.3, -4	1.1, -1
7	1.5, $+4$	1.6, -3	1.1, -2	1.1, -4	6.8, -2
8			2.1, -2	1.4, -5	3.8, -1
9			3.3, -2	2.0, -6	1.3, 0
10			1.6, -1	1.8, -6	2.1, 0
11			1.2, 0	1.7, -6	
12			1.4, $+1$		

conclusion supported further by Lewis), both methods will fail for large values of N. Thus in practice it is necessary to add extra terms cautiously and to stop when maximum achievable accuracy has been reached. This is easy when the exact solution is known but less easy without such knowledge. Lewis discusses an apparently effective but very expensive method for estimating the 'optimal' cutoff value for N.

It is clear from the examples that the eigenfunction expansion method can work well. It has however one other drawback. This is that, not only is it relatively expensive to find all the eigenvectors of an $N \times N$ matrix, but the value of N required may be quite large, because eigenfunction expansions such as (1) converge in general only rather slowly.

12.2.2 Non-Hermitian kernels

If K is not Hermitian then the eigenvalues may not be real and the eigenfunctions are not orthogonal. We may however extend the method to this case by using an expansion in the singular functions of K. Recall from Chapter 6 that the singular values μ_i and singular functions $\{u_i, v_i\}$ satisfy the relations

$$u_i = \mu_i K v_i, \tag{5}$$

$$v_i = \mu_i K^\dagger u_i,$$

and that $(u_i, u_j) = (v_i, v_j) = \delta_{ij}$.

Further,

$$\lim_{i \to \infty} \mu_i^{-1} = 0. \tag{6}$$

Now suppose we make the expansions

$$x = \sum_{i=1}^{\infty} \beta_i v_i = \sum_{i=1}^{\infty} (v_i, x) v_i, \tag{7}$$

$$y = \sum_{i=1}^{\infty} \alpha_i u_i = \sum_{i=1}^{\infty} (u_i, y) u_i.$$

Then it follows from (5) that $\beta_i = \mu_i(u_i, y)$, and the counterpart of (3) is

$$x = \sum_{i=1}^{\infty} (u_i, y) \mu_i v_i. \tag{8}$$

This is the singular function expansion of x; it can be used numerically just as the eigenfunction expansion (3) can. As with (3), it is usually necessary to compute the singular functions numerically, and methods for doing this reduce the problem to that of finding the singular vectors of a finite matrix. This is a well conditioned problem, so that the main advantage of the eigenfunction method is retained.

12.2.3   Deficient kernels

As described above, the eigenfunction or singular function methods appear to break down if the set of eigenfunctions or singular functions is not complete. This is the case, for example, with a kernel of finite rank, which has only a finite set of eigenfunctions. However, such problems, although very illconditioned, can still be solved by the methods of this section *if they have a solution*, since the solution then exists in the space spanned by the (finite set of) eigenfunctions. Consider for simplicity the kernel of finite rank (12.1.7). We saw from Theorem 12.1.1 that if (12.1.1) has a solution x, it lies in the space spanned by $\{a_1, \ldots, a_n\}$. But this is also the space spanned by the eigenfunctions, as was shown in Chapter 6. If we know the representation (12.1.7) of the finite kernel, the numerical solution of (12.1.1) with a finite eigenfunction expansion (or directly from (12.1.9)) is straightforward. If we do not have this information then the eigenfunction or singular function expansion we use will involve the eigenfunction space corresponding to zero eigenvalues (see Section 6.2.2) and then equation (3) becomes completely unstable numerically, since it contains terms $(v_i, y)\lambda_i^{-1} = 0/0$. These terms reflect the lack of uniqueness of the solution for a finite rank kernel.

12.3 The method of regularisation

The mathematical basis for truncating the computed eigenfunction expression (12.2.3) or singular function expansion (12.2.8) is that once the terms in the series start to diverge they cannot represent the $\mathscr{L}^2$ solution which the problem formulation demands. This also corresponds closely to the physical basis: we expect the solution to be 'smooth' in some sense and the presence of large high-order terms in the expansion will violate this expectation, so we cut them off. Provided we can formulate precisely what we mean by 'smooth', the requirement can be used to provide a method of solution which automatically leads to the required property. One physically satisfying model which we can construct is the following. In many practical applications the driving term $y(s)$ represents some measured function, the integral operator K represents a model of the instrument used to do the measuring and $x(s)$ represents the 'true' measured quantity, $y(s)$ being a 'smeared' version of $x(s)$ as seen through the instrument being used. Now $y(s)$ will be known only to some finite accuracy ε; we should therefore only expect to find $\|Kx - y\| \leq \varepsilon$. Of all the functions x satisfying this relation, we seek that which is the smoothest, in the sense that for some linear operator L, $\|Lx\|$ has the minimum value. This yields the constrained minimisation problem

$$\underset{x}{\text{minimise}} \|Lx\| \qquad (1)$$

subject to

$$\|Kx - y\| \leq \varepsilon.$$

This problem can be solved directly, in any given norm. It is however relatively complicated to do so (but see Section 12.4 for a related approach).

It is therefore usual to solve, not (1), but a related problem, which we develop as follows. We note that the minimum value of $\|Lx\|$ in (1) will decrease as ε increases, that is as the constraints weaken. Therefore, at the minimum of (1), the constraints will be binding, that is, $\|Kx - y\| = \varepsilon$. Now if we solve the unconstrained problem (for fixed α)

$$\underset{x}{\text{minimise}} \|Kx - y\|^2 + \alpha \|Lx\|^2 \qquad (2)$$

we will find at the minimum some value η for $\|Kx - y\|$. As $\alpha \to 0$, $\eta \to 0$ provided that a solution of (12.1.1) exists, and for some value of α, $\eta = \varepsilon$. This makes plausible the following result, which we quote without proof:

For some α (which depends on ε) the solution of problem (2) is identical to that of problem (1).

Because (2) is an unconstrained problem it is easier to solve than (1). It is referred to as the *regularised problem*, and the method based on a numerical solution of (2), as a *regularisation method*. The choice of operator L must be made on qualitative grounds; the usual choices are:

$$L = I; \frac{d}{ds}; \frac{d^2}{ds^2}.$$

If the kth derivative is chosen for L, the method is referred to as kth order regularisation. We consider further only the case $L = I$ and the natural L_2 norm. Then (2) has the form

$$\underset{x}{\text{minimise}} \; F(x) = (Kx - y, \, Kx - y) + \alpha(x, x). \tag{2a}$$

But, reordering, we find

$$F(x) = (x, (K^\dagger K + \alpha I)x) - (x, K^\dagger y) - (K^\dagger y, x) + (y, y). \tag{2b}$$

The minimum of this is at the point x such that for any function p,

$$\frac{\partial F}{\partial \varepsilon}(x + \varepsilon p)|_{\varepsilon = 0} = 0.$$

Carrying out this differentiation, we find that x is a minimum point if

$$[K^\dagger K + \alpha I]x = K^\dagger y. \tag{3}$$

Writing this equation out in full, we see that it has the form

$$\int_a^b \hat{K}(s, t)x(t)dt + \alpha x(s) = \hat{y}(s) \tag{3a}$$

where

$$\hat{K}(s, t) = \int_a^b K^*(\xi, s)K(\xi, t)ds,$$

$$\hat{y}(s) = \int_a^b K^*(t, s)y(t)dt. \tag{3b}$$

That is, the regularised solution can be obtained by solving a Fredholm equation of the second kind, although one with a so-called iterated kernel. Provided that a suitable value of α is known, this makes the method of (zero order) regularisation very convenient in practice; solving (3a) is much cheaper than finding the singular vectors required by the singular value method. The two methods are however closely connected. Recall that the singular functions v_i in equation (12.2.5) are eigenfunctions of the operator $K^\dagger K$:

$$K^\dagger K v_i = \mu_i^{-2} v_i. \tag{4}$$

Now if we insert the expansions (12.2.7) for x and y and use (12.2.5), we find from (3)

$$\sum_{i=1}^{\infty} (\beta_i \mu_i^{-2} + \alpha \beta_i)v_i = \sum_{i=1}^{\infty} \alpha_i K^\dagger u_i = \sum_{i=1}^{\infty} (u_i, y)\mu_i^{-1} v_i. \tag{5}$$

Equating coefficients of v_i gives the explicit solution

$$x = \sum_{i=1}^{\infty} \frac{\mu_i(u_i, y)}{(1 + \alpha\mu_i^2)} v_i. \tag{6}$$

We now compare this with the formally exact expansion (12.2.8). If α is small, $(1 + \alpha\mu_i^2)$ is close to 1 for small i (that is, small μ_i), and the terms of (6) agree closely with those of (12.2.8). For large i, $\mu_i \to \infty$; hence the expansion (6) cuts off, in a gradual manner, the usually unwanted high components which in the singular function method we were forced to truncate. We can say that zero order regularisation inserts a *filter factor* $(1 + \alpha\mu_i^2)^{-1}$ into the expansion (12.2.8). Other regularisation choices have similar effects, but with different filter factors. We now see that the choice of α is a compromise; we must not choose α too small, or the method will become illconditioned; nor must we choose α too large, or the expansion (6) will become distorted in the early terms with the result that accuracy will be lost. The choice of α is possibly the most difficult part of a regularisation calculation and many attempts have been made to provide an algorithm for determining an 'optimal' choice for a given equation; see, for example, Ribière (1967), Wahba (1977). As a rule of thumb, note that the minimum value of the function $\mu/(1 + \alpha\mu^2)$ is $-1/2\alpha^{\frac{1}{2}}$. It follows (see the discussion in Section 12.2) that an error δy in y may make a change δx in x with $\|\delta x\| \leq \|\delta y\|/2\alpha^{\frac{1}{2}}$. We must therefore at least keep this quantity small, and this sets a lower limit on the value of α: $\alpha \gtrsim \|\delta y\|^2$.

12.4   The Augmented Galerkin method
12.4.1   The method

We now turn to the use of expansion methods for first kind equations. We find that within an expansion framework, and in particular within the Fast Galerkin scheme (see Chapters 8, 11), the necessary smoothness conditions can be imposed in a rather direct way. Consider again the equation

$$\int_a^b K(s,t)x(t)\mathrm{d}t = y(s), \quad a \leq s \leq b, \tag{1}$$

and let $\{h_i(s)\}_{i=0}^{\infty}$ be a basis for $\mathscr{L}^2[a,b]$. Then we may approximate the solution x to (1) by

$$x_N(s) = \sum_{i=0}^{N} a_i h_i(s). \tag{2}$$

The (weighted) Galerkin equations for the coefficients a_i are

$$\check{B}a = \bar{y} \tag{3a}$$

where

$$\bar{B}_{ij} = \int_a^b w(s)h_i(s) \int_a^b K(s,t)h_j(t)\mathrm{d}t\,\mathrm{d}s, \quad i,j = 0, 1, \dots, N,$$

$$\bar{y}_i = \int_a^b w(s)y(s)h_i(s)\mathrm{d}s, \qquad\qquad i = 0, 1, \dots, N, \qquad (3b)$$

$$\mathbf{a} = [a_0, a_1, \dots, a_N]^T,$$

and $w(s)$ is a positive weight function on $[a,b]$. As with the Nystrom method, the matrix $\bar{\mathbf{B}}$ in (3) is in general quite illconditioned, and any attempt to solve (3) directly leads to numerical nonsense for large values of N; the computed solution vector $\mathbf{a}$ typically defines a solution x_N which for large N oscillates wildly. If we now choose the set $\{h_i\}$ to be orthonormal with respect to $w(s)$ on $[a, b]$, then the assumption that (1) has an $\mathscr{L}^2$ solution implies that the representation

$$x(s) = \sum_{i=0}^{\infty} b_i h_i(s) \qquad\qquad (4)$$

is convergent and that hence

$$\sum_{i=0}^{\infty} |b_i{}^2| < \infty\,;\, |b_n| = \mathcal{O}(n^{-\frac{1}{2}}). \qquad\qquad (5)$$

Hence there exist constants $C_b > 0$, $r \geq \frac{1}{2}$ such that

$$|b_i| \leq C_b \hat{\imath}^{-r} = \delta_i, \quad i = 0, 1, \dots, \qquad\qquad (6)$$

where

$$\hat{\imath} = i, \quad i > 0,$$
$$= 1, \quad i = 0. \qquad\qquad (6a)$$

The maximum exponent r for which (6) can be satisfied with C_b finite reflects the smoothness properties of the solution; we would normally expect $r > 1$, while if x is analytic and $\{h_i\}$ a set of orthogonal (normalised) polynomials, (6) may be valid for all finite r. For a discussion of the convergence rates of orthogonal expansions, see Chapter 9.

Given an estimate of C_b, r we may regularise the solution by imposing (6) as a constraint on the computed solution vector $\mathbf{a}$. That is, rather than solving (3) we solve the problem

$$\underset{\mathbf{a}}{\text{minimise}} \, \| \bar{\mathbf{B}}\mathbf{a} - \bar{\mathbf{y}} \| \qquad\qquad (7)$$

subject to

$$|a_i| \leq \delta_i, \quad i = 0, 1, \dots, N.$$

This problem may be formulated in any norm; we discuss in detail only the L_1 norm.

In this case (7) has the form

$$\text{minimise} \quad v = \sum_{i=0}^{N} \left| \sum_{j=0}^{N} B_{ij} a_j - y_i \right| \tag{8}$$

subject to

$$|a_i| \leq \delta_i = C_b i^{-r}.$$

Here, $\mathbf{B}$, $\mathbf{y}$ are numerical approximations to the Galerkin matrix $\mathbf{\check{B}}$ and vector $\mathbf{\bar{y}}$. We discuss the numerical calculation of $\mathbf{\check{B}}$, $\mathbf{\bar{y}}$ in Section 12.4.3.

The constrained optimisation problem (8) may be transformed to a standard linear programming problem (L.P.P.) as follows. Let

$$\beta_i = \text{sign}(y_i),$$

$$a_i = b_i - c_i, \quad i = 0, 1, \ldots, N,$$

$$\sum_{j=0}^{N} \beta_i B_{ij} a_j - |y_i| = u_i - v_i,$$

$$u_i, v_i, b_i, c_i \geq 0.$$

Then problem (8) is equivalent to

$$\text{minimise} \quad v = \sum_{i=0}^{N} (u_i + v_i) \tag{9a}$$

subject to

$$\sum_{j=0}^{N} \beta_i B_{ij} (b_j - c_j) - u_i + v_i = |y_i|,$$

$$b_i + c_i + z_i = \delta_i, \quad i = 0, 1, \ldots, N, \tag{9b}$$

$$b_i, c_i, u_i, v_i, z_i \geq 0,$$

provided that at the optimal solution of (9)

$$u_i v_i = b_i c_i = 0, \quad \text{for} \quad i = 0, 1, \ldots, N.$$

These relations are automatically satisfied by the 'simplex procedure' (see Barrodale and Roberts, 1973).

Alternatively, problem (8) is equivalent to the problem:

$$\text{maximise} \quad \xi = \sum_{j=0}^{N} \left[\sum_{i=0}^{N} \beta_i B_{ij} (b_j - c_j) - 2u_j \right] \tag{10}$$

subject to constraints (9b).

If we define $\mathbf{A}$ such that $A_{ij} = \beta_i B_{ij}$ and $\mathbf{I}_n$ as the identity matrix of order n then the condensed tableau of (10) has the matrix form

$$\mathbf{X} = \begin{pmatrix} \mathbf{A} & -\mathbf{A} & -\mathbf{I}_{N+1} \\ \mathbf{I}_{N+1} & \mathbf{I}_{N+1} & \mathbf{0} \end{pmatrix}.$$

The rank of **X** is exactly $2N + 2$ (full rank) because

$$\det\begin{pmatrix} \mathbf{A} & -\mathbf{I}_{N+1} \\ \mathbf{I}_{N+1} & \mathbf{0} \end{pmatrix} = 1.$$

Let ξ^* be the optimal value of the objective function in (10); then at the optimal solution of (10) we have

$$v = \sum_{i=0}^{N}\left|\sum_{j=0}^{N} B_{ij}a_j - y_i\right| = \sum_{i=0}^{N}|y_i| - \xi^*.$$

In the next section we show that the solution of (10) is stable against perturbations in **B**, **y**. The estimation of the parameters C_b, r is discussed in Section 12.4.3.

12.4.2 Stability of the Augmented Galerkin method

Let **a** be the optimal solution to (7) and at this solution $v = \|\mathbf{Ba} - \mathbf{y}\|_1$. We consider the perturbed problem

$$\text{minimise } \|(\mathbf{B} + \delta\mathbf{B})\mathbf{a}' - (\mathbf{y} + \delta\mathbf{y})\| \tag{11}$$

subject to

$$|a_i'| \le \delta_i, \quad i = 0, 1, \ldots, N,$$

under the assumption

$$\|\delta\mathbf{B}\|_1 < \varepsilon_B; \quad \|\delta\mathbf{y}\|_1 < \varepsilon_y. \tag{12}$$

The following theorems show that provided $\varepsilon_B, \varepsilon_y$ are independent of N, the residual norm v and the solution vector **a** are stable as $N \to \infty$; that is, for large N the computed solution will remain a satisfactory approximation for $\varepsilon_B, \varepsilon_y$ not too large.

Theorem 12.4.1. Let (1) have an $\mathcal{L}^2$ solution x satisfying (4), (6) with $r > 1$ and let the optimal solution of the perturbed system (11) have residual norm v'. Then for sufficiently small $\varepsilon_B, \varepsilon_y$ there exists m, independent of N, such that

$$|v' - v| < \varepsilon_B m + \varepsilon_y. \tag{13}$$

Proof. For $\varepsilon_B, \varepsilon_y$ sufficiently small **a** is a feasible solution to (11). Hence

$$v' \le \|(\mathbf{B} + \delta\mathbf{B})\mathbf{a} - (\mathbf{y} + \delta\mathbf{y})\|_1 \le v + \|\delta\mathbf{Ba} - \delta\mathbf{y}\|_1$$
$$\le v + \varepsilon_B\|\mathbf{a}\|_1 + \varepsilon_y$$

and similarly $v \le v' + \|\delta\mathbf{Ba}' - \delta\mathbf{y}\|_1 \le v' + \varepsilon_B\|\mathbf{a}'\|_1 + \varepsilon_y$.

But if $r > 1$, it follows from (8) that

$$\|\mathbf{a}\|_1 = \sum_{i=0}^{N}|a_i| \le ((2r-1)/(r-1))C_b = m, \text{ say,}$$

and similarly for $\|\mathbf{a}'\|_1$. Hence the result follows.

It is clear from the proof that the theorem remains valid in other norms.

Theorem 12.4.2. Let $r > 1$. Then for any $\varepsilon_B, \varepsilon_y$ there exists a constant K such that for all N:

$$\|\boldsymbol{\delta}\mathbf{a}\|_1 = \|\mathbf{a} - \mathbf{a}'\|_1 \le K. \tag{14}$$

Further, provided that problem (8) has an unique solution,

$$\lim_{\varepsilon_y, \varepsilon_B \to 0} \|\boldsymbol{\delta}\mathbf{a}\|_1 = 0. \tag{15}$$

Proof. Inequality (14) follows directly from the constraints imposed. We have

$$|a_i - a_i'| \le 2\delta_i = 2C_b i^{-r}$$

whence if $r > 1$ we may take $K = 2C_b((2r - 1)/(r - 1))$.

Now, by Theorem 12.4.1,

$$\lim_{\varepsilon_B, \varepsilon_y \to 0} v' = v$$

and for $\varepsilon_B, \varepsilon_y$ sufficiently small, $\mathbf{a}'$ is a feasible solution to (8). Then if the solution to (8) is unique, it follows that

$$\lim_{\varepsilon_B, \varepsilon_y \to 0} \mathbf{a}' = \mathbf{a}.$$

Theorems 12.4.1 and 12.4.2 show that it is not necessary to artificially truncate the expansion (2) to avoid instability; hence the method avoids the main difficulty associated with the eigenfunction/singular-function expansion approach.

12.4.3 Computational details

We now make the formalism concrete with a specific choice of the various parameters.

Choice of basis and evaluation of $\bar{\mathbf{B}}$, $\bar{\mathbf{y}}$

As with the Fast Galerkin method, computational economies result if we map $[a, b]$ onto $[-1, 1]$, and take

$$w(s) = (1 - s^2)^{-\frac{1}{2}}; \quad h_i(s) = T_i(s)$$

where $T_i(s)$ is the ith Chebyshev polynomial. Then the $(N + 1) \times (N + 1)$ matrix $\mathbf{B}$ and $(N + 1)$-vector $\mathbf{y}$ can be evaluated in $\mathcal{O}(N^2 \ln N)$ operations, using the techniques described in Chapter 8 for regular kernels and in Chapter 11 for singular kernels and/or driving terms.

Estimation of the parameters C_b and r

As with the regularisation method, the Augmented Galerkin method contains parameters which must be adjusted to give good results.

In fact, we seem to be worse off in that there are two parameters, C_b and r, rather than one. However, whereas the user normally has little information available to him *a priori* as to suitable values of α, the parameters C_b, r are directly related to the form of the solution and it is very common that at least some prior knowledge exists on this. The parameter C_b is a measure of the norm of the solution and r a measure of its smoothness.

The form of the constraints suggests that we should not choose r too large, nor C_b too small, since if the chosen values are incompatible with the behaviour of the exact solution $x(s)$, the computed solution cannot well approximate x. Numerical experience suggests however that the results are insensitive to the exact values of C_b and r chosen. As a result, it proves possible to make the choice of values of C_b, r either automatic (the algorithm itself chooses both C_b and r) or at least semi-automatic (a default choice of r is made, or specified by the user, and C_b is estimated). We consider three methods for estimating these parameters:

Method 1. Choose r arbitrary but small ($r = 2, 3$). From (3) and (6) we have

$$C_b > \| \mathbf{a} \|_\infty \geq \frac{\| \bar{\mathbf{y}} \|_\infty}{\| \bar{\mathbf{B}} \|_\infty}$$

and we choose

$$C_b = \frac{\lambda \| \bar{\mathbf{y}} \|_\infty}{\| \bar{\mathbf{B}} \|_\infty}$$

where λ must now be set heuristically; the values $\lambda = 2 - 4$ seem to work quite well.

Method 2. For a wide class of kernels, matrix $\bar{\mathbf{B}}$ in (3) is well conditioned for small values of N. Hence, we choose N_0 small and solve system (3) directly to obtain $\mathbf{a}$. Now, taking logarithms of the constraint equations (8), we obtain

$$\ln |a_i| \leq \ln C_b - r \ln \hat{i}, \quad i = 0, 1, \ldots, N_0. \tag{16}$$

We may therefore estimate C_b, r from a least squares solution to the overdetermined system of equations

$$\ln |a_i| = \ln C_b - r \ln i, \quad i = 1, \ldots, N_0. \tag{17}$$

However, we cannot take these equations as they stand. For r to be positive, the sequence $\{|a_i|\}$ should be decreasing (or should contain a decreasing subsequence). If $\{|a_i|\}$ is not monotone decreasing we smooth it by replacing the ith equation by the $(i + 1)$st in (17) whenever $|a_i| < |a_{i+1}|$. We then solve equation (17) for $\ln C_b, r$ and finally multiply the computed C_b by a heuristic safety factor λ, $1 < \lambda < 4$.

Method 3. This method is a compromise between methods 1 and 2. We evaluate **a** as described in method 2. We then assign a value to r and calculate C_b from the formula

$$C_b = \max_{0 \le i \le N_0} (|a_i| \hat{i}^r). \tag{18}$$

Solution of the constrained system

Since the constrained problem (7) can be formulated as a linear programming problem, in principle any linear programming routine may be used for its solution. However, it is advantageous to provide a routine tailored to this context; in particular the routine can and should take cognisance of:

(i) The illconditioning of the matrix $\bar{\mathbf{B}}$

(ii) The presence of the inequality constraints; these can be treated rather more efficiently than the direct reduction to a standard L.P.P. would suggest.

A suitable procedure is that given in Barrodale and Roberts (1977), which directly solves an overdetermined set of linear equations in the L_1 norm, with added inequality (and/or equality) constraints. It uses the simplex method and takes precautions to recognise and treat illconditioning, so that degeneracy, whether exact or (as here) numerical, yields no difficulties.

Examples

We compare these methods on a number of examples.

1. *Examples* 12.1.1 *and* 12.2.1

Results for these examples using the Nystrom and eigenfunction expansion methods were given earlier; the (very poor) Nystrom results are identical (except from rounding errors) to those which would be obtained using the Galerkin method with Chebyshev basis but no additional regularity constraints; that is, solving the Galerkin equations directly. Tables 12.4.1 to 12.4.3 show the results obtained with the Augmented Galerkin method, using methods 1, 2 and 3 to estimate C_b, r.

2. *Examples* **12.4.1** *and* **12.4.2**

$$\int_0^1 K(s,t)x(t)dt = y(s),$$

$$K(s,t) = t(s-1), \quad t < s,$$
$$ = s(t-1), \quad s \le t.$$

Table 12.4.1. *The computed maximum error* $\| e_N \|_\infty$ *for test problem* 12.1.1 *with* $\beta = 1$, *using the Augmented Galerkin methods* 1, 2, 3

	Method 1		Method 2		Method 3	
N	$r=3, \lambda=2$	$r=5, \lambda=4$	$N_0=3$	$N_0=4$	$N_0=3, r=3$	$N_0=3, r=5$
2	4.0×10^{-1}	2.3, −1	8.0, −1	3.5, −1	1.9, −1	3.8, −1
3	1.9, −1	8.0, −2	4.7, −1	3.3, −1	9.7, −2	1.3, −1
4	1.3, −1	2.7, −2	3.3, −1	2.4, −1	8.7, −2	6.5, −2
5	3.2, −2	1.1, −2	3.2, −2	3.2, −2	3.5, −2	2.2, −2
6	1.6, −3	1.6, −3	1.6, −3	1.6, −3	6.8, −3	2.6, −3
7	7.7, −5	7.7, −5	7.7, −5	7.7, −5	7.7, −3	7.7, −5
8	7.5, −5	7.5, −5	7.5, −5	7.5, −5	7.5, −3	7.7, −5
9	6.4, −5	6.4, −5	6.4, −5	6.4, −5	7.7, −3	6.4, −5
10	6.0, −5	6.0, −5	6.0, −5	6.0, −5	7.7, −3	6.0, −5
15	5.6, −5	5.6, −5	5.6, −5	5.6, −5	7.7, −3	5.6, −5
20	6.7, −5	6.7, −5	6.7, −5	6.7, −5	7.7, −3	6.7, −5

Table 12.4.2. *The computed maximum error* $\| e_N \|_\infty$ *for test problem* 12.1.1 *with* $\beta = 2$, *using the Augmented Galerkin methods* 1, 2, 3

	Method 1		Method 2		Method 3	
N	$r=3, \lambda=2$	$r=5, \lambda=4$	$N_0=3$	$N_0=4$	$N_0=3, r=3$	$N_0=3, r=5$
2	4.0×10^{-1}	2.3, −1	8.3, −1	3.4, −1	1.8, −1	3.3, −1
3	2.0, −1	8.2, −2	4.5, −1	3.3, −1	9.5, −2	1.3, −1
4	1.3, −1	2.8, −2	3.4, −1	2.9, −1	8.3, −2	5.5, −2
5	2.9, −2	1.1, −2	2.9, −2	2.9, −2	3.3, −2	1.9, −2
6	2.7, −3	2.7, −3	2.7, −3	2.7, −3	4.4, −3	2.7, −3
7	9.1, −5	9.1, −5	9.1, −5	9.1, −5	5.3, −3	9.1, −5
8	6.3, −5	6.3, −5	6.3, −5	6.3, −5	5.3, −3	6.3, −5
9	6.2, −5	6.1, −5	6.2, −5	6.2, −5	5.3, −3	6.2, −5
10	6.4, −5	6.4, −5	6.4, −5	6.4, −5	5.3, −3	6.4, −5
15	6.4, −5	6.4, −5	6.5, −5	6.4, −5	5.3, −3	6.4, −5
20	1.1, −4	1.1, −4	1.1, −4	1.1, −4	5.3, −3	1.1, −4

Exact solution: $x(t) = y''(t)$, provided that $y''(t)$ is continuous on $[0, 1]$ and $y(0) = y(1) = 0$. We consider the cases

***Example* 12.4.1.** $y(s) = (s^3 - s)/6$.

***Example* 12.4.2.** $y(s) = e^s + (1 - e)s - 1$.

Results for these examples are given in Table 12.4.4.

We draw several interesting conclusions from these results:

(i) *Stability*. The method clearly has very satisfactory stability properties. As N increases, the error reduces initially and then finally stabilizes; increasing N further neither gains nor (significantly) loses accuracy.

For most of the problems the constraints (6) are binding for large i, N, as would be expected. However, for some examples, these constraints are not binding for any value of N. This behaviour stems from the solution

Table 12.4.3. *The computed maximum error* $\|e_N\|_\infty$ *for test problem* 12.2.1 *using the Augmented Galerkin method.*

Method 1		Method 2		Method 3	
N $r=3, \lambda=2$	$r=5, \lambda=4$	$N_0=3$	$N_0=4$	$N_0=3, r=3$	$N_0=3, r=5$
2 1.5×10^{-1}	8.7, -2	4.2, -1	4.2, -1	2.1, -1	4.2, -1
3 1.1, -1	6.5, -2	1.1, -1	1.1, -1	1.1, -1	1.1, -1
4 5.3, -2	1.3, -2	2.1, -1	2.1, -1	6.6, -2	6.2, -2
5 3.5, -2	1.0, -2	1.3, -1	1.3, -1	4.0, -2	3.3, -2
6 2.7, -2	4.1, -3	1.1, -1	1.1, -1	3.5, -2	2.3, -2
7 2.4, -2	3.7, -3	2.6, -2	2.6, -1	2.9, -2	2.8, -2
8 1.6, -2	4.2, -3	2.1, -2	2.1, -2	2.6, -2	8.9, -3
9 2.4, -2	2.0, -3	1.8, -2	1.8, -2	1.9, -2	1.5, -2
10 1.1, -2	2.5, -3	1.9, -2	1.9, -2	3.8, -2	1.4, -2
15 1.1, -2	1.8, -3	6.8, -2	5.8, -2	4.0, -2	8.0, -3
20 1.2, -2	4.7, -4	1.0, -1	6.0, -2	3.0, -2	4.0, -3

procedure used (CL1: Barrodale and Roberts, 1977), which for these cases has decided that the matrix **B** is numerically singular for $N > N_0$. In these circumstances it sets the remaining expansion coefficients a_i, $i = N_0 + 1, \ldots, N$, to zero while returning a valid solution to the (numerically degenerate) problem (8); the inequality constraints are then unlikely to be binding. Thus CL1 is quite a reasonable solver of first kind equations in its own right.

(ii) *Estimation of C_b and r.* Satisfactory results are obtained with all three methods for estimating these parameters, although method 1 performs significantly better than the other two and method 3 seems better than the 'fully automated' method 2.

(iii) *Ease of use.* Factors (i) and (ii) imply that the method is easy to use in practice. The relative insensitivity of the results to the parameters C_b, r means that these can be set automatically, with no effort required by the user, who can then run the method with increasing N until the computed results settle down; the fact that they indeed do settle down is crucial in this respect.

(iv) *Comparison with other methods.* For problem 12.1.1 the maximum accuracy achieved is comparable with that of method (*d*) of Lewis. For problem 12.2.1, method (*d*) achieves a very high accuracy – much higher than that obtained by Lewis's other algorithms, which typically returned accuracies comparable to that of the current method. These results are almost certainly due to the singular behaviour of the kernel in Example 12.2.1 The mild singularity at $s = t = 0$ makes the equation less ill-posed than those of Example 12.1.1 with $\beta = 1$ and $\beta = 2$, so that high accuracy is achievable, while making the integrals required by a Galerkin or an eigenfunction method difficult to evaluate. Lewis overcomes this difficulty

Table 12.4.4. *The computed error* $\|e_N\|_\infty$
using the Augmented Galerkin method 1
$(r = 3, \lambda = 2)$ *for test problems* 12.4.1
and 12.4.2

N	Problem 12.4.1	Problem 12.4.2
2	4.9×10^{-1}	$6.6, -1$
3	$5.0, -2$	$1.9, -1$
4	$1.6, -9$	$7.7, -2$
5	$9.2, -10$	$4.9, -2$
6	$1.8, -8$	$3.4, -3$
7	$4.1, -8$	$1.6, -4$
8	$1.8, -8$	$7.0, -6$
9	$1.2, -8$	$1.4, -7$
10	$5.1, -8$	$2.1, -7$
15	$2.5, -7$	$8.7, -7$
20	$3.2, -7$	$9.4, -7$

by specifying that every integral be evaluated (independently) by an automatic quadrature routine. The Fast Galerkin algorithm evaluates the $(N + 1) \times (N + 1)$ matrix **B** in $\mathcal{O}(N^2 \ln N)$ operations using an $(N + 1)$-point Gauss–Chebyshev quadrature rule. Singular kernels are handled effectively within this scheme by storing Chebyshev expansions of the singular part in a library; the particular singularity displayed by the kernel of problem 12.2.1 was not in the library used, and in these circumstances the quadrature rule used converges only slowly. The effectiveness of the Fast Galerkin scheme on singular kernels when these *are* of standard form is shown by Examples 12.4.1 and 12.4.2. Here the kernel has a discontinuous first derivative across the line $s = t$, but very rapid convergence is obtained as well as very high final accuracy. This high achievable accuracy suggests that the problems 12.4.1, 12.4.2 are 'less illconditioned' than problem 12.1.1. This reflects a general observation on such first kind problems: the more singular the kernel, the less illconditioned the problem, with 'smooth' kernels such as that in Example 12.1.1 leading to the hardest numerical problems.

12.4.4   A numerical criterion for the existence of a solution

We noted in Section 12.2.1 that equation (12.2.3) could be used to give a numerical test for the existence of a solution of equation (12.1.1). Such tests can be of obvious value. We show here that a similar, although more complicated, numerical criterion for existence can be given for the Augmented Galerkin method, and that in practice it works rather well. It is based on an analysis of the structure of the matrix $\bar{\mathbf{B}}$ and vector $\bar{\mathbf{y}}$ in equation (12.4.3b). On the interval $[-1, 1]$ these have the form given by (8.5.5,6) which, as we noted there, identify $\bar{B}_{ij}, \bar{y}_i$ as coefficients in the orthogonal expansions

$$K(s,t)(1-t^2)^{\frac{1}{2}} = \frac{4}{\pi^2} \sum_{i=0}^{\infty}{}' \sum_{j=0}^{\infty}{}' \bar{B}_{ij} T_i(s) T_j(t),$$

$$y(s) = \frac{2}{\pi} \sum_{i=0}^{\infty}{}' \bar{y}_i T_i(s).$$

This in turn implies (see Chapter 9) that there exist constants $p, q, p', q' > \frac{1}{2}$; $c_k > 0$ such that

$$|\bar{B}_{ij}| \leq c_k \hat{i}^{-p'} \hat{j}^{-q'}, \quad i > j = 0, 1, \ldots,$$
$$\leq c_k \hat{i}^{-p} \hat{j}^{-q}, \quad j \geq i = 0, 1, \ldots. \tag{19a}$$

The coefficients p, p', q, q' depend on the analyticity properties of $K(s,t)(1-t^2)^{\frac{1}{2}}$; because of the square root singularity we expect that $p' > p, q' > q$. Following (19a) we introduce the representation for $\bar{B}_{ij}$:

$$\bar{B}_{ij} = C_{ij} i^{-p'} \hat{j}^{-q'}, \quad i > j, \tag{19b}$$

so that

$$|C_{ij}| \leq c_k.$$

Similarly there exist constants C_y, u such that

$$|\bar{y}_i| \leq C_y \hat{i}^{-u}. \tag{20a}$$

The existence of a solution of (1) requires that the analyticity properties of $K(s,t)$ and of $y(s)$ are 'compatible'; the following theorem relates this compatibility to the exponents p, q, p', q', u.

Theorem 12.4.3.
Let $\bar{B}, \bar{y}$ satisfy (19), (20a), and in addition let there exist a constant $c_y' > 0$ and an infinite sequence of integers P such that

$$c_y' \hat{i}^{-u} \leq |\bar{y}_i|, \quad i \in P. \tag{20b}$$

Then a necessary condition for (1) to have an integrable solution is that:

$$\text{Case 1: if } p' \leq p + q - \tfrac{1}{2}, \quad \text{then} \quad u \geq p'$$

or

$$\text{Case 2: if } p' > p + q - \tfrac{1}{2}, \quad \text{then} \quad u > p + q - \tfrac{1}{2}.$$

Proof. Assume that (1) has an integrable solution x, with expansion (4) satisfying (6) with $r > \frac{1}{2}$. Then the expansion coefficients **b** satisfy the Galerkin equations

$$\sum_{j=0}^{\infty} \bar{B}_{ij} b_j = \bar{y}_i, \quad i = 0, 1, \ldots. \tag{21}$$

Using (19) and (6a) we can write, for $i > 0$:

$$\left| \sum_{j=0}^{\infty} \bar{B}_{ij} b_j \right| \le c_k \left[i^{-p'} \left(\sum_{j=0}^{i-1} \hat{j}^{-q'} |b_j| \right) \right.$$

$$\left. + C_b i^{-p} \left(i^{-(q+r)} + \sum_{j=i+1}^{\infty} j^{-(q+r)} \right) \right]$$

$$\le \left(c_k \sum_{j=0}^{\infty} \hat{j}^{-q'} |b_j| \right) i^{-p'} + \left(\frac{c_k C_b (q+r)}{q+r-1} \right) i^{-(p+q+r-1)}$$

whence using (20b) we obtain

$$\frac{\left| \sum_{j=0}^{\infty} \bar{B}_{ij} b_j \right|}{\bar{y}_i} \le C_1 i^{-p'+u} + C_2 i^{-(p+q+r-1)+u}, \quad i = 1, 2, \ldots,$$

where

$$C_1 = \frac{\left(c_k \sum_{j=0}^{\infty} (j^{-q'} |b_j|) \right)}{c_v'}, \quad C_2 = \frac{c_k C_b (q+r)}{c_y' (q+r-1)}.$$

According to equation (21) the left hand side of the above inequality is unity for all $i \ge 1$, so that the right hand side should not tend to zero as i tends to infinity. A sufficient condition for this is given by

$$u \ge \alpha = \min \{ p', p+q+r-1 \}. \tag{22}$$

But the assumption that the solution is square integrable implies $r > \frac{1}{2}$; hence the theorem follows.

Discussion

Theorem 12.4.3 yields a cheaply computable existence criterion. The exponent u can be estimated from the vector $\bar{y}$ by following the procedure described in method 2 of Section 12.4.3; that is, we smooth $\bar{y}$, take logarithms and carry out a least squares fit to the functional form (20a). Similarly, p' can be estimated from (19a) and the first column of $\bar{B}$, and $(p+q)$ from the diagonal elements of $\mathbf{B}$. With appropriate precautions (to detect and allow for odd-even effects, for example) the criterion proves very effective, as the following examples illustrate.

(a) Example 12.4.2.

For this example, $p' = \infty$ and hence case 2 applies. The estimated exponents are given in Table 12.4.5 and correctly predict the existence of a solution.

Table 12.4.5. *Calculated values of* p', $p + q - \frac{1}{2}$, u *and* r (*Theorem 12.4.3*) *for problem 12.4.2*

N	p'	$p + q - \frac{1}{2}$	u	r
4	25	1.84	6.53	2.24
5	25	1.79	8.06	2.36
6	25	1.74	9.43	3.83
7	25	1.72	10.75	5.36
8	25	1.70	12.05	6.85
9	25	1.68	13.36	8.46
10	25	1.67	12.27	8.23
15	25	1.63	12.28	6.25
20	25	1.60	12.28	5.04

The estimate of 25 for p' represents a formal reply of 'very large' from the fitting routine.

Table 12.4.6. *Computed exponents for problems 12.1.3, 12.1.4*

	Problem 12.1.3			Problem 12.1.4		
N	p'	$p + q - \frac{1}{2}$	u	p'	$p + q - \frac{1}{2}$	u
4	3.07	1.81	1.00	5.32	5.06	2.32
5	3.49	2.17	1.00	6.52	6.24	2.32
6	3.76	2.52	1.00	7.71	7.49	2.24
7	3.97	2.52	1.00	8.90	8.77	2.24
8	4.11	2.63	1.00	10.09	8.95	2.20
9	4.23	2.74	1.00	10.93	8.95	2.20
10	4.31	2.74	1.00	11.26	8.95	2.17
15	4.55	2.89	1.00	11.07	8.94	2.13
20	4.66	2.96	1.00	8.54	8.94	2.10

We find that in both problems, $p' > p + q - \frac{1}{2}$ but $u < p + q - \frac{1}{2}$; that is, the necessary condition for a solution to exist does not hold, in agreement with the analytic conclusion.

(*b*) *Examples* 12.1.3 *and* 12.1.4.

For these examples it was noted in Section 12.1 that no solution exists. Table 12.4.6 gives the computed exponents found for these examples.

12.5   First kind Volterra equations

First kind Volterra equations of the form

$$\text{nonlinear:} \quad \int_a^s K(s, t, x(t)) \mathrm{d}t = y(s) \tag{1}$$

$$\text{linear:} \quad \int_a^s K(s, t) x(t) \mathrm{d}t = y(s) \tag{2}$$

can formally be handled by the same methods as first kind Fredholm equations; however, they are often in practice rather less illconditioned than their Fredholm counterparts. The reason for this can be seen informally by the following manipulations. Restricting attention to the linear case, and differentiating the equation (2), we find

$$\frac{d}{ds}\left[\int_a^s K(s,t)x(t)dt\right] = y'(s).$$

That is,

$$K(s,s)x(s) + \int_a^s \frac{\partial K(s,t)}{\partial s}x(t)dt = y'(s)$$

which may be rewritten as

$$x(s) = \bar{y}(s) + \int_a^s \bar{K}(s,t)x(t)dt \qquad (3)$$

with

$$\bar{y}(s) = \frac{y'(s)}{K(s,s)},$$

$$\bar{K}(s,t) = -\frac{\dfrac{\partial K(s,t)}{\partial s}}{K(s,s)}.$$

Equation (3) is a Fredholm equation of the second kind.

Provided $\bar{y}(s)$, $\bar{K}(s,t)$ are well behaved (which usually requires that $K(s,s)$ be nonzero for all s in the range of interest), the standard Fredholm theory is applicable and (3) is (usually) well posed. This implies that the solution of (3) is not pathologically sensitive to changes in $y'(s)$ and here we will expect the solutions of (2) not to be pathologically sensitive to changes in $y(s)$.

This expectation is borne out in practice; while (3) *can* be used as the basis for a numerical solution (and sometimes is, because of its comfortingly second kind form) it is in fact not particularly convenient for this because of the derivatives which it contains. A direct solution of (2) is perhaps more sensible, armed with the knowledge that it is not after all likely to be too illconditioned.

Examples 12.5.1 *and* 12.5.2.

$$\int_0^s \cos(s-t)x(t)dt = \sin s. \qquad (4)$$

Solution: $x(s) = 1$.

Range: $0 \leqslant s \leqslant 1$. (Problem 12.5.1) $\qquad (5)$

Range: $0 \leqslant s \leqslant 2$. (Problem 12.5.2) $\qquad (6)$

Table 12.5.1. *Computed maximum error*
$\|e_N\|_\infty$ *using the Augmented Galerkin*
method for Examples 12.5.1 and 12.5.2

N	Example 12.5.1	Example 12.5.2
2	1.0×10^{-1}	3.2, -1
3	9.3, -2	2.4, -1
4	2.1, -3	2.8, -2
5	7.5, -4	9.2, -3
6	6.4, -6	3.5, -4
7	1.8, -6	8.8, -5
8	8.9, -9	2.2, -6
9	1.1, -9	2.0, -7
10	2.9, -9	1.3, -8
15	2.6, -9	2.3, -9
20	6.6, -9	6.4, -9

Table 12.5.1 gives computed results for these two examples using the Augmented Galerkin method 1 with $r = 3$, $\lambda = 2$. The high accuracy achieved reflects the only moderate degree of illconditioning of these problems; the accuracy is harder to achieve as the range increases and we comment again that 'global' methods of this type are not really suited to Volterra problems if the range involved is large.

Exercises

1. An expansion method for the solution of $Kx = y$ seeks an approximation x_N for x of the form

$$x_N(s) = \sum_{i=1}^{N} a_i^{(N)} h_i(s).$$

Show that the method of least squares approximation is defined by the following set of linear equations

$$\mathbf{L}_{LS}^{(N)} \mathbf{a}^{(N)} = \mathbf{y}_{LS}^{(N)} \tag{A}$$

where

$$(\mathbf{L}_{LS}^{(N)})_{ij} = \int_a^b Kh_i^*(s) Kh_j(s) \, ds, \quad i, j = 1, \ldots, N,$$

$$(\mathbf{y}_{LS})_i = \int_a^b y(s) Kh_i(s) \, ds, \quad i = 1, \ldots, N,$$

$$Kh_i(s) = \int_a^b K(s, t) h_i(t) \, dt.$$

The first kind Fredholm equation

$$\int_0^\pi \sin(s+t)x(t)dt = \frac{\pi}{2}\sin s$$

has the exact solution $x(s) = \cos s$. Find the least squares solution of the form

$$x_2(s) = a_1{}^{(2)} + a_2{}^{(2)}s^2$$

by performing all the required integrals in (A) analytically.

2. Decide whether the following equations have a solution, giving reasons for your answer. Where a solution exists, use the results of Theorem 12.1.1 to find such a solution.

(a) $\displaystyle\int_0^\pi \sin(s+t)x(t)dt = e^{2s}$, $\qquad\qquad 0 \le s \le \pi$

(b) $\displaystyle\int_{-\pi}^\pi \sin(s+t)x(t)dt = 3\sin s + 2\cos s$, $\quad -\pi \le s \le \pi$

(c) $\displaystyle\int_{-1}^1 e^{(s+1)(t+1)/4}x(t)dt = (1-s^2)^{\frac{1}{2}}$, $\qquad -1 \le s \le 1$

3. The method of zero order regularisation replaces

$$\int_a^b K(s,t)x(t)dt = y(s)$$

by the problem

$$\underset{x}{\text{minimise }} F(x)$$

where

$$F(x) = (Kx - y, Kx - y) + \alpha(x, x).$$

Show, by reordering, that

$$F(x) = (x,(K^\dagger K + \alpha I)x) - (K^\dagger y, x) - (x, K^\dagger y) + (y, y) + \alpha(x, x).$$

Hence show that the defining equations for x are given by equation (12.3.3).

4. (a) Show that the first kind Volterra equation

$$\int_0^s \cos(s-t)x(t)dt = \sin s$$

can be reformulated as the second kind Volterra equation

$$x(s) = \cos s + \int_0^s \sin(s-t)x(t)dt.$$

(b) Explain the difficulty encountered when a similar attempt is made to reformulate

$$\int_0^s \sin(s-t)x(t)\mathrm{d}t = s - \sin s$$

as a second kind equation. Suggest a way of overcoming this problem and show that the second kind formulation may be given by

$$x(s) = \sin s + \int_0^s \sin(s-t)x(t)\mathrm{d}t.$$

13

Integro-differential equations

13.1 Introduction

An integro-differential equation is an equation involving one (or more) unknown functions $x(s)$, together with both differential and integral operations on x. Such a description covers a very broad class of functional relations and we restrict discussion here to the simplest types of one-dimensional integro-differential equations, which form a natural generalisation of Volterra and Fredholm integral equations. In particular we shall consider *nonlinear first order ordinary Volterra integro-differential equations* of the form

$$\left. \begin{array}{l} x'(s) = g(s, x(s)) + \lambda \displaystyle\int_a^s K(s, t, x(t)) \mathrm{d}t, \\[2mm] x(a) = \alpha, \end{array} \right\} \tag{1}$$

and *linear first and second order ordinary Fredholm integro-differential equations* of the form

$$\left. \begin{array}{l} P(s)x''(s) + Q(s)x'(s) + R(s)x(s) + \lambda \displaystyle\int_a^b K(s,t)x(t)\mathrm{d}t = g(s), \\[2mm] \mathbf{C}x(\mathbf{r}) + \mathbf{D}x'(\mathbf{r}) = \mathbf{e}, \end{array} \right\} \tag{2}$$

where

$$\begin{aligned} \mathbf{r} &= (r_1, \ldots, r_m)^T, \quad a \le r_i \le b, \\ x(\mathbf{r}) &= (x(r_1), \ldots, x(r_m))^T, \\ x'(\mathbf{r}) &= (x'(r_1), \ldots, x'(r_m))^T, \end{aligned}$$

and where for a pth order problem, $\mathbf{C}, \mathbf{D}$ are $p \times m$ matrices; $\mathbf{e}$ is a $p \times 1$ matrix.

In equations (1) and (2) g, K, P, Q, R are known functions but x is not. Note the appearance in (1) and (2) of 'boundary condition' equations. These are necessary to help prove that an unique solution exists; that such conditions are needed is evident by analogy with first order initial value and second order boundary value problems (set $\lambda = 0$ in (1) and (2)). It is the presence of these additional boundary condition equations which makes the treatment of integro-differential equations significantly different from that of integral equations.

We have seen in previous chapters that numerical methods for the solution of integral equations fall into two major classes, namely quadrature or Nystrom methods and expansion methods. Both these methods generalise readily to the solution of integro-differential equations.

Consider equation (1) with $a = 0$, $\lambda = 1$, $\alpha = x_0$. Then

$$\left.\begin{aligned} x'(s) &= g(s, x(s)) + \int_0^s K(s, t, x(t))\mathrm{d}t, \\ x(0) &= x_0. \end{aligned}\right\} \tag{3}$$

If we set

$$z(s) = \int_0^s K(s, t, x(t))\mathrm{d}t$$

then (3) may be expressed as

$$x'(s) = f(s, x(s), z(s)), \quad x(0) = x_0, \tag{3a}$$

where

$$f(s, x(s), z(s)) = g(s, x(s)) + z(s).$$

In the discussion which follows, we consider the numerical solution of first order nonlinear Volterra integro-differential equations with reference to equation (3a) which we usually write as

$$\begin{aligned} x'(s) &= f(s, x(s), z(s)), \quad x(0) = x_0, \\ z(s) &= \int_0^s K(s, t, x(t))\mathrm{d}t. \end{aligned} \tag{3b}$$

We may consider equation (3b) to be a generalisation of the corresponding problem in ordinary differential equations

$$x'(s) = f(s, x(s)), \quad x(0) = x_0,$$

in which the right hand side is now dependent on all values $x(t)$, $0 \leqslant t \leqslant s$, instead of on $x(s)$ alone. This suggests that we might solve equations (3b) by adapting the methods of solution for first order differential equations after replacing the integral by a quadrature rule of suitable accuracy; we discuss this approach in Section 13.2.1 where we solve equations (3) in the form (3b).

Note that equation (3b) may be expressed as a pair of coupled integral equations:

$$
\left.
\begin{aligned}
x(s) &= x_0 + \int_0^s f(t, x(t), z(t))dt, \\
z(s) &= \int_0^s K(s, t, x(t))dt,
\end{aligned}
\right\}
\tag{3c}
$$

and this formulation suggests an integral equation approach to the solution of (3). In Section 5.6 we outlined a solution scheme for such a system, in the particular case when f and K are linear functions of their arguments, and each integral is replaced by the same quadrature rule. We discuss in Section 13.2.2 the solution of the above system in more general terms: f and K are assumed to be nonlinear and each integral is approximated by a different quadrature rule.

We observe that if K is linearly dependent upon x, that is if

$$
K(s, t, x(t)) = A(s, t) + Bx(t)
$$

then (3b) may be expressed as a pair of differential equations

$$
x'(s) = f(s, x(s), z(s)),
$$

$$
z'(s) = A(s, s) + Bx(s) + \int_0^s \frac{\partial}{\partial s} A(s, t)dt.
\tag{3d}
$$

This first order system may be solved by generalising the methods used to solve a single first order differential equation (see Lambert, 1973); we do not discuss this formulation further.

Equation (2) reduces, on setting

$$
P(s) = 1, \quad Q(s) = R(s) = 0, \quad a = 0, \quad b = 1,
$$
$$
\mathbf{C} = \mathbf{I}, \mathbf{D} = \mathbf{0}, \mathbf{r} = (0, 1)^T, \mathbf{e} = (0, 0)^T
$$

to

$$
\left.
\begin{aligned}
x''(s) + \lambda \int_0^1 K(s, t)x(t)dt &= g(s), \\
x(0) = x(1) &= 0.
\end{aligned}
\right\}
\tag{4}
$$

By analogy with methods for the solution of linear two-point boundary value problems in ordinary differential equations, an obvious way to proceed in the solution of (4) is to replace derivatives by finite difference approximations and the integral by a quadrature rule of suitable accuracy. We shall discuss this approach further in Section 13.2.3.

Alternatively we may rephrase equation (4) as a Fredholm integral equation of the second kind. Introducing

$$
f(s, x(s)) = g(s) - \lambda \int_0^1 K(s, t)x(t)dt
$$

equation (4) reduces to

$$\left. \begin{array}{l} x''(s) = f(s, x(s)), \\ x(0) = x(1) = 0. \end{array} \right\} \qquad (4a)$$

Now if we integrate equation (4a) twice and in addition we require that the boundary conditions are satisfied we find that x satisfies the integral equation (see Section 0.3):

$$x(s) = -s \int_0^1 (1 - t)f(t, x(t))\mathrm{d}t + \int_0^s (s - t)f(t, x(t))\mathrm{d}t. \qquad (5)$$

Substituting for f gives

$$x(s) + \lambda \int_0^1 G(s, t) \int_0^1 K(t, u)x(u)\mathrm{d}u \, \mathrm{d}t = \int_0^1 G(s, t)g(t)\mathrm{d}t \qquad (6)$$

where

$$\begin{array}{ll} G(s, t) = -t(1 - s), & 0 \leqslant t \leqslant s, \\ = -s(1 - t), & s \leqslant t \leqslant 1, \end{array}$$

is the Green's function for the problem. Thus x satisfies the Fredholm integral equation

$$x(s) + \lambda \int_0^1 H(s, u)x(u)\mathrm{d}u = \gamma(s) \qquad (7)$$

where

$$H(s, u) = \int_0^1 G(s, t)K(t, u)\mathrm{d}t, \qquad (7a)$$

$$\gamma(s) = \int_0^1 G(s, t)g(t)\mathrm{d}t, \qquad (7b)$$

and the solution of this equation is characterised by the Fredholm alternative of Chapter 6. Clearly the methods of Chapter 4 are directly applicable to the solution of (7), having first replaced the integrals in (7a, b) by quadrature rules of suitable accuracy. In general, the existence of a Green's function may be known but its form may be difficult to obtain, especially for a general boundary condition such as in (2), so that this approach provides theoretical insight rather than a practical general numerical technique.

Expansion methods may also be used to solve Fredholm and Volterra integro-differential equations, either directly or in their integrated forms, and we discuss this approach in Section 13.3.

13.2   Quadrature methods for the numerical solution of integro-differential equations
In the previous section we saw that different formulations of the integro-differential equation accordingly suggest solution by different methods of approach. We give below a brief review of quadrature methods applied to these formulations.

13.2.1   Differential equation methods for Volterra-type equations
We consider the first order nonlinear Volterra integro-differential equation of Section 13.1 in the form

$$x'(s) = f(s, x(s), z(s)), \quad x(0) = x_0, \tag{1a}$$

$$z(s) = \int_0^s K(s, t, x(t)) \, dt, \tag{1b}$$

over the finite interval $[0, A]$. We assume that f and K are uniformly continuous functions of their arguments, that K satisfies a Lipschitz condition in x and that f satisfies a Lipschitz condition in both x and z. Then with these assumptions, equations $(1a, b)$ have an unique solution in $0 \le s \le A$. (See Linz, 1969*b*).

Clearly we can obtain a step-by-step solution of the differential equation $(1a)$ if we approximate $z(s)$ numerically and Linz (1969*b*) discusses an extension of multistep methods for first order differential equations to deal with $(1a, b)$. We present this extension below, using the notation of Section 5.1.6 to present the new multistep method for $(1a)$:

$$x_{n+1} = \sum_{i=0}^{p} a_i x_{n-i} + h \sum_{i=-1}^{p} b_i f(s_{n-i}, x_{n-i}, z_{n-i}) \tag{2a}$$

where the a_i, b_i are constants and x_r, z_r are approximations to $x(s_r), z(s_r)$ respectively, with $s_r = rh, r = 0, 1, \ldots, N$, $h = A/N$. Equation $(2a)$ is a $(p+1)$-step method which is explicit or implicit according as b_{-1} is zero or nonzero, and which requires $p + 1$ starting conditions. Clearly if equation $(2a)$ is implicit we shall need to employ a predictor formula (which is necessarily explicit) to correspond to the role of $(2a)$ as the corrector formula. Equation $(2a)$ also requires the values $z_{n-i}, i = -1, 0, \ldots, p$ which follow on replacing the integral in $(1b)$ by a standard quadrature rule of suitable accuracy:

$$z_{n-i} = \sum_{j=0}^{n-i} w_{n-i, j} K(s_{n-i}, t_j, x_j), \quad i = -1, 0, \ldots, p, \tag{2b}$$

where $s_i = t_i, i = 0, 1, \ldots, N$. This formula also requires starting values and we return later to the question of provision of starting values for $(2a)$ and $(2b)$. Usually the weights $w_{n-i, j}$ correspond to those of an appropriate repeated quadrature rule or alternatively to those of a single application of the Gregory rule (see Section 4.3.4) which has the form:

$$\int_0^{s_r} \phi(s)ds = h[\tfrac{1}{2}\phi_0 + \phi_1 + \cdots + \phi_{r-1} + \tfrac{1}{2}\phi_r] + \Delta \tag{3}$$

where

$$\Delta = -h(\tfrac{1}{12}\nabla + \tfrac{1}{24}\nabla^2 + \tfrac{19}{720}\nabla^3 + \tfrac{3}{160}\nabla^4 + \cdots)\phi_r$$
$$+ h(\tfrac{1}{12}\Delta - \tfrac{1}{24}\Delta^2 + \tfrac{19}{720}\Delta^3 - \tfrac{3}{160}\Delta^4 + \cdots)\phi_0.$$

The order of the multistep method in (2a) will usually determine the number of differences to be retained in (3) if a Gregory rule is used to approximate the integral in (1b). If the kernel function in (1b) is badly behaved then the integral may be replaced by an appropriate product rule (see Section 5.5).

Example **13.2.1.** Suppose we use Euler's method in equation (2a) and the Trapezoid rule in (2b). Then equations (2a, b) take the simple form

$$x_{n+1} = x_n + hf(s_n, x_n, z_n),$$

$$z_n = h\sum_{j=0}^{n}{}'' K(s_n, s_j, x_j), \quad n \geq 1.$$

This is a one-step method ($p = 0$) which requires only the values of x_0 and z_0 in order to proceed; x_0 is given as the initial condition and z_0 is clearly zero.

For $p \geq 1$ starting values for the multistep method (2a, b) must be supplied. One starting procedure, given by Brunner and Lambert (1974), uses a power series expansion of the form

$$\tilde{x}(s) = \sum_{v=0}^{w} \frac{a_v s^v}{v!}, \quad w \geq 1, \tag{4}$$

where w is an integer value determined by the order of the multistep method (2a, b). (See Brunner and Lambert for further details.) The coefficients $\{a_v\}$ are determined from the following conditions:

   (i) $\tilde{x}(0) = x_0$

   (ii) $\tilde{x}'(0) = f(0, x_0, 0)$

   (iii) $\tilde{x}'(s_j) = f(s_j, \tilde{x}(s_j), \tilde{z}(s_j)), \quad j = 1, \ldots, w - 1,$ (5)

where

$$\tilde{z}(s_j) = \int_0^{s_j} K(s_j, t, \tilde{x}(t))dt;$$

this integral is approximated by numerical quadrature. For example a Gregory rule of suitable order could be used. It follows from (i) and (ii) that

$$a_0 = x_0,$$
$$a_1 = f(0, x_0, 0).$$

Condition (iii) gives rise to a set of $(w-1)$ nonlinear equations for a_2, $a_3, \ldots, a_w$ which under suitable conditions (sufficiently small steps) may be solved by fixed point iteration to provide an unique solution of (5) (see Brunner and Lambert, 1974). This starting procedure seems preferable to a direct Taylor series expansion method, which for methods of order greater than two would require finding derivatives of the kernel function K. The above starting procedure does require however that we solve a system of nonlinear equations.

Linz (1969b) has proposed an alternative starting procedure which uses appropriate combinations of interpolation and quadrature rules to solve problem (1a,b) in the equivalent form

$$x(s_0) = x_0, \tag{6a}$$

$$x(s_{n+r}) = x(s_n) + \int_{s_n}^{s_{n+r}} f(t, x(t), z(t))\mathrm{d}t, \tag{6b}$$

$$z(t) = \int_0^t K(t, u, x(u))\mathrm{d}u. \tag{6c}$$

Now if Simpson's rule is used to approximate the integrals in (6) we have for $n = 0, 2, 4, \ldots,$

$$\begin{aligned} x_{n+1} = x_n + \tfrac{1}{6}h[f(s_n, x_n, z_n) + 4f(s_{n+\frac{1}{2}}, x_{n+\frac{1}{2}}, z_{n+\frac{1}{2}}) \\ + f(s_{n+1}, x_{n+1}, z_{n+1})], \end{aligned} \tag{7a}$$

$$\begin{aligned} x_{n+2} = x_n + \tfrac{1}{3}h[f(s_n, x_n, z_n) + 4f(s_{n+1}, x_{n+1}, z_{n+1}) \\ + f(s_{n+2}, x_{n+2}, z_{n+2})], \end{aligned} \tag{7b}$$

$$\begin{aligned} z_{n+1} = \tfrac{1}{3}h\sum_{j=0}^{n} w_j K(s_{n+1}, s_j, x_j) + \tfrac{1}{6}h[K(s_{n+1}, s_n, x_n) \\ + 4K(s_{n+1}, s_{n+\frac{1}{2}}, x_{n+\frac{1}{2}}) + K(s_{n+1}, s_{n+1}, x_{n+1})], \end{aligned} \tag{7c}$$

where

$$w_0 = w_n = 1; \quad w_j = 3 - (-1)^j, \quad 1 \le j \le n-1,$$

$$z_{n+2} = \tfrac{1}{3}h\sum_{j=0}^{n+2} w_j K(s_{n+2}, s_j, x_j), \tag{7d}$$

where

$$w_0 = w_{n+2} = 1; \quad w_j = 3 - (-1)^j, \quad 1 \le j \le n+1.$$

Thus provided that we have some means of approximating the values $x_{n+\frac{1}{2}}$ and $z_{n+\frac{1}{2}}$, equations (6a) and (7a)–(7d) may be used to provide starting values for the multistep method (2a,b). Let $P_2(s)$ be the quadratic polynomial interpolating the solution $x(s)$ at the points s_n, s_{n+1}, s_{n+2}; then if we set $x_{n+\frac{1}{2}} = P_2(s_{n+\frac{1}{2}})$ it follows that

$$x_{n+\frac{1}{2}} = \tfrac{3}{8}x_n + \tfrac{3}{4}x_{n+1} - \tfrac{1}{8}x_{n+2}. \tag{8a}$$

Similarly quadratic interpolation of $z(s)$ at the points s_n, s_{n+1}, s_{n+2} yields

$$z_{n+\frac{1}{2}} = \tfrac{3}{8}z_n + \tfrac{3}{4}z_{n+1} - \tfrac{1}{8}z_{n+2}. \tag{8b}$$

Thus for $n = 0, 2, 4, \ldots$ equations $(7a)$–$(7d)$, $(8a,b)$ are an implicit set of equations for x_{n+1} and x_{n+2}, which can be solved by iteration provided h is sufficiently small. Indeed the repeated use of these equations defines a new method for solving the complete problem, with the advantage that no starting values are required. We illustrate, via an example taken from Linz (1969b), the use of equations $(7a)$–$(7d)$ and $(8a,b)$ both as a method of solution over the whole interval and as a starting procedure used in conjunction with the multistep method $(2a, b)$.

***Example* 13.2.2.** The equation

$$\left. \begin{array}{l} x'(s) = 1 + x(s) - se^{-s^2} - 2\displaystyle\int_0^s ste^{-x^2(t)}dt, \quad 0 \le s \le 2, \\[2mm] x(0) = 0, \end{array} \right\} \tag{9}$$

whose exact solution is $x(s) = s$, is solved using

   (i) equations $(7a)$–$(7d)$ and $(8a,b)$ over the interval $[0,2]$, and
  (ii) the third order Adams–Bashforth/Adams–Moulton method which uses the predictor–corrector pair

$$x_{n+3}^{(p)} = x_{n+2} + \tfrac{1}{12}h(23f_{n+2} - 16f_{n+1} + 5f_n), \tag{10a}$$

$$x_{n+3}^{(c)} = x_{n+2} + \tfrac{1}{12}h(5f_{n+3}^{(p)} + 8f_{n+2} - f_{n+1}), \tag{10b}$$

where $f_{n+3}^{(p)} = f(s_{n+3}, x_{n+3}^{(p)}, z_{n+3})$, with the third order Gregory rule

$$z_r = \sum_{j=0}^{r} w_{rj}K(s_r, t_j, x_j) \tag{10c}$$

for $r \ge 3$, with weights defined by

$$w_{r0} = w_{rr} = \tfrac{5}{12},$$
$$w_{r1} = w_{r,r-1} = \tfrac{13}{12},$$
$$w_{ri} = 1, \quad i = 2, 3, \ldots, r - 2.$$

(These weights follow on neglecting second and higher differences in (3)). The required starting values are found using equations $(7a)$–$(7d)$ and $(8a, b)$.

Table 13.2.1 shows the modulus of the maximum error $\varepsilon_{\max}$ obtained on solution of equation (9) over the interval $[0, 2]$, for each of the two methods with a steplength h.

Method (i) clearly gives better accuracy; as Linz points out, this is to be expected because of the high accuracy of Simpson's rule which is used

Table 13.2.1. *These results are taken from Table II of Linz (1969b)*

Method	h	$|\varepsilon_{max}|$
(i)	0.1	1.7×10^{-5}
(i)	0.05	$1.2, -6$
(ii)	0.1	$4.5, -4$
(ii)	0.05	$7.0, -5$

throughout. Although this method has the advantage of being self-starting it requires additional work to solve the nonlinear systems of equations for $\{x_{n+1}, x_{n+2}\}_{n=0,2,4,\ldots}$ which arise. The user must therefore decide on an order of priorities and choose the method of solution accordingly. Method (i) is in fact an example of a step-by-step quadrature method and these methods are discussed more fully in Section 13.2.2 where the integro-differential equation is solved in the form of two coupled integral equations.

Finally we discuss briefly the use of Runge–Kutta methods to provide starting values or to compute the entire numerical solution of equations $(1a, b)$. Runge–Kutta methods for the solution of ordinary differential equations have been discussed at length in Section 5.1.3 and these methods are normally easily extended to solve integro-differential equations of the form $(1a, b)$; for convenience we write equations $(1a, b)$ equivalently as

$$\left.\begin{array}{l} x'(s) = F(s, x(s)), \\ x(0) = x_0, \end{array}\right\} \tag{11}$$

where F depends on s and on $x(t)$ for $0 \le t \le s$. The general p-stage Runge–Kutta method is given by (see Section 5.1.3)

$$x_{i+1} = x_i + h \sum_{l=0}^{p-1} A_{pl} F(s_i + \theta_l h, x_{i+\theta_l}) \tag{12}$$

where

$$0 = \theta_0 \le \theta_1 \le \cdots \le \theta_{p-1} \le 1.$$

Clearly we shall require approximations (in terms of $x_0, x_1, \ldots, x_i$) of appropriate order to the integrals

$$\int_0^{s_i + \theta_l h} K(s_i + \theta_l h, t, x(t)) \mathrm{d}t$$

which we treat as

$$\int_0^{s_i} K(s_i + \theta_l h, t, x(t)) \mathrm{d}t + \int_{s_i}^{s_i + \theta_l h} K(s_i + \theta_l h, t, x(t)) \mathrm{d}t.$$

Suppose that the Runge–Kutta method has order k, that is the local truncation error per step is $\mathcal{O}(h^{k+1})$. Then we may use quadrature formulae of order $k-1$ to approximate each of the integral terms; the quadrature errors of these integrals are $\mathcal{O}(h^k)$ and this is consistent with the use of a kth order Runge–Kutta method.

Example 13.2.3. Consider an extension of the improved Euler method (see Example 5.1.2) which we write as:

$$x_{i+1} = x_i + \tfrac{1}{2}h(k_0 + k_1) \tag{13}$$

with

$$k_0 = F(s_i, x_i), \tag{13a}$$

$$k_1 = F(s_{i+1}, x_i + hk_0). \tag{13b}$$

This is a second order method whose local truncation error is $\mathcal{O}(h^3)$. This suggests that we use the repeated Trapezoid rule to estimate the integral in (13a):

$$\int_0^{s_i} K(s_i, t, x(t))\mathrm{d}t \approx h \sum_{j=0}^{i}{}'' K(s_i, s_j, x_j).$$

We write the integral in (13b) as

$$\int_0^{s_{i+1}} K(s_{i+1}, t, x(t))\mathrm{d}t = \int_0^{s_i} K(s_{i+1}, t, x(t))\mathrm{d}t$$
$$+ \int_{s_i}^{s_{i+1}} K(s_{i+1}, t, x(t))\mathrm{d}t$$

and treat each integral by a separate quadrature rule of order one. We approximate the first and second integrals respectively by the repeated Trapezoid rule and Euler's rule, that is

$$\int_0^{s_{i+1}} K(s_{i+1}, t, x(t))\mathrm{d}t \approx h \sum_{j=0}^{i}{}'' K(s_{i+1}, s_j, x_j)$$
$$+ hK(s_{i+1}, s_i, x_i).$$

Hence this second order Runge–Kutta method for ordinary differential equations can be easily extended to provide a self-starting method for the solution of (1a, b).

13.2.2 Integral equation methods for Volterrra-type equations

We treat the first order nonlinear Volterra integro-differential equation of Section 13.1 in the form

$$x(s) = x_0 + \int_0^s f(t, x(t), z(t))\mathrm{d}t, \tag{14a}$$

$$z(s) = \int_0^s K(s, t, x(t))\mathrm{d}t, \tag{14b}$$

using a step-by-step quadrature method which we describe below. As in the previous section we solve the problem over the finite interval $[0, A]$ returning the solution at the equally spaced points $s_r = rh, r = 0, 1, \ldots, N$ where

$$0 = s_0 < s_1 < \cdots < s_N = A.$$

We proceed by approximating each integral by a quadrature rule of suitable accuracy. Thus when $s = s_r$, we obtain

$$x_r = x_0 + h \sum_{i=0}^{r} w_{ri} f(s_i, x_i, z_i), \quad r = k, k+1, \ldots, N, \tag{15a}$$

$$z_i = h \sum_{j=0}^{i} w_{ij}' K(s_i, s_j, x_j), \quad i = 1, 2, \ldots, r. \tag{15b}$$

If $x_0, x_1, \ldots, x_{r-1}$ are known then the quadrature method given by (15a) can be used to solve for x_r; in general the first k values of x_r are computed via special starting procedures. If $w_{rr} = 0$ then (15a) represents an explicit system of equations, but in general w_{rr} is nonzero. In this case (15a) defines the x_r implicitly and the equations must be solved by an iterative method, for example by simple iteration or by Newton's method. The former iteration has been described in Section 5.1.2 and we present here the solution of equations (15a, b) using the Newton iteration. Let ϕ_1, ϕ_2 be functions of x_r, z_r defined by

$$\phi_1(x_r, z_r) = x_r - hw_{rr} f(s_r, x_r, z_r) - \psi_{1,r},$$

$$\phi_2(x_r, z_r) = z_r - hw_{rr}' K(s_r, s_r, x_r) - \psi_{2,r},$$

where

$$\psi_{1,r} = x_0 + h \sum_{i=0}^{r-1} w_{ri} f(s_i, x_i, z_i),$$

$$\psi_{2,r} = h \sum_{j=0}^{r-1} w_{rj}' K(s_r, s_j, x_j),$$

are known values. Then equations (15a, b) are equivalent to the following system of equations:

$$\phi(\mathbf{x}_r) = \begin{pmatrix} \phi_1(x_r, z_r) \\ \phi_2(x_r, z_r) \end{pmatrix} = \begin{pmatrix} 0 \\ 0 \end{pmatrix} \tag{16}$$

with $\mathbf{x}_r = (x_r, z_r)^T$. Equation (16) can now be solved using the Newton iteration:

$$\mathbf{x}_r^{(k+1)} = \mathbf{x}_r^{(k)} - \mathbf{J}^{-1}(\mathbf{x}_r^{(k)}) \phi(\mathbf{x}_r^{(k)}) \tag{17}$$

where $\mathbf{J}(\mathbf{x}_r)$ is the Jacobian matrix given by

$$\mathbf{J}(\mathbf{x}_r) = \begin{pmatrix} \dfrac{\partial \phi_1}{\partial x_r}(x_r, z_r) & \dfrac{\partial \phi_1}{\partial z_r}(x_r, z_r) \\ \dfrac{\partial \phi_2}{\partial x_r}(x_r, z_r) & \dfrac{\partial \phi_2}{\partial z_r}(x_r, z_r) \end{pmatrix}.$$

The Newton iteration is normally solved as a system of linear equations for the vector $\delta_r^{(k)} = x_r^{(k+1)} - x_r^{(k)}$, that is

$$J(x_r^{(k)})\delta_r^{(k)} = -\phi(x_r^{(k)}). \tag{18}$$

A convenient starting value is $x_r^{(0)} = x_{r-1}$ and sufficient conditions for the convergence of this iteration are provided by the Newton–Kantorovich theorem (see Ortega, 1972). We illustrate the method for the problem in equation (9).

Example 13.2.4.
  The equation

$$x'(s) = 1 + x(s) - se^{-s^2} - 2\int_0^s ste^{-x^2(t)}dt, \quad x(0) = 0, \tag{19}$$

may be written equivalently as

$$\left.\begin{aligned} x(s) &= \int_0^s f(t, x(t), z(t))dt, \\ z(s) &= \int_0^s ste^{-x^2(t)}dt, \end{aligned}\right\} \tag{20}$$

where

$$f(s, x(s), z(s)) = 1 + x(s) - se^{-s^2} - 2z(s).$$

Suppose we use the Trapezoid rule to approximate each integral in (20), that is,

$$w_{r0} = w_{rr} = w_{r0}' = w_{rr}' = \tfrac{1}{2}; \quad r \geq 1,$$
$$w_{ri} = w_{ri}' = 1, \quad i = 1,\ldots,r-1; \quad r \geq 2,$$
$$w_{00}' = 0.$$

The starting value $x_0 = x(0)$ is required and this is given by the initial condition $x(0) = 0$. Then the Newton iteration is given by

$$J(x_r^{(k)})\delta_r^{(k)} = -\phi(x_r^{(k)})$$

where

$$J(x_r) = \begin{pmatrix} 1 - \dfrac{h}{2} & h \\ hs_r^2 x_r e^{-x_r^2} & 1 \end{pmatrix},$$

$$\phi(x_r) = \begin{pmatrix} x_r - \dfrac{h}{2}(1 + x_r - s_r e^{-s_r^2} - 2z_r) - h\displaystyle\sum_{i=0}^{r-1}{}'(1 + x_i - s_i e^{-s_i^2} - 2z_i) \\ z_r - \dfrac{h}{2}(s_r^2 e^{-x_r^2}) - h\displaystyle\sum_{j=0}^{r-1}{}'(s_r s_j e^{-x_j^2}) \end{pmatrix}.$$

Table 13.2.2. *Results for Example* 13.2.4

s	x(s)	x̃(s), h = 0.1	N	x̃(s), h = 0.01	N
0	0	0.000 000 00	0	0.000 000 00	0
0.1	0.1	0.100 000 29	2	0.100 000 00	2
0.2	0.2	0.200 002 93	2	0.200 000 02	2
0.3	0.3	0.300 011 55	3	0.300 000 10	2
0.4	0.4	0.400 034 17	3	0.400 000 32	2
0.5	0.5	0.500 079 46	3	0.500 000 77	2
0.6	0.6	0.600 157 11	3	0.600 001 56	2
0.7	0.7	0.700 279 67	4	0.700 002 81	2
0.8	0.8	0.800 458 07	4	0.800 004 48	3
0.9	0.9	0.900 693 19	5	0.900 006 86	3
1.0	1.0	1.001 001 03	7	1.000 010 00	3

Continuing the iteration process until the convergence condition

$$\| \delta_r^{(k)} \|_\infty < 5 \times 10^{-5}$$

is satisfied for each value of r, Table 13.2.2 records, at equally spaced intervals, the computed and exact solutions $\tilde{x}(s)$ and $x(s)$ obtained on solution of problem (19) over the interval $[0, 1]$. The starting value is taken as $x_1^{(0)} = x_0 = (0, 0)^T$ and the number of iterations N required to achieve the specified accuracy at each step, with two different steplengths h, is also recorded.

Note that for this particular case, where the Trapezoid rule is used to approximate both integrals in (15a, b), equation (15a) may be differenced for successive values of r to give

$$x_{r+1} - x_r = \frac{h}{2}(f_r + f_{r+1}) \tag{21}$$

where

$$f_j = f(jh, x_j, z_j),$$

$$z_j = h \sum_{i=0}^{j} {}'' K(jh, ih, x_i).$$

This procedure reduces the method to a multistep method of the type discussed in the previous section. In particular we note that the use of equations (21) involves considerably less work than the corresponding step-by-step quadrature method applied to equations (20). Clearly if such a reduction is possible, be it to a multistep formula or to some other scheme which proves to be computationally more advantageous than the use of equation (15a) in the quadrature method, then this method of approach should always be adopted. Baker (1977) remarks that those examples where equations (15a, b) reproduce a multistep method of the form (2a, b) are provided by choosing the weights w_{ri} to correspond to those of the Gregory rules (see equation (3)) with a fixed number of differences; then subtraction of the equation for x_r from that for x_{r+1} in (15a) yields an Adams–Moulton corrector formula.

Before a step-by-step quadrature method can proceed, starting values for equation (15a) must be supplied. We describe a starting procedure (see Day, 1970) which provides third order approximations (with error proportional to h^4) for these values; this procedure could in fact be used to generate the entire numerical solution of the problem. For ease of presentation we consider equations (14a, b) in the equivalent form

$$\left.\begin{aligned} x'(s) &= g(s, x(s)) + \int_0^s K(s, t, x(t))dt, \\ x(0) &= x(s_0) = x_0. \end{aligned}\right\} \tag{22}$$

The first stage of the procedure produces a third order approximation x_1 to $x(s_1)$ using the explicit Runge–Kutta formula (with $n = 0$):

$$x_{n+1} = x_n + \frac{h}{4}[F(s_n, x_n) + 3F(s_n + \tfrac{2}{3}h, x_{n+\frac{2}{3}})] \tag{23}$$

where $F(s, x(s))$ corresponds to the right hand side of equation (22). A first order approximation to $x(\tfrac{2}{3}h)$ is given by the truncated Taylor series expansion

$$\hat{x}_{\frac{2}{3}} = x_0 + \tfrac{2}{3}hx_0' = x_0 + \tfrac{2}{3}hg(0, x_0) \tag{24}$$

where $x_0' \approx x'(s_0)$. On integrating equation (22) we obtain

$$x(s_0 + bh) = x(s_0 + ah) + \int_{s_0 + ah}^{s_0 + bh} g(s, x(s))ds$$
$$+ \int_{s_0 + ah}^{s_0 + bh} \int_0^s K(s, t, x(t))ds\,dt \tag{22a}$$

and if we set $a = 0$, $b = \tfrac{2}{3}$ and approximate each integral by the Trapezoid rule we produce a second order approximation to $x(\tfrac{2}{3}h)$:

$$\tilde{x}_{\frac{2}{3}} = x_0 + \tfrac{1}{3}h[g(0, x_0) + g(\tfrac{2}{3}h, \hat{x}_{\frac{2}{3}})]$$
$$+ \tfrac{1}{9}h^2[K(\tfrac{2}{3}h, 0, x_0) + K(\tfrac{2}{3}h, \tfrac{2}{3}h, \hat{x}_{\frac{2}{3}})]. \tag{25}$$

Then we have (from (23), again using the Trapezoid rule to approximate any integrals)

$$x_1 = x_0 + \tfrac{1}{4}h[g(0, x_0) + 3g(\tfrac{2}{3}h, \tilde{x}_{\frac{2}{3}})]$$
$$+ \tfrac{1}{4}h^2[K(\tfrac{2}{3}h, 0, x_0) + K(\tfrac{2}{3}h, \tfrac{2}{3}h, \tilde{x}_{\frac{2}{3}})]. \tag{26}$$

The second stage of the procedure delivers a third order approximation x_2

to $x(s_2)$. Substituting $s = s_1 = h$ in (22) we find

$$x'(s_1) = g(s_1, x_1) + \int_0^{s_1} K(s_1, t, x(t))dt$$

and (using the Trapezoid rule to approximate the integral) we have

$$x_1' = g(s_1, x_1) + \frac{h}{2}[K(s_1, 0, x_0) + K(s_1, s_1, x_1)] \tag{27}$$

where $x_1' \approx x'(s_1)$. Now if we set $a = 1$, $b = 2$ in (22a) it follows that

$$x(s_2) = x(s_1) + \int_{s_1}^{s_2} g(s, x(s))ds + \int_{s_1}^{s_2} \int_0^s K(s, t, x(t))ds\,dt$$

and using the Trapezoid rule to approximate the first integral and the outer integral of the second we obtain

$$\hat{x}_2 = x_1 + \tfrac{1}{2}h[g(s_1, x_1) + g(s_2, x_1 + hx_1')]$$
$$+ \tfrac{1}{2}h\left[\int_0^{s_1} K(s_1, t, x(t))dt + \int_0^{s_2} K(s_2, t, x(t))dt\right].$$

Using the Trapezoid rule and Simpson's rule where appropriate we have

$$\hat{x}_2 = x_1 + \tfrac{1}{2}h[g(s_1, x_1) + g(s_2, x_1 + hx_1')]$$
$$+ \tfrac{1}{4}h^2[K(s_1, 0, x_0) + K(s_1, s_1, x_1)]$$
$$+ \tfrac{1}{6}h^2[K(s_2, 0, x_0) + 4K(s_2, s_1, x_1)$$
$$+ K(s_2, s_2, x_1 + hx_1')] \tag{28}$$

which is a second order approximation to $x(2h)$. Finally we set $a = 0$, $b = 2$ in (22a) and approximate each integral by the Trapezoid or Simpson's rule according as the range of integration is $[s_0, s_1]$ or $[s_0, s_2]$. Then a third order approximation to $x(s_2)$ is given by

$$x_2 = x_0 + \tfrac{1}{3}h[g(0, x_0) + 4g(s_1, x_1) + g(s_2, \hat{x}_2)]$$
$$+ \tfrac{2}{3}h^2[K(s_1, 0, x_0) + K(s_1, s_1, x_1)]$$
$$+ \tfrac{1}{9}h^2[K(s_2, 0, x_0) + 4K(s_2, s_1, x_1) + K(s_2, s_2, \hat{x}_2)]. \tag{29}$$

The third and subsequent stages of the procedure deliver third order approximations x_r to $x(s_r)$, $r \geq 3$, using the two-step implicit Adams–Moulton formula

$$x_r = x_{r-1} + \tfrac{1}{12}h[5F(s_r, x_r) + 8F(s_{r-1}, x_{r-1}) - F(s_{r-2}, x_{r-2})].$$

Each integral is approximated by the repeated Trapezoid rule, giving

$$x_r' = x_{r-1} + \tfrac{1}{12}h[5g(s_r, x_r) + 8g(s_{r-1}, x_{r-1}) - g(s_{r-2}, x_{r-2})]$$
$$+ \tfrac{5}{12}h^2 \sum_{i=0}^{r}{}'' K(s_r, s_i, x_i) + \tfrac{2}{3}h^2 \sum_{i=0}^{r-1}{}'' K(s_{r-1}, s_i, x_i)$$
$$- \tfrac{1}{12}h^2 \sum_{i=0}^{r-2}{}'' K(s_{r-2}, s_i, x_i). \tag{30}$$

Equation (30) is a nonlinear equation for x_r which must be solved iteratively; a good first guess follows on replacing x_r on the right hand side of (30) by $(x_{r-1} + hx_{r-1}')$ where x_{r-1}' is an approximation to $x'(s_{r-1})$ with

$$x_{r-1}' = g(s_{r-1}, x_{r-1}) + h\sum_{i=0}^{r-1}{}'' K(s_{r-1}, s_i, x_i). \qquad (31)$$

Thus the method is defined by equations (24)–(31). Clearly algorithms of higher order can be generated in this fashion; that is, by producing a sequence of intermediate approximations of suitable accuracy. (See, for example Goldfine, 1972.)

13.2.3 Differential equation methods for Fredholm-type equations

We next consider the solution of the Fredholm integro-differential equation

$$x''(s) + \lambda \int_0^1 K(s,t)x(t)dt = g(s),$$
$$x(0) = x(1) = 0, \qquad (32)$$

in the form of a two-point boundary value problem in ordinary differential equations:

$$x''(s) = f(s,x),$$
$$x(0) = x(1) = 0, \qquad (33)$$

where $f(s,x) = g(s) - \lambda \int_0^1 K(s,t)x(t)dt$. Let the interval $[0,1]$ be divided into N equal intervals of length h with

$$0 = s_0 < s_1 < \cdots < s_{N-1} < s_N = 1,$$
$$s_r = rh, \quad r = 0, 1, \ldots, N.$$

We can use central differences to approximate the second derivative term in (33) at the point s_i. We have

$$x''(s_i) = \frac{1}{h^2}(\delta^2 x_i - \tfrac{1}{12}\delta^4 x_i + \tfrac{1}{90}\delta^6 x_i - \cdots), \quad i = 1, 2, \ldots, N-1, \quad (34)$$

where δ is the central difference operator. In particular, if we neglect fourth and higher differences, the second derivative at $s = s_i$ is approximated by

$$x_i'' \approx \frac{x_{i+1} - 2x_i + x_{i-1}}{h^2}, \quad i = 1, 2, \ldots, N-1, \qquad (35)$$

with an error of $\mathcal{O}(h^2)$. Then if we impose the boundary conditions

$$x_0 = x_N = 0 \qquad (36)$$

we have the following set of equations to solve:

$$\mathbf{Ax} = \mathbf{f} \qquad (37)$$

where

$$\mathbf{A} = \frac{1}{h^2}\begin{pmatrix} -2 & 1 & & & 0 \\ 1 & -2 & 1 & & \\ & 1 & -2 & \cdot & \\ & & \cdot & \cdot & \cdot & 1 \\ 0 & & & \cdot & 1 & -2 \end{pmatrix}$$

is an $(N-1) \times (N-1)$ tridiagonal matrix and $\mathbf{x} = (x_1, \ldots, x_{N-1})^T$, $\mathbf{f} = (f_1, \ldots, f_{N-1})^T$. Equation (37) may be solved using Gauss elimination; in particular the special structure of $\mathbf{A}$ results in a greatly simplified algorithm for the solution of (37). To obtain good accuracy with this low-order approximation, h must be small, or in other words N must be large. A higher order approximation is obtained if sixth and higher differences are neglected in (34); although this may allow us to choose a larger value of h the resulting coefficient matrix $\mathbf{A}$ is less sparse (its bandwidth is increased by one) and difficulties arise at $i = 0$ and $i = N$. The two main methods for improving the accuracy of low-order approximations to the solution of boundary value problems in ordinary differential equations are the deferred approach to the limit and the difference correction technique. These methods have been described in Chapter 4 in connection with the solution of boundary value problems in the form of Fredholm integral equations of the second kind.

Returning now to the solution of the integro-differential equation (32), a similar approach requires that we approximate the integral in (32) numerically by a quadrature rule of suitable accuracy. Consider for example, the use of Gregory's rule to approximate the integral in (32) at the point $s = s_i$:

$$\int_0^1 K(s_i, t)x(t)\mathrm{d}t = h\sum_{j=0}^{N}{}'' K(s_i, s_j)x_j + \Delta_i \tag{38}$$

where

$$\Delta_i = -h(\tfrac{1}{12}\nabla + \tfrac{1}{24}\nabla^2 + \tfrac{19}{720}\nabla^3 + \tfrac{3}{160}\nabla^4 + \cdots)K(s_i, s_N)x_N$$
$$+ h(\tfrac{1}{12}\Delta - \tfrac{1}{24}\Delta^2 + \tfrac{19}{720}\Delta^3 - \tfrac{3}{160}\Delta^4 + \cdots)K(s_i, s_0)x_0.$$

If we use (35) as an approximation to the second derivative term at $s = s_i$, then the repeated Trapezoid rule (obtained by neglecting Δ_i in (38)) is a natural choice to replace the integral term since the order of the associated error of the quadrature rule matches that of the difference approximation.

At $s = s_i$ we have

$$\frac{x_{i+1} - 2x_i + x_{i-1}}{h^2} + \lambda h \sum_{j=0}^{N} {}'' K(s_i, s_j) x_j = g_i, \quad i = 1, \ldots, N-1, \quad (39)$$

with

$$x_0 = x_N = 0.$$

This leads to the following set of equations:

$$(\mathbf{A} + \lambda \mathbf{B})\mathbf{x} = \mathbf{g} \tag{40}$$

where

$$\mathbf{B} = h \begin{pmatrix} K(h, h) & K(h, 2h) & \cdots & K(h, 1-h) \\ K(2h, h) & K(2h, 2h) & \cdots & K(2h, 1-h) \\ \vdots & \vdots & & \vdots \\ K(1-h, h) & K(1-h, 2h) & \cdots & K(1-h, 1-h) \end{pmatrix}$$

$$\mathbf{g} = (g_1, \ldots, g_{N-1})^T,$$

and $\mathbf{A}, \mathbf{x}$ are defined in equation (37). Notice that the coefficient matrix $\mathbf{A} + \lambda \mathbf{B}$ is a full matrix so that the solution cost increases rapidly with N. Equation (40) may be solved by a standard Gauss elimination algorithm and there is therefore a considerable advantage to be gained, in terms of cost, from the use of high-order approximations for both the integral and differential operators. This in turn should lead to more accurate approximations for $x(s_i)$.

***Example* 13.2.5.** The well known method of Numerov to solve the differential equation (33) is obtained by operating on (33), at $s = s_i$, with the central difference operator $(1 + \frac{1}{12}\delta^2)$, using equation (34) to provide an approximation to $x''(s_i)$.
Numerov's method is then:

$$x_{i+1} - 2x_i + x_{i-1} = \frac{h^2}{12}(f_{i+1} + 10f_i + f_{i-1}), \quad i = 1, \ldots, N-1$$

with leading term in the local truncation error given by $-\frac{1}{240}\delta^6 x_i$. Adapting this method to the solution of (32) is straightforward; we use Gregory's rule, neglecting fifth and higher differences, to provide an appropriate quadrature rule (with error $\mathcal{O}(h^6)$) for the integral in (32). Then the method is

$$x_{i+1} - 2x_i + x_{i-1} = \frac{h^2}{12}(g_{i+1} + 10g_i + g_{i-1})$$

$$- \frac{\lambda h^3}{12} \left\{ \sum_{j=0}^{N} {}'' [K(s_{i+1}, s_j) + 10K(s_i, s_j) + K(s_{i-1}, s_j)] x_j \right\}$$

$$- \frac{\lambda h^2}{12} \{ \Delta_{i+1}^{(4)} + 10\Delta_i^{(4)} + \Delta_{i-1}^{(4)} \}, \quad i = 1, \ldots, N-1,$$

where

$$\Delta_r^{(4)} = -h(\tfrac{1}{12}\nabla + \tfrac{1}{24}\nabla^2 + \tfrac{19}{720}\nabla^3 + \tfrac{3}{160}\nabla^4)K(s_r, s_N)x_N$$
$$+ h(\tfrac{1}{12}\Delta - \tfrac{1}{24}\Delta^2 + \tfrac{19}{720}\Delta^3 - \tfrac{3}{160}\Delta^4)K(s_r, s_0)x_0.$$

If the error in the computed solution has an asymptotic expansion in increasing powers of h (or h^2 as is the case above) then the methods of deferred approach to the limit and difference correction are certainly applicable.

13.3   Expansion methods

These methods approximate the solution $x(s)$ of the integro-differential equation by

$$x(s) \approx x_N(s) = \sum_{i=0}^{N} a_i^{(N)}h_i(s) \tag{1}$$

where the set $\{h_i(s)\}$ is complete in $\mathscr{L}^2(a, b)$. The expansion methods of Chapters 7 and 8 may be readily adapted to the solution of Volterra and Fredholm integro-differential equations. Consider the Fredholm integro-differential equation of the previous section:

$$x''(s) + \lambda \int_0^1 K(s, t)x(t)\mathrm{d}t = g(s), \tag{2a}$$

$$x(0) = x(1) = 0, \tag{2b}$$

and introduce the residual $r_N(s)$ which satisfies

$$r_N(s) = x_N''(s) + \lambda \int_0^1 K(s, t)x_N(t)\mathrm{d}t - g(s). \tag{3}$$

The $h_i(s)$ are usually chosen so that $x_N(s)$ automatically satisfies the boundary conditions (2b); in this case we obtain a so-called 'interior method' (see Delves and Walsh, 1974). Assuming that $h_i''(s)$ exists for $i = 0, 1, \ldots, N$ we can substitute for x_N, x_N'' in (2) to obtain

$$r_N(s) = \sum_{i=0}^{N} a_i^{(N)}\{h_i''(s) + \lambda k_i(s)\} - g(s) \tag{4}$$

where

$$k_i(s) = \int_0^1 K(s, t)h_i(t)\mathrm{d}t.$$

Thus equation (4) can be used to determine the coefficients $a_i^{(N)}$, $i = 0, 1, \ldots, N$, in (1) by requiring that some measure of $r_N(s)$ is small. In this way we can obtain approximations $x_N(s)$ to the solution of the integro-differential equation which satisfy the boundary conditions.

For example, if we choose $\{s_j\}_{j=0,1,\ldots,N}$ in $[0,1]$ such that

$$0 \le s_0 < s_1 < \cdots < s_N \le 1$$

and

$$r_N(s_j) = 0, \quad j = 0, 1, \ldots, N \tag{4a}$$

we obtain the well known collocation method.

Example 13.3.1.

$$x''(s) - 60 \int_0^1 (s-t)x(t)dt = s - 2,$$

$$x(0) = x(1) = 0,$$

has the solution $x(s) = s(s-1)^2$. Let $x_1(s) = \sum_{i=0}^1 a_i h_i(s)$ with $h_0(s) = s(s-1)$, $h_1(s) = s^2(s-1)$. Now $h_0(0) = h_0(1) = 0$ and $h_1(0) = h_1(1) = 0$ so $x_1(s)$ clearly satisfies the boundary conditions. The collocation equations are:

$$\sum_{i=0}^1 a_i \{h_i''(s_j) + \lambda k_i(s_j)\} - g(s_j) = 0, \quad j = 0, 1,$$

where $\lambda = -60$, $g(s) = s - 2$, $k_i(s) = \int_0^1 (s-t)h_i(t)dt$. It follows that $k_0(s) = \frac{1}{12}(1 - 2s)$, $k_1(s) = \frac{1}{60}(3 - 5s)$. Hence if we use the end points of the interval $[0,1]$ as collocation points we have the following set of equations to solve:

$$3a_0 + 5a_1 = 2,$$

$$7a_0 + 6a_1 = -1.$$

The solution of this system is $a_0 = -1$, $a_1 = 1$ and we see that $x_1(s)$ agrees with the exact solution of the integro-differential equation.

If instead of (4a) we require that $r_N(s)$ be orthogonal with weight function $w(s)$ to each of the functions $h_0, h_1, \ldots, h_N$, that is,

$$\int_a^b w(s)h_i^*(s)r_N(s)ds = 0, \quad i = 0, 1, \ldots, N, \tag{5}$$

we obtain the (symmetric) Galerkin method. For example, the Galerkin equations for $(2a,b)$ are

$$L_G^{(N)}a^{(N)} = g^{(N)} \tag{6}$$

where

$$(L_G^{(N)})_{ij} = \int_0^1 w(s)h_i^*(s)Lh_j(s)ds$$

$$= \int_0^1 w(s)h_i^*(s)h_j''(s)ds$$

$$+ \lambda \int_0^1 w(s) \int_0^1 h_i^*(s)K(s,t)h_j(t)dt\,ds,$$

$$(g^{(N)})_i = \int_0^1 w(s)h_i^*(s)g(s)ds, \quad i,j = 0,1,\ldots,N.$$

A popular choice of expansion set $\{h_i(s)\}$ is the Chebyshev polynomials $\{T_i(s)\}$ which are orthogonal on the interval $[-1,1]$ with weight function $(1-s^2)^{-\frac{1}{2}}$. We make the following expansion for $x_N(s)$:

$$x(s) \approx x_N(s) = \sum_{i=0}^{N}{}'' a_i T_i(s) \tag{7}$$

and consider the solution of first order linear Fredholm integro-differential equations of the type (see Section 13.1)

$$\left.\begin{aligned} x'(s) + R(s)x(s) + \lambda \int_{-1}^{1} K(s,t)x(t)dt = g(s), \\ c^T x(r) = e, \quad r = (r_1,\ldots,r_m)^T, \quad -1 \leq r_i \leq 1. \end{aligned}\right\} \tag{8}$$

The methods which we discuss are equally applicable to first order Volterra integro-differential equations and we assume, without loss of generality, that the independent variable s has domain $-1 \leq s \leq 1$.

13.3.1   El-gendi's method

The method of El-gendi (1969) for the solution of the Fredholm integral equation

$$x(s) = y(s) + \lambda \int_{-1}^{1} K(s,t)x(t)dt \tag{9}$$

has been discussed in Chapter 4. There we saw that on approximating the solution of (9) by a finite Chebyshev expansion and the integral by the Clenshaw–Curtis quadrature rule at $s = s_i = \cos(i\pi/N)$, that is,

$$\int_{-1}^{1} K(s_i,t)x(t)dt \approx \sum_{j=0}^{N}{}'' w_j K(s_i,s_j)x(s_j), \quad i = 0,1,\ldots,N, \tag{10}$$

$$w_j = \frac{4}{N} \sum_{\substack{r=0 \\ r\,\text{even}}}^{N}{}'' \frac{1}{1-r^2} \cos\left(\frac{rj\pi}{N}\right), \quad j = 0,1,\ldots,N,$$

we obtain a set of linear equations for the solution of (9) at the quadrature points s_i, $i = 0,1,\ldots,N$. The solution of this system enables the coefficients a_i in (7) to be directly determined (from (13)) and the method can be extended to solve equation (8) with a single point boundary condition (that is, $m = 1$); a preliminary integration yields

$$x(s) + \int_{-1}^{s} R(t)x(t)dt + \lambda \int_{-1}^{s}\int_{-1}^{1} K(s,t)x(t)ds\,dt$$

$$= e + \int_{-1}^{s} g(t)dt. \tag{11}$$

Thus at $s = s_i$ we have

$$x(s_i) + \int_{-1}^{s_i} R(t)x(t)dt + \lambda \int_{-1}^{s_i} \int_{-1}^{1} K(s_i, t)x(t)ds_i dt$$

$$= e + \int_{-1}^{s_i} g(t)dt, \quad i = 0, 1, \ldots, N. \tag{12}$$

We consider first the evaluation of the right side of (12). Using (7) and the discrete orthogonality properties of the Chebyshev polynomials (see Chapter 2, Exercise 4) we see that

$$a_i = \frac{2}{N} \sum_{j=0}^{N}{}'' x_N(s_j) \cos\left(\frac{ij\pi}{N}\right), \quad i = 0, 1, \ldots, N, \tag{13}$$

so that the coefficients a_i can be readily computed once the approximate solution $x_N(s_i)$, $i = 0, 1, \ldots, N$, is known. We compute $x_N(s_i)$ in an obvious way by approximating the integrals in (12).

Suppose $g(t)$ has the approximate Chebyshev expansion

$$g(t) \approx g_N(t) = \sum_{j=0}^{N}{}'' \alpha_j T_j(t). \tag{14}$$

Then we may write

$$\int_{-1}^{s} g_N(t)dt = \sum_{j=0}^{N}{}'' \alpha_j \int_{-1}^{s} T_j(t)dt. \tag{15}$$

The following relations[†] prove useful:

$$\int_{-1}^{s} T_j(t)dt = \begin{cases} \dfrac{T_{j+1}(s)}{2(j+1)} - \dfrac{T_{j-1}(s)}{2(j-1)} + \dfrac{(-1)^{j+1}}{j^2 - 1}, & j \geq 2, \\[2mm] \frac{1}{4}\{T_2(s) - 1\} & , \quad j = 1, \\[2mm] T_1(s) + 1 & , \quad j = 0, \end{cases} \tag{16}$$

and we obtain (see Exercise 4)

$$\int_{-1}^{s} g_N(t)dt = \sum_{r=0}^{N+1} \beta_r T_r(s) \tag{17}$$

where

$$\beta_0 = \sum_{\substack{j=0 \\ j \neq 1}}^{N}{}'' \frac{(-1)^{j+1}\alpha_j}{j^2 - 1} - \tfrac{1}{4}\alpha_1,$$

$$\beta_k = \frac{\alpha_{k-1} - \alpha_{k+1}}{2k}, \quad k = 1, 2, \ldots, N-2,$$

[†] These relations follow on generalising the results of Exercise 2, Chapter 8.

$$\beta_{N-1} = \frac{\alpha_{N-2} - \frac{1}{2}\alpha_N}{2(N-1)},$$

$$\beta_N = \frac{\alpha_{N-1}}{2N},$$

$$\beta_{N+1} = \frac{\frac{1}{2}\alpha_N}{2(N+1)}.$$

Now we have (from (14) and the discrete orthogonality conditions of Chapter 2, Exercise 4)

$$\alpha_j = \frac{2}{N} \sum_{k=0}^{N}{}'' g_N(s_k) T_j(s_k), \quad j = 0, 1, \ldots, N, \tag{18}$$

where $s_k = \cos(k\pi/N)$. Thus if we set $s = s_i$ in (17) and substitute for β_r in terms of the $\alpha_r, r = 0, 1, \ldots, N$, and subsequently for α_r in terms of $g_N(s_k)$, $k = 0, 1, \ldots, N$, we obtain a set of equations of the form

$$\mathbf{G}\mathbf{g}_N = \bar{\mathbf{g}}_N \tag{19}$$

where $\mathbf{G}$ is an $(N+1) \times (N+1)$ matrix and $\bar{\mathbf{g}}_N, \mathbf{g}_N$ are vectors of length $N+1$ with

$$(\bar{\mathbf{g}}_N)_i = \int_{-1}^{s_i} g_N(t) \mathrm{d}t,$$

$$(\mathbf{g}_N)_i = g_N(s_i),$$

$$G_{ij} = w_{ij}, \quad i, j = 0, 1, \ldots, N,$$

where w_{ij} are the weights associated with a quadrature rule for the interval $[-1, s_i]$:

$$\int_{-1}^{s_i} g_N(t) \mathrm{d}t = \sum_{j=0}^{N} w_{ij} g_N(s_j), \quad i = 0, 1, \ldots, N. \tag{19a}$$

Clearly when $i = N$, equation (19a) reduces to

$$\int_{-1}^{1} g_N(t) \mathrm{d}t = \sum_{j=0}^{N} w_{Nj} g_N(s_j)$$

or

$$\int_{-1}^{1} g_N(t) \mathrm{d}t = \sum_{j=0}^{N}{}'' \bar{w}_{Nj} g_N(s_j)$$

and the $\bar{w}_{Nj}$ correspond to the weights w_j of the Clenshaw–Curtis rule of equation (10). Now so far we have not imposed any special conditions on the approximation (14); we make (19) usable by imposing the interpolating conditions

$$g_N(s_i) \equiv g(s_i), \quad i = 0, 1, \ldots, N. \tag{19b}$$

***Example* 13.3.2.** Suppose we approximate $g(s)$ by

$$g_2(s) = \sum_{j=0}^{2} {}'' \alpha_j T_j(s).$$

Then it follows from (17) that

$$\int_{-1}^{s} g_2(t)dt = \sum_{r=0}^{3} \beta_r T_r(s) \tag{20}$$

where

$$\beta_0 = \tfrac{1}{2}\alpha_0 - \tfrac{1}{4}\alpha_1 - \tfrac{1}{6}\alpha_2,$$
$$\beta_1 = \tfrac{1}{2}\alpha_0 - \tfrac{1}{4}\alpha_2,$$
$$\beta_2 = \tfrac{1}{4}\alpha_1,$$
$$\beta_3 = \tfrac{1}{12}\alpha_2.$$

Also we have from (18) and (19b):

$$\alpha_j = \sum_{k=0}^{2} {}'' g(s_k)T_j(s_k), \quad j = 0, 1, 2.$$

Hence substituting for β_r in terms of $g(s_k), k = 0, 1, 2$, in equation (20), with $s_k = \cos(k\pi/2)$, yields the following quadrature rule:

$$\int_{-1}^{s_i} g_2(t)dt = \sum_{j=0}^{2} w_{ij}g(s_j)$$

where

$$w_{00} = \tfrac{1}{3}, \quad w_{01} = \tfrac{4}{3}, \quad w_{02} = \tfrac{1}{3},$$
$$w_{10} = -\tfrac{1}{12}, \quad w_{11} = \tfrac{2}{3}, \quad w_{12} = \tfrac{5}{12},$$
$$w_{20} = w_{21} = w_{22} = 0.$$

The remaining integrals in equation (12) can be treated in an analogous way (see El-gendi, 1969) by approximating each integrand by a finite Chebyshev expansion of order N, and we state below the results we shall need:

(i) $$\int_{-1}^{s_i} \int_{-1}^{1} K(s_i, t)x(t)ds_i\,dt$$

$$= \sum_{j=0}^{N} \bar{K}_{ij}x_N(s_j), \quad i = 0, 1, \ldots, N \tag{21}$$

where
$$\bar{\mathbf{K}} = \mathbf{GC} \tag{21a}$$

with
$$C_{ij} = G_{Nj}K_{ij}, \tag{21b}$$
$$K_{ij} = K(s_i, s_j), \quad i, j = 0, 1, \ldots, N.$$

(ii) $\displaystyle\int_{-1}^{s_i} R(t)x(t)\mathrm{d}t = \sum_{j=0}^{N} \bar{R}_{ij}x_N(s_j)$ (22)

where

$$\bar{\mathbf{R}} = \mathbf{GR}$$ (22a)

with

$$R_{ij} = \begin{cases} R(s_i), & i=j, \\ 0, & i\neq j. \end{cases}$$ (22b)

These results enable us to approximate each integral in (12) using the appropriate quadrature rule at the points $s_i = \cos(i\pi/N)$, $i = 0, 1,\ldots,N$, and the following set of equations

$$(\mathbf{I} + \bar{\mathbf{R}} + \lambda\bar{\mathbf{K}})\mathbf{x}_N = \mathbf{e} + \mathbf{Gg}$$ (23)

where

$$\mathbf{x}_N = (x_N(s_0), x_N(s_1),\ldots,x_N(s_N))^T, \quad \mathbf{e} = (e,e,\ldots,e)^T,$$

may be solved to yield the solution at the quadrature points s_i.

This method, although (as perhaps the reader has decided) complicated to describe, can also be applied to second order integro-differential equations with a two-point boundary condition; the extension to treat integro-differential equations with the general boundary condition (13.1.2) is however less straightforward, and we shall describe in Section 13.3.3 a method, based on Galerkin techniques, which overcomes this difficulty.

13.3.2 Wolfe's method

Wolfe's method for the solution of integro-differential equations differs from that of El-gendi in that it treats the equation directly, rather than first converting the equation to an integral equation. The method uses a generalisation of the technique of Clenshaw and Norton (1963) to solve first for $x'(s)$, recovering $x(s)$ at the end by a final integration. With reference to equation (8) the method generates the iterative sequence $\{x_i(s)\}$ where

$$x_{i+1}'(s) + R(s)x_i(s) + \lambda\int_{-1}^{1} K(s,t)x_i(t)\mathrm{d}t = g(s)$$ (24a)

with

$$\mathbf{c}^T x(\mathbf{r}) = c, \quad m = 1,$$ (24b)

where (24b) is a single point boundary condition. Then given finite Chebyshev expansions for the functions $x_i(s)$ and $K(s,t)x_i(t)$, of degree $N+1$ and M respectively, that is,

$$x_i(s) = \sum_{j=0}^{N+1}{}' a_{ij}T_j(s),$$ (25a)

$$K(s,t)x_i(t) = \sum_{k=0}^{M}{}' b_{ik}(s)T_k(t), \tag{25b}$$

the method proceeds to find the unknown Chebyshev coefficients in the expansion for $x_{i+1}{}'(s)$:

$$x_{i+1}{}'(s) = \sum_{l=0}^{N}{}' d_{i+1,l}T_l(s). \tag{25c}$$

It follows that

$$\int_{-1}^{1} K(s,t)x_i(t)dt = \sum_{k=1}^{M+1} c_{i,k}(s)[1 - (-1)^k] \tag{26}$$

where

$$c_{ik}(s) = \frac{b_{i,k-1}(s) - b_{i,k+1}(s)}{2k}, \quad k = 1,\dots,M-1,$$

$$c_{iM}(s) = \frac{b_{i,M-1}(s)}{2M},$$

$$c_{i,M+1}(s) = \frac{b_{iM}(s)}{2(M+1)}.$$

Now if we set $s = s_r = \cos(r\pi/N)$, $r = 0, 1, \dots, N$, in (24a) we can determine the $x_{i+1}{}'(s_r)$ from the relation

$$x_{i+1}{}'(s_r) = g(s_r) - R(s_r)x_i(s_r)$$
$$- \lambda \sum_{k=1}^{M+1} c_{ik}(s_r)[1 - (-1)^k], \quad r = 0,1,\dots,N, \tag{27}$$

where $c_{ik}(s_r)$ is determined from the relations in (26) and the expansion

$$b_{ik}(s_r) \approx \frac{2}{N}\sum_{l=0}^{N}{}'' K(s_r, s_l)x_i(s_l)\cos\left(\frac{kl\pi}{N}\right), \quad r = 0,1,\dots,N. \tag{28}$$

It follows from (25c) that

$$d_{i+1,l} \approx \frac{2}{N}\sum_{r=0}^{N}{}'' x_{i+1}{}'(s_r)\cos\left(\frac{lr\pi}{N}\right), \quad l = 0,1,\dots,N, \tag{29}$$

and it only remains to find the Chebyshev coefficients of $x_{i+1}(s)$:

$$x_{i+1}(s) = \sum_{j=0}^{N+1}{}' a_{i+1,j}T_j(s). \tag{30}$$

These are directly related to those of the derivative function $x_{i+1}{}'(s)$ whose Chebyshev expansion is given by (25c).

Thus a relation analogous to (26) and the boundary condition (24b) yields the coefficients $a_{i+1,j}$, $j = 0, 1, \dots, N+1$. We note that equation (24) is a generalisation of Picard iteration for ordiary differential equations, which may or may not converge. Assuming convergence of the method we may continue the procedure until

$$\frac{|x_i(s_r) - x_{i-1}(s_r)|}{|x_i(s_r)|} \leq \varepsilon, \quad r = 0, 1, \dots, N, \tag{31}$$

for some small real $\varepsilon > 0$. Wolfe recommends that the iteration be started with a low value of $N = M$ and that with each iteration, N (and M) be increased by one until the condition (31) is met. Clearly the method requires that a large number of sets of Chebyshev coefficients be computed at each iteration. The method could be recast as a method for setting up simultaneous equations, in which case the iteration can be considered as an iterative method for solving these equations.

Although described in the context of expansion methods, the methods of both El-gendi and Wolfe are in fact 'hybrid' methods in that they solve directly for $x(s)$, $x'(s)$ at a set of quadrature points. We describe in the next section an alternative expansion scheme for linear integro-differential equations which solves directly for the Chebyshev coefficients of the unknown function $x(s)$, and which has the advantage of being easy and efficient to use as well as yielding rapid convergence for increasing values of N.

13.3.3   The Fast Galerkin scheme for linear integro-differential equations (Babolian and Delves, 1981)

This method is based on the Fast Galerkin scheme (see Chapters 8 and 9) for integral equations. Many generalisations of this scheme are possible but the one presented here has the following advantages:

(i) singularities in the kernel, driving terms or coefficients of the differential operators are treated efficiently provided that they are of 'standard' type;

(ii) the rapid convergence and $\mathcal{O}(N^2 \ln N)$ operation count of the Fast Galerkin scheme are retained;

(iii) equations of both Fredholm and Volterra type are handled, although the latter are perhaps more naturally handled by the step-by-step methods of Section 13.2.1;

(iv) second and higher order differential operators, with a wide class of single point or multipoint boundary conditions (corresponding respectively to $m = 1$ or $m > 1$ in (13.1.2)) can be treated in a uniform manner;

(v) the defining equations for the method are well conditioned.

These points, especially (iv) and (v) depend upon the details of the method, and in particular on the way that the differential operator and the boundary conditions are incorporated. As in Wolfe's method, the method first solves for the highest occurring derivative of $x(s)$ and the function itself is recovered in a subsidiary operation.

We present the method for the solution of the following first order linear integro-differential equation with linear boundary condition:

$$Q(s)x'(s) + R(s)x(s) + \lambda \int_{-1}^{1} K(s,t)x(t)dt = g(s), \quad -1 \le s \le 1, \tag{32a}$$

$$\mathbf{c}^T x(\mathbf{r}) + \mathbf{d}^T x'(\mathbf{r}) = e, \tag{32b}$$

where

$$\mathbf{c} = (c_1,\ldots,c_m)^T, \quad \mathbf{d} = (d_1,\ldots,d_m)^T,$$
$$\mathbf{r} = (r_1,\ldots,r_m)^T, \quad -1 \le r_i \le 1,$$

and $x(\mathbf{r}), x'(\mathbf{r})$ are defined in equation (13.1.2). We shall assume that

$$Q(s) \text{ does not change sign on } [-1,1]. \tag{32c}$$

If (32c) does not hold the integro-differential equation may or may not have an unique solution and the numerical method may or may not encounter difficulties, (see Babolian and Delves, 1981). Equations (32a, b) correspond to equation (13.1.2) with $P(s) \equiv 0$, $a = -1$, $b = 1$, and we note that in most cases, for a first order equation, $\mathbf{d} = \mathbf{0}$. However, its inclusion causes no complications and yields a uniform treatment for first and second order problems. Suppose $x(s)$ and $x'(s)$ have the Chebyshev expansions

$$x(s) = \sum_{j=0}^{\infty}{}' a_j T_j(s), \tag{33a}$$

$$x'(s) = \sum_{j=0}^{\infty}{}' a_j' T_j(s), \tag{33b}$$

where the coefficients a_j' and a_j are to be determined by the method in that order. Suppose that Q, R and g have the known expansions

$$Q(s) = \sum_{j=0}^{\infty}{}' q_j T_j(s), \tag{34a}$$

$$R(s) = \sum_{j=0}^{\infty}{}' r_j T_j(s), \tag{34b}$$

$$g(s) = \sum_{j=0}^{\infty}{}' g_j T_j(s). \tag{34c}$$

(In practice numerical approximations to the coefficients q_j, r_j, g_j can be computed using Fast Fourier Transform techniques as in Section 8.5.) Then if we substitute these expansions into (32a) and apply the weighted Galerkin technique we obtain the infinite system of equations

$$
\sum_{j=0}^{\infty}{}' a_j' \left(\sum_{k=0}^{\infty}{}' q_k \int_{-1}^{1} \frac{T_k(s)T_j(s)T_i(s)\,ds}{(1-s^2)^{\frac{1}{2}}} \right)
$$
$$
+ \sum_{j=0}^{\infty}{}' a_j \left\{ \sum_{k=0}^{\infty}{}' r_k \int_{-1}^{1} \frac{T_k(s)T_j(s)T_i(s)\,ds}{(1-s^2)^{\frac{1}{2}}} \right.
$$
$$
\left. + \lambda \int_{-1}^{1} \int_{-1}^{1} \frac{K(s,t)T_j(t)T_i(s)\,ds\,dt}{(1-s^2)^{\frac{1}{2}}} \right\}
$$
$$
= \sum_{k=0}^{\infty}{}' g_k \int_{-1}^{1} \frac{T_k(s)T_i(s)\,ds}{(1-s^2)^{\frac{1}{2}}}. \tag{35}
$$

Using the orthogonality properties of the Chebyshev polynomials and the relation

$$
T_i(s)T_j(s) = \tfrac{1}{2}[T_{i+j}(s) + T_{|i-j|}(s)]
$$

it follows that

$$
\int_{-1}^{1} \frac{T_k(s)T_j(s)T_i(s)}{(1-s^2)^{\frac{1}{2}}}\,ds =
\begin{cases}
\pi & \text{if } i=j=k=0, \\[2mm]
\dfrac{\pi}{2}\delta_{ij} & \text{if } i+j>0 \text{ and } k=0, \\[2mm]
\dfrac{\pi}{2}(\delta_{k,i+j} + \delta_{k,|i-j|}) & \text{if } k>0.
\end{cases}
$$

Thus equation (35) may be written in the form

$$
\mathbf{Q}\mathbf{a}' + (\mathbf{R} + \lambda\mathbf{B})\mathbf{a} = \mathbf{g} \tag{36}
$$

where

$$
\left.
\begin{aligned}
&Q_{ij} = \frac{q_{i+j} + q_{|i-j|}}{2}, \quad R_{ij} = \frac{r_{i+j} + r_{|i-j|}}{2}, \\[3mm]
&B_{ij} = \frac{2}{\pi}\int_{-1}^{1}\int_{-1}^{1} \frac{K(s,t)T_j(t)T_i(s)\,ds\,dt}{(1-s^2)^{\frac{1}{2}}}, \quad \begin{array}{l} i=0,1,\ldots, \\ j=1,2,\ldots, \end{array}
\end{aligned}
\right\} \tag{36a}
$$

and for $i \geq 0$

$$
Q_{i0} = \frac{q_i}{2}, \quad R_{i0} = \frac{r_i}{2}, \quad B_{i0} = \frac{1}{\pi}\int_{-1}^{1}\int_{-1}^{1} \frac{K(s,t)T_i(s)\,ds\,dt}{(1-s^2)^{\frac{1}{2}}}. \tag{36b}
$$

Now the coefficients a_j' are related to the a_j by

$$
a_j = \frac{a_{j-1}' - a_{j+1}'}{2j}, \quad j=1,2,\ldots, \tag{37}
$$

or (in the notation of Babolian and Delves)

$$
\mathbf{a}^{(1)} = \mathbf{A}\mathbf{a}' \tag{37a}
$$

where

$$
A_{ij} = \begin{cases} \dfrac{1}{2(j+1)}, & j = i \geq 0, \\[2mm] \dfrac{-1}{2(j-1)}, & j = i+2, \\[2mm] 0, & \text{otherwise}, \end{cases}
$$

$$
\mathbf{a}^{(1)} = (a_1, a_2, \ldots)^T,
$$
$$
\mathbf{a}' = (a_0', a_1', \ldots)^T.
$$

In addition an expression for a_0 can be obtained by considering the boundary condition (32b), which we write equivalently as

$$
\sum_{i=1}^{m} [c_i x(r_i) + d_i x'(r_i)] = e.
$$

Now if we substitute the Chebyshev expansions for $x(r_i)$ and $x'(r_i)$ in the above equation we obtain

$$
\sum_{i=1}^{m} \left[c_i \sum_{j=0}^{\infty}{}' a_j T_j(r_i) + d_i \sum_{j=0}^{\infty}{}' a_j' T_j(r_i) \right] = e
$$

or

$$
\mathbf{c}^T \mathbf{T} \mathbf{a} + \mathbf{d}^T \mathbf{T} \mathbf{a}' = e, \tag{38}
$$

where

$$
T_{ij} = T_j(r_i), \quad i = 1, 2, \ldots, m; \quad j = 1, 2, \ldots;
$$
$$
T_{i0} = \tfrac{1}{2}, \quad i = 1, 2, \ldots, m.
$$

We may write the term $\mathbf{c}^T \mathbf{T} \mathbf{a}$ as

$$
\mathbf{c}^T \mathbf{T} \mathbf{a} = (\mathbf{c}^T \mathbf{l}) a_0 + \mathbf{c}^T \mathbf{T}^{(1)} \mathbf{a}^{(1)}
$$

where $\mathbf{l} = (\tfrac{1}{2}, \ldots, \tfrac{1}{2})^T$, $T_{ij}^{(1)} = T_{i,j+1}$, $i = 1, 2, \ldots, m; j = 0, 1, \ldots$.

Then rearranging the above equation for a_0 and using equation (37a), we obtain

$$
a_0 = \frac{e - (\mathbf{d}^T \mathbf{T} + \mathbf{c}^T \mathbf{T}^{(1)} \mathbf{A}) \mathbf{a}'}{\Delta_1} \tag{39}
$$

where $\Delta_1 = \mathbf{c}^T \mathbf{l}$ is assumed to be nonzero. Now if we set

$$
\mu = \frac{e}{\Delta_1}, \quad \mathbf{k}^T = -\frac{(\mathbf{d}^T \mathbf{T} + \mathbf{c}^T \mathbf{T}^{(1)} \mathbf{A})}{\Delta_1}
$$

equations (37a) and (39) may be combined to give

$$
\mathbf{a} = \mathbf{A}' \mathbf{a}' + \boldsymbol{\mu} \tag{40}
$$

where

$$
\mathbf{A}' = \begin{pmatrix} \mathbf{k}^T \\ \mathbf{A} \end{pmatrix}, \quad \boldsymbol{\mu} = \begin{pmatrix} \mu \\ \mathbf{0} \end{pmatrix}.
$$

The method thus consists of solving first for $\mathbf{a}'$ from the equation

$$[\mathbf{Q} + (\mathbf{R} + \lambda\mathbf{B})\mathbf{A}']\mathbf{a}' = \mathbf{g}_1 \qquad (41)$$

where

$$\mathbf{g}_1 = \mathbf{g} - (\mathbf{R} + \lambda\mathbf{B})\boldsymbol{\mu}$$

and we have substituted for $\mathbf{a}$ in equation (36), and then solving for $\mathbf{a}$ from equation (40). Clearly, in order to proceed numerically, the infinite system (41) must be truncated. Babolian and Delves (1981) have adopted the simple strategy of truncating every expansion after the term involving T_N; then all vectors are replaced by $(N+1)$-vectors and infinite square matrices by their $(N+1) \times (N+1)$ leading submatrices. Thus equation (41) reduces to the finite system

$$\mathbf{L}_1^{(N)}\mathbf{a}'^{(N)} = \mathbf{g}_1^{(N)} \qquad (42)$$

where $\mathbf{L}_1^{(N)}$ is the leading submatrix of order $(N+1) \times (N+1)$ obtained from the infinite matrix $\mathbf{L}_1 = \mathbf{Q} + (\mathbf{R} + \lambda\mathbf{B})\mathbf{A}'$. This procedure is justified by the error analysis of Babolian and Delves where it is shown that the cost of setting up the defining equations is $\mathcal{O}(N^2 \ln N)$. Babolian and Delves make the additional assumption

$$m \le 2; \quad |r_i| = 1, \quad 1 \le i \le m; \quad \mathbf{d} = \mathbf{0}, \qquad (43)$$

along with some smoothness requirements on the functions Q, R and K. They show that the coefficient matrix $\mathbf{L}_1^{(N)}$ is asymptotically diagonal of type B (see Babolian, 1980, Section 6.0) thus implying that the matrix problem (42) is well conditioned. Imposing the condition (43) merely serves to simplify the error analysis for the method, although numerical evidence suggests that the stability and convergence properties of the method do not depend on (43). A further attraction of the method over those of the two previous sections is that it provides a cheap and effective numerical error estimate; the interested reader is referred to Babolian and Delves.

Example 13.3.3. We illustrate the method for the following example taken from Babolian (1980):

$$s^2 x'(s) + e^s x(s) + \int_{-1}^{1} e^{(s+1)t}x(t)\,dt$$

$$= (s^2 + e^s)e^s + \frac{(e^{s+2} - e^{-s-2})}{s+2}, \quad -1 \le s \le 1, \qquad (44a)$$

$$x(-1) + x(1) = e + \frac{1}{e}. \qquad (44b)$$

Equation (44) has the solution $x(s) = e^s$ and we note that $Q(s) = s^2$ is

Table 13.3.1 *Computed error* $\|e_N\|_\infty$
and estimated error E_{est} *for problem*
(44)

N	$\|e_N\|_\infty$	E_{est}
3	4.2×10^{-1}	8.4, 0
4	8.3, −2	1.4, 0
5	2.4, −2	2.4, −1
6	2.4, −3	3.5, −2
7	3.0, −4	4.3, −3
8	3.3, −5	4.7, −4
9	3.3, −6	4.4, −5
10	3.0, −7	3.8, −6
11	2.7, −8	2.8, −7
12	2.0, −9	1.9, −8
13	2.0, −10	1.9, −9
14	2.2, −10	9.7, −10
15	2.3, −10	2.3, −10

These results are taken from Table 6.1
of Babolian (1980).

unsigned on the interval $[-1, 1]$. The error norm $\|e_N\|_\infty$, given by

$$\|e_N\|_\infty = \max_{-1 \le s \le 1} |x(s) - x_N(s)|$$

where $x_N(s)$ is the computed solution

$$x_N(s) = \sum_{j=0}^{N}{}' a_j^{(N)} T_j(s) \tag{45}$$

has been computed for a range of values of N. The Chebyshev coefficients $\mathbf{a}^{(N)}$ in (45) have been evaluated from $\mathbf{a}'^{(N)}$, the numerical solution of the system (42). Table 13.3.1 shows the value $\|e_N\|_\infty$ along with the modulus of the estimated error E_{est}.

These results show the rapid convergence which can be achieved using the Fast Galerkin method. The estimated error is clearly reliable for this problem.

The application of the method to second and higher order integro-differential equations is straightforward. Consider the second order equation of Section 13.1:

$$P(s)x''(s) + Q(s)x'(s) + R(s)x(s) + \lambda \int_{-1}^{1} K(s, t)x(t)dt = g(s),$$
$$-1 \le s \le 1, \tag{46a}$$

$$Cx(\mathbf{r}) + Dx'(\mathbf{r}) = \mathbf{e}, \tag{46b}$$

where we have set $a = -1$, $b = 1$ in (13.1.2). We assume that

$$P(s) \text{ does not change sign on } [-1, 1]. \tag{46c}$$

If we introduce the further expansions

$$x''(s) = \sum_{j=0}^{\infty}{}' a_j'' T_j(s),$$ (47)

$$P(s) = \sum_{j=0}^{\infty}{}' p_j T_j(s),$$ (48)

and apply the weighted Galerkin technique to equation (46) we obtain the infinite system of equations

$$\mathbf{P}\mathbf{a}'' + \mathbf{Q}\mathbf{a}' + (\mathbf{R} + \lambda\mathbf{B})\mathbf{a} = \mathbf{g},$$ (49)

where

$$P_{ij} = \frac{p_{i+j} + p_{|i-j|}}{2}, \quad i = 0, 1, \ldots; \quad j = 0, 1, 2, \ldots;$$

$$P_{i0} = \frac{p_i}{2}, \qquad\qquad i = 0, 1, \ldots,$$

and the other matrices are as defined in (36). The coefficients a_j'' are related to the a_j' by

$$\mathbf{a}' = \mathbf{A}''\mathbf{a}'' + \boldsymbol{\eta}$$ (50)

where $\mathbf{A}'', \boldsymbol{\eta}$ are defined in Babolian and Delves. Thus we can use equations (40) and (50) to write (49) as

$$\mathbf{L}_2\mathbf{a}'' = \mathbf{g}_2$$ (51)

where

$$\mathbf{L}_2 = \mathbf{P} + [\mathbf{Q} + (\mathbf{R} + \lambda\mathbf{B})\mathbf{A}']\mathbf{A}'',$$

$$\mathbf{g}_2 = \mathbf{g} - [\mathbf{Q} + (\mathbf{R} + \lambda\mathbf{B})\mathbf{A}']\boldsymbol{\eta} - (\mathbf{R} + \lambda\mathbf{B})\boldsymbol{\mu}.$$

The method proceeds to solve the finite system

$$\mathbf{L}_2^{(N)}\mathbf{a}''^{(N)} = \mathbf{g}_2^{(N)}$$ (52)

for the $(N + 1)$ Chebyshev coefficients of the finite expansion

$$x_N''(s) = \sum_{j=0}^{N}{}' a_j'' T_j(s)$$ (53)

where $\mathbf{L}_2^{(N)}$ is the leading submatrix of order $(N + 1) \times (N + 1)$ obtained from the infinite matrix $\mathbf{L}_2$. $\mathbf{a}'^{(N)}$ and $\mathbf{a}^{(N)}$ are then evaluated from equations (50) and (40) respectively. The error analysis is simplified by making the assumptions

$$m \leq 2; |r_i| = 1, \quad 1 \leq i \leq m;$$

$$\mathbf{D} \text{ has one zero row, say } d_{1i} = 0, \ 1 \leq i \leq m,$$ (54)

along with some smoothness requirements on the functions P, Q, R and K, and $\mathbf{L}_2^{(N)}$ is shown to be asymptotically diagonal of type B.

The following example shows the behaviour of the method for two problems, taken from Babolian and Delves. These examples have been particularly chosen to show how essential (or non-essential) are some of the restrictions made in the error analysis on the boundary conditions (both on points and coefficients) and on the functions under consideration.

***Example* 13.3.4.** Consider the following two problems:

Problem 1: Equation (46a) with $K(s, t) = e^{(s+1)t}$, $\lambda = 1$, $P(s) = e^s$,

$$Q(s) = \cos s, \quad R(s) = \sin s,$$

$$g(s) = (e^s + \cos s + \sin s)e^s + \frac{(e^{s+2} - e^{-s-2})}{s+2},$$

with boundary conditions:

$$(a)\begin{cases} x(1) + x(-1) = e + \dfrac{1}{e}, \\ x(1) + x(-1) - x'(-1) = e; \end{cases}$$

$$(b)\begin{cases} x(1) + x(-1) + x'(1) = 2e + \dfrac{1}{e}, \\ x(1) + x(-1) - x'(-1) = e. \end{cases}$$

This problem, whose exact solution is $x(s) = e^s$, has smooth coefficients with $P(s)$ unsigned on $[-1, 1]$. Note that the associated boundary conditions (a) satisfy condition (54) while in case (b) $d_{12} \neq 0$ for $\mathbf{r} = (-1, 1)^T$. Results for problems 1(a) and 1(b) are shown in Table 13.3.2 where the maximum and estimated errors are given for a range of values of N.

Problem 2: Equation (46a) with $K(s, t) = (s^3 + t^3 + 3)^{\frac{1}{2}}$, $\lambda = 1$,

$$P(s) = \ln(1 + s), \quad Q(s) = \cos s, \quad R(s) = \sin s,$$
$$g(s) = 2\ln(1 + s) + 2s\cos s + s^2\sin s$$
$$+ \tfrac{2}{9}[(s^3 + 4)^{\frac{3}{2}} - (s^3 + 2)^{\frac{3}{2}}],$$

with boundary conditions

$$x'(-1) = -2, \quad x(1) - x'(1) = -1.$$

The exact solution of this problem is $x(s) = s^2$. Condition (54) is not satisfied, since for $\mathbf{r} = (-1, 1)^T$, $d_{11} = 1$. In addition $P(s) = \ln(1 + s)$ is singular and the smoothness conditions imposed by Babolian and Delves are violated. There is therefore no error estimate available for this problem. The Fast Galerkin technique of Chapter 11, which takes the singularity into account, may be used to set up the coefficient matrix for this problem and the results are shown in the last column of Table 13.3.2.

Problem 1(a) has been chosen to satisfy all those conditions on which the error analysis of Babolian and Delves is based. The numerical results for this problem show that the error estimate is rather reliable, and similarly for

Table 13.3.2. *Computed error* $\|e_N\|_\infty$ *and estimated error* E_{est} *for problems* 1–2.

	Problem 1(a)		Problem 1(b)		Problem 2
N	$\|e_N\|_\infty$	E_{est}	$\|e_N\|_\infty$	E_{est}	$\|e_N\|_\infty$
3	5.3×10^{-2}	1.4, +1	6.1, −1	1.6, +1	2.1, −2
4	1.1, −2	2.4, 0	1.8, −2	2.9, 0	1.0, −2
5	9.0, −4	4.1, −1	7.8, −3	4.7, −1	2.8, −3
6	1.4, −4	6.7, −2	6.2, −4	7.2, −2	7.4, −5
7	1.4, −5	1.0, −2	2.0, −4	1.0, −2	3.5, −5
8	1.9, −6	1.3, −3	1.1, −5	1.4, −3	4.5, −6
9	1.5, −7	1.5, −4	2.4, −6	1.5, −4	1.9, −7
10	1.4, −8	1.2, −5	1.0, −7	1.4, −5	2.8, −7
11	9.0, −10	1.2, −6	1.6, −8	1.1, −6	2.1, −8
12	8.7, −11	8.9, −8	4.9, −10	8.1, −8	4.3, −8
13	8.7, −11	1.8, −8	3.1, −10	9.3, −9	1.2, −9
14	5.8, −11	9.5, −9	3.3, −10	5.9, −9	1.8, −9
15	8.3, −11	5.4, −9	5.7, −10	2.8, −9	4.0, −10

These results are taken from Tables 1 and 2 of Babolian and Delves (1981).

problem 1(b), despite the fact that this problem violates conditions (54). This, together with further numerical evidence based on problems with $m > 2$ or the first row of $\mathbf{D} \neq \mathbf{0}$, suggests that condition (54) is not essential. The results also show that the method works as well for problem 2 as for the others. The application of the Singular Galerkin method of Chapter 11 results in the truncated matrix being accurately computed. The singularities then cancel fortuitously in the solution, yielding a rapidly convergent scheme. (Such cancellations are not taken into account in the error analysis of Babolian and Delves.)

In cases where condition (46c) (or (32c)) fails to hold but the equation (46a, b) (or (32a, b)) has an unique solution, a modification of the numerical method has been investigated by Babolian and Delves (1981). They show that with this modification the satisfactory behaviour of the algorithm is retained and the reader is referred to their paper for details.

Exercises

1. Show that the following integro-differential equations have the (stated) solution:

(a) $x'(s) = x(s) + \displaystyle\int_0^1 \frac{x(t)}{s + e^t} \, dt - \ln\left(\frac{s + e}{s + 1}\right), \quad 0 \leq s \leq 1,$

$x(0) = 1.$

[solution $x(s) = e^s$]

(b) $x''(s) = x(s) - \dfrac{4}{\pi} \displaystyle\int_0^\pi \cos(s-t)\,x(t)\mathrm{d}t, \quad 0 \le s \le \pi,$

$x(0) = 1, \quad x'(\pi) = 0.$

[solution $x(s) = \cos s$]

(c) $\ln(1+s)\,x''(s) + \cos s\,x'(s) + \sin s\,x(s)$

$\qquad + \lambda \displaystyle\int_{-1}^1 (s^3 + t^3 + 3)^{\frac{1}{2}} x(t)\mathrm{d}t = g(s), \quad -1 \le s \le 1,$

$g(s) = 2\ln(1+s) + 2s\cos s + s^2 \sin s + \dfrac{2\lambda}{9}[(s^3+4)^{\frac{3}{2}} - (s^3+2)^{\frac{3}{2}}],$

$x'(-1) = -2, \quad x(1) - x'(1) = -1.$

[solution $x(s) = s^2$]

(d) $x'(s) = 1 + x(s) - s e^{-s^2} - 2\displaystyle\int_0^s s t e^{-x^2(t)}\mathrm{d}t, \quad 0 \le s \le 2,$

$x(0) = 0.$

[solution $x(s) = s$]

2. Show that the integro-differential equation

$$x''(s) + \lambda \int_0^1 K(s,t)x(t)\mathrm{d}t = g(s), \quad 0 \le s \le 1,$$

$x(0) = x(1) = 0,$

may be rephrased as the following Fredholm integral equation of the second kind:

$$x(s) + \lambda \int_0^1 H(s,u)x(u)\mathrm{d}u = \gamma(s) \qquad (A)$$

where

$$H(s,u) = \int_0^1 G(s,t)K(t,u)\mathrm{d}t,$$

$$\gamma(s) = \int_0^1 G(s,t)g(t)\mathrm{d}t,$$

and $G(s,t)$ is the Green's function for the problem, defined in equation (13.1.6). Show further that when the Trapezoid rule is used to replace each integral in (A) the approximate solution at $s = s_i$ is given by

$$x_i + \lambda h^2 \sum_{j,k=0}^{N}{}'' G(s_i,s_k)K(s_k,s_j)x_j = h\sum_{l=0}^{N}{}'' G(s_i,s_l)g_l$$

where $s_i = ih, \ i = 0,1,\dots,N,$ with $Nh = 1.$

3. The first order Volterra integro-differential equation (13.1.3) may be written as

$$x(s) = x_0 + \int_0^s f(t, x(t), z(t))dt, \quad 0 \le s \le l,$$

$$z(s) = \int_0^s K(s, t, x(t))dt.$$

This may be approximated at $s = s_r = rh$, $r = 0, 1, \ldots, N$, with $h = l/N$ by replacing each of the integrals with the repeated Simpson's rule and the single step Trapezoid rule applied to the upper end of the range of integration when r is odd. Show that the equations of the resulting step-by-step quadrature rule may be differenced to give:

$$x_{2r+1} - x_{2r} = \tfrac{1}{2}h(f_{2r+1} + f_{2r}),$$
$$x_{2r} - x_{2r-2} = \tfrac{1}{3}h(f_{2r} + 4f_{2r-1} + f_{2r-2}),$$

where x_r is an approximation to $x(s_r)$ and $f_r = f(s_r, x_r, z_r)$ with z_r an approximation to $z(s_r)$.

4. If $x(s)$ and $\int_{-1}^s x(t)dt$ have the following finite Chebyshev expansions

$$x(s) = \sum_{j=0}^{N}{}'' \alpha_j T_j(s),$$

$$\int_{-1}^s x(t)dt = \sum_{j=0}^{N+1} \beta_j T_j(s),$$

show, using the relations in equation (13.3.16), that the coefficients β_j are given by

$$\beta_0 = \sum_{\substack{j=0\\j\ne1}}^{N}{}'' \frac{(-1)^{j+1}\alpha_j}{j^2-1} - \tfrac{1}{4}\alpha_1,$$

$$\beta_k = \frac{\alpha_{k-1} - \alpha_{k+1}}{2k}, \quad k = 1, 2, \ldots, N-2,$$

$$\beta_{N-1} = \frac{\alpha_{N-2} - \tfrac{1}{2}\alpha_N}{2(N-1)},$$

$$\beta_N = \frac{\alpha_{N-1}}{2N},$$

$$\beta_{N+1} = \frac{\tfrac{1}{2}\alpha_N}{2(N+1)}.$$

5. The Fast Galerkin scheme applied to the solution of second order

Fredholm integro-differential equations makes use of relations between the Chebyshev coefficients of the solution $x(s)$ and those of the derivative functions $x'(s)$, $x''(s)$. In the notation of Section 13.3.3 show that if $x(s)$, $x'(s)$, $x''(s)$ have the following expansions:

$$x(s) = {\sum_{j=0}^{\infty}}' a_j T_j(s), \quad x'(s) = {\sum_{j=0}^{\infty}}' a_j' T_j(s),$$

$$x''(s) = {\sum_{j=0}^{\infty}}' a_j'' T_j(s),$$

then the Chebyshev coefficients of $x(s)$ and $x''(s)$ are related by

$$\mathbf{a}^{(2)} = \mathbf{B}\mathbf{a}''$$

where

$$B_{ij} = \begin{cases} \dfrac{1}{4(j+1)(j+2)}, & j = i \geq 0, \\[2mm] \dfrac{-1}{2(j^2-1)}, & j = i+2, \\[2mm] \dfrac{1}{4(j-1)(j-2)}, & j = i+4, \\[2mm] 0, & \text{otherwise,} \end{cases}$$

$$\mathbf{a}^{(2)} = (a_2, a_3, \dots)^T,$$

$$\mathbf{a}'' = (a_0'', a_1'', \dots)^T.$$

(*HINT*: Delete the first row and column of the matrix $\mathbf{A}$ in equation (3.3.37a) to obtain the new system $\mathbf{a}^{(2)} = \bar{\mathbf{A}}\mathbf{a}^{(1)'}$. Now substitute for $\mathbf{a}^{(1)'}$, an expression for which is obtained on differentiating the system in (3.3.37a).)

APPENDIX

Singular expansions

We give without proof a number of Chebyshev expansions for 'singular', that is nonsmooth, functions for use with the algorithm described in Section 11.6.2.

A.1 Expansions for singular driving terms

We consider expansions of the form

$$h(s) = \sum_{i=0}^{\infty}{}' h_i T_i(s), \quad -1 \leq s \leq 1, \tag{1}$$

$$h_i = \frac{2}{\pi} \int_{-1}^{1} \frac{T_i(s)h(s)ds}{(1-s^2)^{\frac{1}{2}}}. \tag{2}$$

Table A.1 lists the expansion coefficients h_i for a number of singular Chebyshev expansions of some functions $h(s)$.

A.2 Expansions for singular kernels

For a given singular factor $Q(s,t)$ we consider here the three associated functions $H_I(s,t)$ defined as follows:

$$H_I(s,t) = Q(s,t)(1-t^2)^{\frac{1}{2}}, \quad I = \text{Fredholm}, \tag{1}$$

$$= \begin{cases} Q(s,t)(1-t^2)^{\frac{1}{2}}, & t \leq s, \\ 0, & t > s, \end{cases} \bigg\} I = \text{Volterra}, \tag{2}$$

$$= \begin{cases} Q(s,t)(1-t^2)^{\frac{1}{2}}, & t > s, \\ 0, & t \leq s, \end{cases} \bigg\} I = \text{Inverse Volterra}. \tag{3}$$

The suffix indicates the applicability of these factors; note that a kernel of Green's function type may be written as the sum of a Volterra and an Inverse Volterra kernel, and that the expansions for the Inverse Volterra case follow directly from those for Fredholm and Volterra operators.

We introduce the expansions

$$H_I(s,t) = \sum_{i}^{\infty}{}' \sum_{j}^{\infty}{}' H_{ij} T_i(s) T_j(t), \quad -1 \leq s,t \leq 1. \tag{4}$$

Table A.1.

$h(s)$	h_i
$(1-s)^\alpha : \alpha > -\frac{1}{2}$	$\begin{cases} h_0 = 2^{\alpha+1}\Gamma(\alpha+\frac{1}{2})/(\sqrt{\pi}\,\Gamma(\alpha+1)) \\ h_i = [(i-1-\alpha)/(i+\alpha)]h_{i-1}, \quad i \ge 1 \end{cases}$
$\alpha = \frac{1}{2}$	$h_i = -\sqrt{2/(\pi(i^2-\frac{1}{4}))}, \quad i \ge 0$
$(1+s)^\alpha : \alpha > -\frac{1}{2}$	$\begin{cases} h_0 = 2^{\alpha+1}\Gamma(\alpha+\frac{1}{2})/(\sqrt{\pi}\,\Gamma(\alpha+1)) \\ h_i = [(\alpha-i+1)/(\alpha+i)]h_{i-1}, \quad i \ge 1 \end{cases}$
$\alpha = \frac{1}{2}$	$h_i = \sqrt{2}(-1)^{i+1}/(\pi(i^2-\frac{1}{4})), \quad i \ge 0$
$\lvert s \rvert^\alpha, \alpha > -1$	$h_0 = 2\Gamma\left(\frac{\alpha}{2}+\frac{1}{2}\right)\Big/\left(\sqrt{\pi}\,\Gamma\left(\frac{\alpha}{2}+1\right)\right)$
	$h_1 = 0$
	$h_{i+1} = [(\alpha+1-i)/(\alpha+1+i)]h_{i-1}, \quad i \ge 1$
$\ln\lvert s \rvert$	$h_0 = -2\ln 2$
	$h_{2i} = (-1)^{i+1}/i, \quad i \ge 1$
	$h_{2i+1} = 0, \quad i \ge 0$
$\ln(1+s)$	$h_0 = -2\ln 2$
	$h_i = 2(-1)^{i+1}/i, \quad i > 0$
$\ln(1-s)$	$h_0 = -2\ln 2$
	$h_i = -2/i, \quad i > 0$
$(1-s^2)^{\frac{1}{2}}$	$h_0 = 4/\pi; h_1 = 0$
	$h_i = -2(1+(-1)^i)/(\pi(i^2-1)), \quad i \ge 2$
$(1-s^2)^{\frac{1}{2}}(1+s)^\alpha : \alpha > -1$	$h_0 = 2^{\alpha+2}/(\pi(\alpha+1))$
	$h_1 = 2^{\alpha+3}/(\pi(\alpha+2)) - h_0$
	$h_2 = 2[2^{\alpha+1}/\pi - 2h_1 - h_0]/(\alpha+3)$
	$h_i = i[-2^{\alpha+3}/(\pi i(i-2)) - 2h_{i-1}$
	$\qquad + h_{i-2}(\alpha-i+3)/(i-2)]/(i+\alpha+1), \quad i > 2$

Then the following results can be derived.

    1. $Q(s,t) = 1$

    (a) *Fredholm*

$$H_{ij} = 0, \quad i > 0$$

$$H_{0,0} = \frac{8}{\pi}$$

$$H_{0,1} = 0$$

$$H_{0,j} = \frac{-4[1+(-1)^j]}{\pi(j^2-1)}, \quad j > 1$$

(b) *Volterra*

$$H_{0,0} = \frac{4}{\pi}; \quad H_{1,0} = \frac{2}{\pi}; \quad H_{i,0} = 0, \quad i > 1$$

$$H_{0,1} = -\frac{1}{\pi}; \quad H_{2,1} = \frac{1}{(2\pi)}; \quad H_{i,1} = 0, \quad i \neq 0,2$$

For $j \geq 2$:

$$H_{0,j} = \frac{4(-1)^{j-1}}{\pi(j^2 - 1)}; \quad H_{j-1,j} = \frac{-1}{\pi(j-1)}; \quad H_{j+1,j} = \frac{1}{\pi(j+1)}$$

$H_{i,j} = 0$ (other values of i)

(c) *Inverse Volterra*

$H_{i,j}(\text{Inverse Volterra}) = H_{i,j}(\text{Fredholm}) - H_{i,j}(\text{Volterra})$

2. $Q(s,t) = |s - t|^\alpha; \quad \alpha > -1$

In all three cases H_{ij} may be computed from the recurrence relations:

$$H_{i,0} = \frac{-4}{(\alpha + 1)} \frac{I_{i,0}}{\pi^2}, \quad i \geq 0$$

$$\begin{vmatrix} H_{0,1} = \frac{-1}{(\alpha + 2)} \left[\frac{4}{\pi^2} I_{0,1} - H_{1,0} \right] \\[2mm] H_{i,1} = \frac{-1}{2(\alpha + 2)} \left[\frac{8 I_{i,1}}{\pi^2} - (H_{i+1,0} + H_{i-1,0}) \right], \quad i \geq 1 \\[2mm] H_{0,2} = \frac{2}{(\alpha + 3)} \left[\frac{-2 I_{0,2}}{\pi^2} + (2 H_{1,1} - H_{0,0}) \right] \\[2mm] H_{i,2} = \frac{2}{(\alpha + 3)} \left[\frac{-2 I_{i,2}}{\pi^2} + (H_{i+1,1} + H_{i-1,1} - H_{i,0}) \right], \quad i \geq 1 \end{vmatrix}$$

For $j > 2$:

$$H_{0,j} = \frac{1}{(j - 2)(j + \alpha + 1)} \left[\frac{8 I_{0,j}}{\pi^2} + 2j(j - 2)H_{1,j-1} \right.$$
$$\left. + j(\alpha + 3 - j)H_{0,j-2} \right]$$

$$H_{i,j} = \frac{1}{(j - 2)(j + \alpha + 1)} \left[\frac{8 I_{i,j}}{\pi^2} + j(j - 2)(H_{i+1,j-1} + H_{i-1,j-1}) \right.$$
$$\left. + j(\alpha + 3 - j)H_{i,j-2} \right], \quad i \geq 1$$

Here, the auxiliary function $I_{i,j}$ has the following definitions:

(a) *Volterra*

$$I_{i,j} = (-1)^{j+1} \int_{-1}^{1} \frac{T_i(s)}{(1-s^2)^{\frac{1}{2}}} (1+s)^{\alpha+1} ds$$

and may be evaluated from the recurrence relation

$$I_{0,j} = (-1)^{j+1} \sqrt{\pi} \, 2^{\alpha+1} \frac{\Gamma(\alpha+\frac{3}{2})}{\Gamma(\alpha+2)}$$

$$I_{i,j} = \frac{(\alpha-i+2)}{(\alpha+i+1)} I_{i-1,j}, \quad i \geq 1.$$

For $\alpha = \pm\frac{1}{2}$ the recurrence relation yields

$$I_{ij} = \frac{(-1)^{i+j}}{\sqrt{2}(i^2-\frac{1}{4})}, \quad \alpha = -\frac{1}{2}$$

and

$$I_{i,j} = \frac{3(-1)^{i+j+1}}{\sqrt{2}(i^2-\frac{1}{4})(i^2-\frac{9}{4})}, \quad \alpha = \frac{1}{2}$$

(b) *Inverse Volterra*

$$I_{i,j} = -\int_{-1}^{1} \frac{T_i(s)}{(1-s^2)^{\frac{1}{2}}} (1-s)^{\alpha+1} ds$$

$$I_{0,j} = \frac{-\sqrt{\pi} \, 2^{\alpha+1} \Gamma(\alpha+\frac{3}{2})}{\Gamma(\alpha+2)}$$

$$I_{i,j} = \frac{(i-2-\alpha)}{(i+\alpha+1)} I_{i-1,j}, \quad i \geq 1$$

For $\alpha = \pm\frac{1}{2}$ this recurrence relation yields

$$I_{i,j} = \frac{1}{\sqrt{2}(i^2-\frac{1}{4})}, \quad \alpha = -\frac{1}{2}$$

$$I_{i,j} = \frac{-3}{\sqrt{2}(i^2-\frac{1}{4})(i^2-\frac{9}{4})}, \quad \alpha = \frac{1}{2}$$

(c) *Fredholm*

$$I_{i,j}(\text{Fredholm}) = I_{i,j}(\text{Volterra}) + I_{i,j}(\text{Inverse Volterra});$$

whence we find

$$I_{0,j} = \frac{-(1+(-1)^j)\sqrt{\pi} \, 2^{\alpha+1} \Gamma(\alpha+\frac{3}{2})}{\Gamma(\alpha+2)}$$

$$I_{i,j} = \frac{(i-\alpha-2)}{(\alpha+i+1)} I_{i-1,j-1}, \quad i \geq 1.$$

For $\alpha = \pm\frac{1}{2}$ the recurrence relation yields

$$I_{i,j} = \frac{((-1)^{i+j}+1)}{\sqrt{2(i^2-\frac{1}{4})}}, \quad \alpha = -\frac{1}{2}$$

$$I_{i,j} = \frac{-3((-1)^{i+j}+1)}{\sqrt{2(i^2-\frac{1}{4})(i^2-\frac{9}{4})}}, \quad \alpha = \frac{1}{2}.$$

3. $Q(s,t) = \ln|s-t|$

   In all three cases, $H_{i,j}$ may be computed from the recurrence relations:

$$H_{i,0} = \frac{4}{\pi^2}[I_{i,0} - J_{i,0}], \quad i \geq 0$$

$$\begin{cases} H_{0,1} = \frac{1}{2}\left[\frac{4}{\pi^2}(I_{0,1} - J_{0,1}) + H_{1,0}\right] \\[2mm] H_{i,1} = \frac{1}{4}\left[\frac{8}{\pi^2}(I_{i,1} - J_{i,1}) + (H_{i+1,0} + H_{i-1,0})\right], \quad i \geq 1 \end{cases}$$

$$\begin{cases} H_{0,2} = \frac{2}{3}\left[\frac{2}{\pi^2}(I_{0,2} - J_{0,2}) + 2H_{1,1} - H_{0,0}\right] \\[2mm] H_{i,2} = \frac{2}{3}\left[\frac{2}{\pi^2}I_{i,2} - \frac{2}{\pi^2}J_{i,2} + (H_{i+1,1} + H_{i-1,1} - H_{i,0})\right], \quad i \geq 1. \end{cases}$$

For $j > 2$:

$$H_{0,j} = \frac{1}{(j+1)(j-2)}\left[\frac{-8I_{0,j}}{\pi^2} - \frac{4}{\pi^2}((j-2)J_{0,j} - jJ_{0,j-2})\right.$$

$$\left. + 2j(j-2)H_{1,j-1} - j(j-3)H_{0,j-2}\right]$$

$$H_{i,j} = \frac{1}{(j+1)(j-2)}\left[\frac{-8I_{i,j}}{\pi^2} - \frac{4}{\pi^2}((j-2)J_{i,j} - jJ_{i,j-2})\right.$$

$$\left. + j(j-2)(H_{i+1,j-1} + H_{i-1,j-1}) - j(j-3)H_{i,j-2}\right], \quad i \geq 1.$$

Here, $I_{i,j}$ and $J_{i,j}$ are given by:

(a) *Fredholm*

$$I_{0,j} = \pi(1 - \ln 2)(1 + (-1)^j)$$

$$I_{1,j} = \frac{\pi}{2}(\ln 2 - \frac{3}{2})(1 + (-1)^{j+1})$$

$$I_{i,j} = \pi(1 + (-1)^{j+i})/i(i^2-1), \quad i \geq 2$$

$$J_{0,0} = 2\pi; \quad J_{i,0} = 0, \quad i \geq 1$$

$$J_{i,1} = 0, \quad \forall i;$$

For $j \geq 2$:

$$J_{i,j} = \begin{cases} 0, & j \quad \text{odd} \\ -2\pi/(j^2-1), & i=0, \; j \quad \text{even} \\ 0, & i>0, \; j \quad \text{even} \end{cases}$$

(b) *Volterra*

$$I_{0,j} = \pi(1 - \ln 2)(-1)^j$$

$$I_{1,j} = \frac{\pi}{2}(\tfrac{3}{2} - \ln 2)(-1)^j$$

$$I_{i,j} = \pi(-1)^{j+i}/i(i^2-1), \quad i \geq 2$$

$$J_{0,0} = \pi; \quad J_{1,0} = \frac{\pi}{2}; \quad J_{i,0} = 0, \quad i > 1$$

$$J_{0,1} = -\frac{\pi}{4}; \quad J_{2,1} = \frac{\pi}{8}; \quad J_{i,1} = 0, \quad i \neq 0, 2$$

For $j \geq 2$:

$$J_{0,j} = \pi(-1)^{j+1}/(j^2-1)$$

$$J_{j-1,j} = \frac{-\pi}{4(j-1)}$$

$$J_{j+1,j} = \frac{\pi}{4(j+1)}$$

$$J_{i,j} = 0, \quad \text{other values of } i$$

(c) *Inverse Volterra*:

$$H_{i,j}(\text{Inverse Volterra}) = H_{i,j}(\text{Fredholm}) - H_{i,j}(\text{Volterra})$$

REFERENCES

Abd-Elal, L.F. & Delves, L.M. (1976). A regularization technique for a class of singular integral problems. *J. Inst. Math. Appl.*, **18**, 37–47.

Atkinson, K.E. (1976). *A Survey of Numerical Methods for the Solution of Fredholm Integral Equations of the Second Kind.* Philadelphia: Society for Industrial and Applied Mathematics.

Babolian, E. (1980). *Galerkin Methods for Integral and Integro-differential Equations.* Ph.D. Thesis, University of Liverpool.

Babolian, E. & Delves, L.M. (1981). A Fast Galerkin scheme for linear integro-differential equations. *IMA J. Numer. Anal.* **1**, 193–213.

Bain, M. & Delves, L.M. (1977). The convergence rates of expansions in Jacobi polynomials. *Numer. Math.*, **27**, 219–25.

Baker, C.T.H. (1977). *The Numerical Treatment of Integral Equations.* Oxford University Press.

Baker, C.T.H., Fox, L., Mayers, D.F. & Wright, K. (1964). Numerical solution of Fredholm integral equations of the first kind. *Comput. J.*, **7**, 141–7.

Baker, C.T.H. & Miller, G.F. (eds.) (1982). *Treatment of Integral Equations by Numerical Methods.* London: Academic Press.

Barrodale, I. (1974). Linear programming solutions to integral equations. In *Numerical Solution of Integral Equations*, eds. L.M. Delves & J. Walsh, pp. 97–107. Oxford University Press.

Barrodale, I. & Phillips, C. (1974). An improved algorithm for discrete Chebyshev linear approximations. In *Proceedings of the Fourth Manitoba Conference on Numerical Mathematics*, ed. R. Stanton, pp. 177–90. Winnipeg: University of Manitoba Press.

Barrodale, I. & Roberts, F.D.K. (1973). An improved algorithm for discrete l_1 linear approximation. *SIAM J. Numer. Anal.*, **10**, 839–48.

Barrodale, I. & Roberts, F.D.K. (1977). Solution of the constrained l_1 linear approximation problem. Research Report, Department of Mathematics, University of Victoria.

Barrodale, I. & Young, A. (1970). Computational experience in solving linear operator equations using the Chebyshev norm. In *Numerical Approximations to Functions and Data*, ed. J.G. Hayes, pp. 115–42. London: Athlone Press.

Brakhage, H. (1960). Über die numerische Behandlung von Integralgleichungen nach der Quadraturformelmethode. *Numer. Math.*, **2**, 183–96.

Brunner, H. & Lambert, J.D. (1974). Stability of numerical methods for Volterra integro-differential equations. *Computing*, **12**, 75–89.

Bückner, H. (1948). A special method of successive approximations for Fredholm integral equations. *Duke Math. J.*, **15**, 197–206.

Casaletto, J., Pickett, M. & Rice, J.R. (1969). A comparison of some numerical integration programs. *SIGNUM Newsletter*, **4**, No. 3, 30–40.

Chan, Y.F. & Fraser, P.A. (1973). S-wave positron scattering by hydrogen atoms. *J. Phys. B*, **6**, 2504–15.

Clenshaw, C.W. & Curtis, A.R. (1960). A method for numerical integration on an automatic computer. *Numer. Math.*, **2**, 197–205.

Clenshaw, C.W. & Norton, H.J. (1963). The solution of nonlinear ordinary differential equations in Chebyshev series. *Comput. J.*, **6**, 88–92.

Cochran, J.A. (1972). *The Analysis of Linear Integral Equations*. New York: McGraw-Hill.

Davis, P.J. (1963). *Interpolation and Approximation*. New York: Blaisdell Publishing Company.

Davis, P.J. & Rabinowitz, P. (1975). *Methods of Numerical Integration*. New York: Academic Press.

Day, J.T. (1968). On the numerical solution of Volterra integral equations. *BIT*, **8**, 134–7.

Day, J.T. (1970). On the numerical solution of integro-differential equations. *BIT*, **10**, 511–14.

Delves, L.M. (1968). The numerical evaluation of principal value integrals. *Comput. J.*, **10**, 389–91.

Delves, L.M. (1977). A fast method for the solution of Fredholm integral equations. *J. Inst. Math. Appl.*, **20**, 173–82.

Delves, L.M. & Abd-Elal, L.F. (1977). The Fast Galerkin algorithm for the solution of linear Fredholm equations. *Comput. J.*, **20**, 374–6.

Delves, L.M., Abd-Elal, L.F. & Hendry, J.A. (1979). A fast Galerkin algorithm for singular integral equations. *J. Inst. Math. Appl.*, **23**, 139–66.

Delves, L.M. & Bain, M. (1977). On the optimum choice of weight functions in a class of variational calculations. *Numer. Math.*, **27**, 209–18.

Delves, L.M. & Freeman, T.L. (1981). *Analysis of Global Expansion Methods: Weakly Asymptotically Diagonal Systems*. London: Academic Press.

Delves, L.M. & Walsh, J. (eds.) (1974). *Numerical Solution of Integral Equations*. Oxford University Press.

Downham, D.Y. & Shah, S.M.M. (1976). The integral equation approach for models of clines. Preprint, Department of S.C.M., University of Liverpool.

El-gendi, S.E. (1969). Chebyshev solution of differential, integral and integro-differential equations. *Comput. J.*, **12**, 282–7.

Enright, W.H., Hull, T.E. & Lindberg, B. (1975). Comparing numerical methods for stiff systems of O.D.E.'s. *BIT*, **15**, 10–48.

Fox, L. (ed.) (1962). *Numerical Solution of Ordinary and Partial Differential Equations*. Oxford: Pergamon Press.

Fox, L. & Goodwin, E.T. (1953). The numerical solution of non-singular linear integral equations. *Phil. Trans. Roy. Soc. A.*, **245**, 501–34.

Freeman, T.L. & Delves, L.M. (1974). On the convergence rates of variational methods. III. Unsymmetric systems. *J. Inst. Math. Appl.*, **14**, 311–23.

Gentleman, W. (1972). Implementing Clenshaw–Curtis quadrature, I Methodology and experience. *C.A.C.M.*, **15**, 337–42.

Gentleman, W. (1972*a*). Implementing Clenshaw–Curtis quadrature. II Computing the cosine transformation. *C.A.C.M.*, **15**, 343–6.

Goldfine, A. (1972). An algorithm for the numerical solution of integro-differential equations. *BIT*, **12**, 578–80.

Green, C.D. (1969). *Integral Equation Methods*. Nelson.

Hanson, R. (1971). A numerical method for solving Fredholm integral equations of the first kind using singular values. *SIAM J. Numer. Anal.*, **8**, 616–22.

Henrici, P. (1962). *Discrete Variable Methods in Ordinary Differential Equations*. New York: Wiley.

Hetherington, J.H. & Schick, L.H. (1965). Exact multiple-scattering analysis of low-energy $K^- - d$ scattering with separable potentials. *Phys. Rev.*, **137B**, 935–48.

Horn, S. & Fraser, P.A. (1975). Low-energy ortho-positronium scattering by hydrogen atoms. *J. Phys. B.*, **8**, 2472–5.

Hull, T.E., Enright, W.H., Fellen. B.M. & Sedgwick, A.E. (1972). Comparing numerical methods for ordinary differential equations. *SIAM J. Numer. Anal.*, **9**, 603–37.

Kahaner, D.K. (1971). Comparison of numerical quadrature formulas. In *Mathematical Software*, ed. J.R. Rice, pp. 229–59. New York: Academic Press.

Kanwal, R.P. (1971). *Linear Integral Equations*. London: Academic Press.

Kershaw, D. (1974). Singular integral and boundary value problems. In *Numerical Solution of Integral Equations*, eds. L.M. Delves & J. Walsh, pp. 258–66. Oxford University Press.

Kreyszig, E. (1978). *Introductory Functional Analysis with Applications*. New York: Wiley.

Lambert, J.D. (1973). *Computational Methods in Ordinary Differential Equations*. London: Wiley.

Lewis, B.A. (1975). On the numerical solution of Fredholm integral equations of the first kind. *J. Inst. Math. Appl.*, **16**, 207–20.

Linz, P. (1968). The numerical solution of Volterra integral equations by finite difference methods. *M.R.C. Tech. Rept.*, **825**

Linz, P. (1969). A method for solving nonlinear Volterra integral equations of the second kind. *Math. Comp.*, **23**, 595–9.

Linz, P. (1969*a*). Numerical methods for Volterra integral equations with singular kernels. *SIAM J. Numer. Anal.*, **6**, 365–74.

Linz, P. (1969*b*). Linear multistep methods for Volterra integro-differential equations. *J.A.C.M*, **16**, 295–301.

Logan, J.E. (1976). *The Approximate Solution of Volterra Integral Equations of the Second Kind*. Ph.D. Thesis, University of Iowa.

Lyness, J.N. (1969). Quadrature methods based on complex function values. *Math. Comp.*, **23**, 601–19.

Lyness, J.N. & Kaganove, J.J. (1976). Comments on the nature of automatic quadrature routines. *ACM Trans. Math. Soft.*, **2**, 65–81.

Lyness, J.N. & Kaganove, J.J. (1977). A technique for comparing quadrature routines. *Comput. J.*, **20**, 170–7.

Mayers, D.F. (1962). Equations of Volterra type. In *Numerical Solution of Ordinary and Partial Differential Equations*, ed. L. Fox, pp. 165–73. Oxford: Pergamon Press.

McKee, S. & Brunner, H. (1980). The repetition factor and numerical stability of Volterra integral equations. *Comp. Math. Appl.*, **6**, 339–47.

Mikhlin, S.G. & Smolitsky, K.L. (1967). *Approximate Methods for the Solution of Differential and Integral Equations*. New York: Elsevier.

Nerinckx, D. (1980). Numerical solution of Volterra integral equations. Rept. TW46, Applied Mathematics & Programming Division, Katholieke Universiteit Leuven.

Noble, B. (1969). Instability when solving Volterra integral equations of the second kind by multistep methods. In *Conference on the Numerical Solution of Differential Equations*, Lecture Notes in Mathematics, No. 109, pp. 23–39. Berlin: Springer-Verlag.

Ortega, J.M. (1972). *Numerical Analysis: A Second Course.* New York: Academic Press.

Petryshyn, W.V. (1963). On a general iterative method for the approximate solution of linear operator equations. *Math. Comp.*, **17**, 1–10.

Phillips, C. (1977). *The Numerical Solution of Volterra Integral Equations of the Second Kind by Step by Step Methods.* Ph.D. Thesis, University of Liverpool.

Rall, L.B. (1955). Error bounds for iterative solution of Fredholm integral equations. *Pacific J. Math.*, **5**, 977–86.

Rhoderick, E.H. & Wilson, E.M. (1962). Current distribution in thin superconducting films. *Nature*, **194**, 1167–9.

Ribière, G. (1967). Regularisation d'opérateurs. *Rev. Franç. Inf. Rech. Opér.*, **1**, 57–79.

Rice, J.R. (1969). *The Approximation of Functions, Volume 2: Advanced Topics.* Reading, Massachusets: Addison-Wesley.

Riddell, I. (1981). *The Numerical Solution of Integral Equations.* Ph.D. Thesis, University of Liverpool.

Riess, R.D. & Johnson, L.W. (1969). Estimating Gauss–Chebyshev quadrature errors. *SIAM J. Numer. Anal.*, **6**, 557–9.

Rumyantsev, I.A. (1965). Programme for solving a system of Volterra integral equations (of the second kind). *USSR Comput. Math. and Math. Phys.*, **5**, 218–24.

Samuelson, P.A. (1953). Rapidly converging solutions to integral equations. *J. Math. Phys.*, **31**, 276–86.

Smithies, F. (1958). *Integral Equations.* Cambridge University Press.

Stroud, A.H. (1971). *Approximate Calculation of Multiple Integrals.* Englewood Cliffs, New Jersey: Prentice–Hall.

Stroud. A.H. & Secrest, D. (1966). *Gaussian Quadrature Formulas.* Englewood Cliffs, New Jersey: Prentice–Hall.

Szego, G. (1939). *Orthogonal Polynomials.* American Mathematical Society Colloquium Publications XXIII. Waverley Press.

Thomas, K.S. (1975). On the approximate solution of operator equations. *Numer. Math.*, **23**, 231–9; Preprint, University of Oxford.

Thomas, K.S. (1976). On the approximate solution of operator equations Part II. Preprint, University of Oxford.

Turchin, V.F., Kozlov, V.P. & Malkevich, M.S. (1971). The use of mathematical statistics methods in the solution of incorrectly posed problems. *Soviet Phys. Usp.*, **13**, 681–702.

Wagner, C. (1951). On the solution of Fredholm integral equations of the second kind by iteration. *J. Math. Phys.*, **30**, 23–30.

Wahba, G. (1977). Practical approximate solutions to linear operator equations when the data are noisy. *SIAM J. Numer. Anal.*, **14**, 651–67.

Wiener, K. (1971). *Beiträge Zur Analysis*, **1**, 109.

Wilkinson, J.H. (1965). *The Algebraic Eigenvalue Problem.* Oxford: Clarendon Press.

Young, A. (1954). Approximate product-integration. *Proc. R. Soc. A.*, **224**, 552–61.

Young, A. (1954a). The application of approximate product–integration to the numerical solution of integral equations. *Proc. R. Soc. A.*, **224**, 561–73.

INDEX

adjoint equation 69
annihilation class 25, 37
asymptotically diagonal matrix 229, 356, 358
  block weakly 226
  point weakly 227
asymptotically lower diagonal matrix 219
asymptotically upper diagonal matrix 219
Augmented Galerkin method 309
  estimation of parameters in 313
  stability of 312

Banach space 72
Battery method 233
Bessel's inequality 18
block methods 136
boundary value problem 5, 341
bounded operator 1, 15

Cauchy integral equations 281
Cauchy sequence 18, 20
Cauchy's theorem 282
central differences 341
characteristic equation 130
characteristic function 70
characteristic polynomial (*see* stability polynomial) 130
characteristic value 4, 70
Chebyshev norm (*see* L_∞-norm) 172
Chebyshev polynomials 27, 32, 201, 216, 347, 354
Christoffel–Darboux identity 30
Clenshaw–Curtis quadrature rule (*see* quadrature rule) 33
collocation 188, 345
complete orthonormal sequence 19
conditioning 212
  inherent 86
  condition number 107, 196
convergence 8
  in the mean (*see also* strong convergence) 18

of quadrature methods 126
of sequences of functions 17
of sequences of operators 20
order of 89, 127
rate of 57, 88
strong 18, 21
weak 18, 21

deferred approach to the limit 99, 124
difference correction 96
difference equation 130, 343
differential equations 1, 4
discontinuities 62

eigenfunction 70
  expansion method 303
eigenvalue 4, 70
  problem 4, 152
El-gendi's method 105, 346
error bounds 9, 50, 51, 81, 221
error estimates 8, 44, 53, 87, 94, 220, 247, 290
error function 83
expansion methods 2, 168
  for approximating the kernel 168
  for eigenvalue problems 180
  for first kind equations 309
  for integro-differential equations 344
extrapolation 97
Euler's rule 335

Fast Galerkin method (for integral equations) 200
  error estimates for 220
  for linear integro-differential equations 352
  quadrature errors for 222
fill-in 144
filter factor 309
finite differences 96, 328